# HP ATP - FlexNetwork Solutions V3
OFFICIAL CERTIFICATION STUDY GUIDE
(EXAMS HP0-Y52 AND HP2-Z37)

**First Edition**

---

**Miriam Allred**

HPE Press
660 4th Street, #802
San Francisco, CA 94107

**HP ATP - FlexNetwork Solutions V3**
**Official Certification Study Guide (Exams HP0-Y52 and HP2-Z37)**
Miriam Allred



Published by:

Hewlett Packard Enterprise Press
660 4th Street, #802
San Francisco, CA 94107



ISBN: 978-942741-20-6

Printed in Mexico

**WARNING AND DISCLAIMER**
This book provides information about the topics covered in the Applying HP FlexNetwork Fundamentals (HP0-Y52) certification exam. Every effort has been made to make this book as complete and as accurate as possible, but no warranty or fitness is implied.

The information is provided on an "as is" basis. The author, and Hewlett Packard Enterprise Press, shall have neither liability nor responsibility to any person or entity with respect to any loss or damages arising from the information contained in this book or from the use of the discs or programs that may accompany it.

The opinions expressed in this book belong to the author and are not necessarily those of Hewlett Packard Enterprise Press.

Note: Books and courses developed prior to the Hewlett-Packard Company separation contain branding, logos, web page links, and other elements/information that has not been updated for each HP Inc. and Hewlett Packard Enterprise. The general knowledge and skills are still considered of value to HP Inc. and Hewlett Packard Enterprise employees (partners/customers) respectively, so these legacy materials are being made available here. Plans are in place for updating the most highly-used content for each HP Inc. and Hewlett Packard Enterprise.

**TRADEMARK ACKNOWLEDGEMENTS**
All third-party trademarks contained herein are the property of their respective owner(s).

**GOVERNMENT AND EDUCATION SALES**
This publisher offers discounts on this book when ordered in quantity for bulk purchases, which may include electronic versions. For more information, please contact U.S. Government and Education Sales 1-855-447-2665 or email sales@hpepressbooks.com.

## Feedback Information

At HPE Press, our goal is to create in-depth reference books of the best quality and value. Each book is crafted with care and precision, undergoing rigorous development that involves the expertise of members from the professional technical community.

Readers' feedback is a continuation of the process. If you have any comments regarding how we could improve the quality of this book, or otherwise alter it to better suit your needs, you can contact us through email at hpepress@epac.com. Please make sure to include the book title and ISBN in your message.

We appreciate your feedback.

**Publisher**: Hewlett Packard Enterprise Press

**HPE Contributors**: Wim Groeneveld, Craig Maxwell, Björn Meyer, Don McCracken, Gerhard Roets

**HPE Press Program Manager**: Michael Bishop

## About the Author

Miriam Allred has spent the last ten years configuring, testing, and troubleshooting HP wired and wireless networks. Miriam combines this wide range of technical expertise with pedagogy and instructional design training, allowing her to create technical training courses for both advanced and entry-level networking professionals. Miriam Allred has a Masters degree from Cleveland State University and a Bachelors degree from Brigham Young University.

## Introduction

This book helps you study for the Applying HP FlexNetwork Fundamentals (HP0-Y52) exam and the HP2-Z37 fast track exam to achieve the HP ATP FlexNetwork Solutions V3 certification. The certification validates that you have the networking skills and expertise to design, implement and manage the modern network, based on the HP FlexNetwork Architecture, for small to mid-size businesses. It also verifies that you understand the converged infrastructure strategy including network virtualization with HP's Intelligent Resilient Framework (IRF).

Areas of study include the ability to:

- Perform basic configuration, access security and setup on HP switches.
- Configure Layer 2 technologies such as STP, Link Aggregation and VLANs
- Configure basic IP Routing with static and dynamic routing technologies (OSPF)
- Manage and monitor networks with HP IMC network management software

## Interactive Embedded Aurasma Links and Videos

Hewlett Packard Enterprise Partner Ready Certification and Learning is driving innovation across our portfolio of technologies to enhance the way you learn. HP Partner Ready Certification has formed a partnership with Aurasma to provide embedded digital content as an additional benefit for customers who purchase this study guide.

Throughout this study guide you will see the HP ExpertOne icon, which indicates that additional digital content is available to you. To view this digital content, you have to download the HP ExpertOne app, powered by Aurasma, open the app and point the viewfinder at the page wherever you see this icon HP ExpertOne to launch your extra content.

Note that the app is currently called HP ExpertOne, and may be renamed to HPE Partner Ready in the future.

## Certification and Learning

Hewlett Packard Enterprise Partner Ready Certification and Learning provides end-to-end continuous learning programs and professional certifications that can help you open doors and succeed in the New Style of Business. We provide:

- continuous learning activities and job-role based learning plans to help you keep pace with the demands of the dynamic, fast paced IT industry;
- professional sales and technical training and certifications to give you the critical skills needed to design, manage and implement the most sought-after IT disciplines; and
- training to help you navigate and seize opportunities within the top IT transformation areas that enable business advantage today.

As a Partner Ready Certification and Learning certified member, your skills, knowledge, and real-world experience are recognized and valued in the marketplace. To continue your professional and career growth, you have access to our large HPE community of world-class IT professionals, trend-makers and decision-makers. Share ideas, best practices, business insights, and challenges as you gain professional connections globally.

To learn more about HPE Partner Ready Certification and Learning, please visit http://certification-learning.hpe.com

## Audience

This book is designed for presales solution architects, consultants, installation technicians and other IT professionals involved in supporting HPE networking solutions and products.

## Minimum Qualifications

To pass the Applying HP FlexNetwork Fundamentals (HP0-Y52) you should have at least a year of experience in designing small and medium-sized networks with intermediate switching, basic routing and wireless technologies. Exams are based on an assumed level of industry standard knowledge that may be gained from training, hands-on experience, or other prerequisites.

## Relevant Certifications

After you pass the exam, your achievement may be applicable toward more than one certification. To determine which certifications can be credited with this achievement, log in to The Learning Center and view the certifications listed on the exam's More Details tab. You might be on your way to achieving additional certifications.

## Preparing for the HP0-Y52 Exam

This self-study guide does not guarantee that you will have all the knowledge you need to pass the exam. It is expected that you will also draw on real-world experience and would benefit from completing the hands-on lab activities provided in the instructor-led training.

## Recommended Training

Recommended training to prepare for each exam is accessible from the exam's page in The Learning Center. See the exam attachment, "Supporting courses," to view and register for the courses.

## Obtain Hands-on Experience

You are not required to take the recommended, supported courses, and completion of training does not guarantee that you will pass the exams. Hewlett Packard Enterprise strongly recommends a combination of training, thorough review of courseware and additional study references, and sufficient on-the-job experience prior to taking an exam.

## Exam Registration

To register for an exam, go to http://certification-learning.hpe.com/tr/certification/learn_more_about_exams.html

# CONTENTS

# Introduction

## Concepts covered in this study guide

This study guide introduces you to:

- Switch management
- HP ProVision command line interface (CLI) and Comware CLI
- Basic networking technologies and protocols, such as link aggregation, virtual LANs (VLANs), Spanning Tree Protocol (STP), HP Intelligent Resilient Framework(IRF), and Simple Network Management Protocol (SNMP)

In addition, you will learn about HP wireless access points (APs) and HP Intelligent Management Center (IMC), which provides a single pane of glass management solution.

At the time this study guide was published, the HP Aruba wireless solutions are not part of the HP0-Y52 exam. For more information about Aruba wireless solutions, please visit http://www.arubanetworks.com/. For more information about training, please visit http://community.arubanetworks.com/t5/Training-Certification-Career/bd-p/training-certification-career.

This study guide is designed to help you pass the related certification exam, HP0-Y52. After you pass this exam, you will obtain the HP Accredited Technical Professional (ATP) FlexNetwork Solutions V3 certification.

For more information about this and other HP certifications, visit: http://certification-learning.hpe.com

Although this study guide is a valuable resource, you should also consider taking the *HP FlexNetwork Fundamentals* instructor-led course because it provides hands-on labs, which you can use to practice configuring HP switches. You will also participate in discussions and lab activities, as you practice configuring these basic technologies on HP ProVision and Comware switches.

## Augmented Reality…what's in it for you?

Hewlett Packard Enterprise Partner Ready Certification and Learning is continuing to drive innovation across our portfolio of technologies to enhance the way you learn. HP Partner Ready Certification and Learning and Aurasma have come together, bringing embedded digital content to this study guide.

Throughout this study guide you will see the HP ExpertOne icon, which indicates that additional digital content is available to you. To view this digital content, you have to download the HP ExpertOne app, powered by Aurasma, open the app and point the viewfinder at the page wherever you see this icon HP ExpertOne to launch your extra content.

Note that the app is currently called HP ExpertOne but maybe renamed as the HPE Partner Ready app in the future.

What do you need to do?

1. First, locate the HP ExpertOne app in the app store you use for your smartphone. For the iPhone locate the HP ExpertOne app in the Mac store. If your Android phone uses Google Play, locate the HP ExpertOne app there. Then follow the standard process for downloading the HP ExpertOne app. (Note that you may be asked to provide a credit card or other payment method. However, you will not actually be charged for the app.)
2. Download the HP ExpertOne, powered by Aurasma, app on your iPhone or Android device.
3. Install the App.
4. Open the App and point your phone on an image in this guide that is marked by the HP ExpertOne icon. Allow the App to focus on the image until the video starts playing.
5. Once the video begins to play, you can move your smartphone or tablet away from the image in this guide. The video will keep playing on your smartphone.

## Example configurations

This study guide provides some example configurations that guide you through the steps required to configure technologies such as STP or HP IRF. Each example configuration includes a network topology such as the one shown in Figure Introduction-1.

These network topologies include four switches, which are labeled ProVision-1, ProVision-2, Comware-1, and Comware-2. The network topology also includes a Windows server and a Windows client.

The network topology shows IP addresses, VLAN assignments, and the ports used to connect to each switch. In Figure Introduction-1, for example, Comware-1 connects to ProVision-1 on port 21 and connects to Comware-2 on port 19.

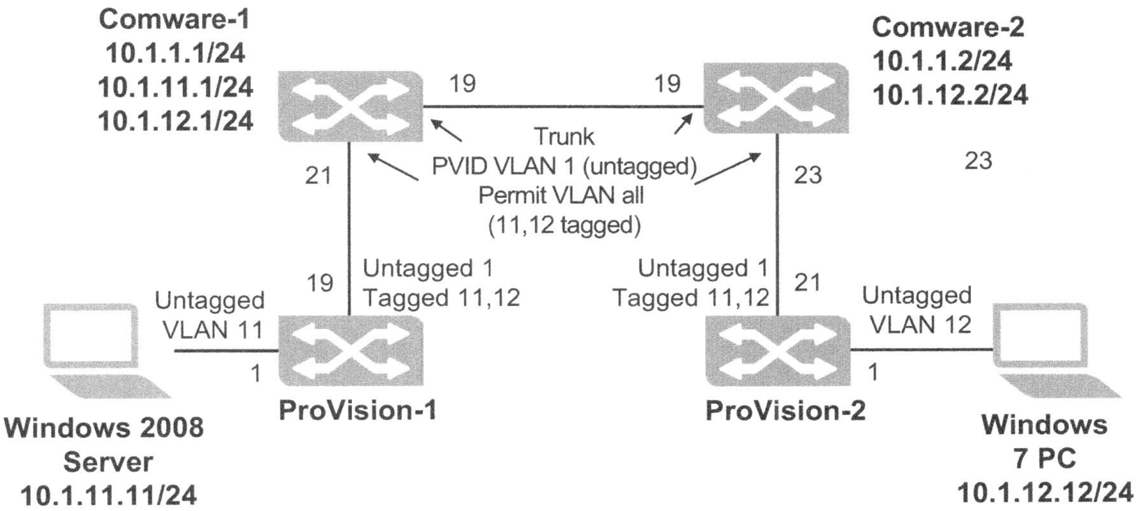


Figure Introduction-1: Example configuration

# 1 Introduction to HP Networking

## EXAM OBJECTIVES

✓ Explain how HP Networking solutions meet companies' requirements

✓ Explain how HP Intelligent Management Center (IMC) simplifies network management

✓ Explain how Software Defined Networking (SDN) revolutionizes networking and encourages innovation

## ASSUMED KNOWLEDGE

Before reading this chapter, you should have a basic understanding of

- Local Area Networks (LANs)
- Wide Area Networks (WANs)
- Data centers
- Campus networks
- Branch networks

## INTRODUCTION

This chapter introduces you to HP networking solutions, outlining the overarching HP architecture called HP FlexNetwork. It also describes HP solutions that work at each part of that architecture. Note that this chapter does not provide a comprehensive list of HP networking solutions; it introduces you to these solutions, allowing you to go to the HP networking website to get up-to-date information about HP networking solutions.

## HP FlexNetwork architecture overview

The HP FlexNetwork Architecture, which is shown in Figure 1-1, forms the basis for HP networking solutions for networks of all sizes, from global enterprise networks to small-to-medium business (SMB) networks. The networking solutions are a key component of the HP Converged Infrastructure. Enterprises can align their networks with their business needs—even as those needs change—by segmenting their networks into the four interrelated modular building blocks that comprise the HP FlexNetwork Architecture: FlexFabric, FlexCampus, FlexBranch, and FlexManagement.

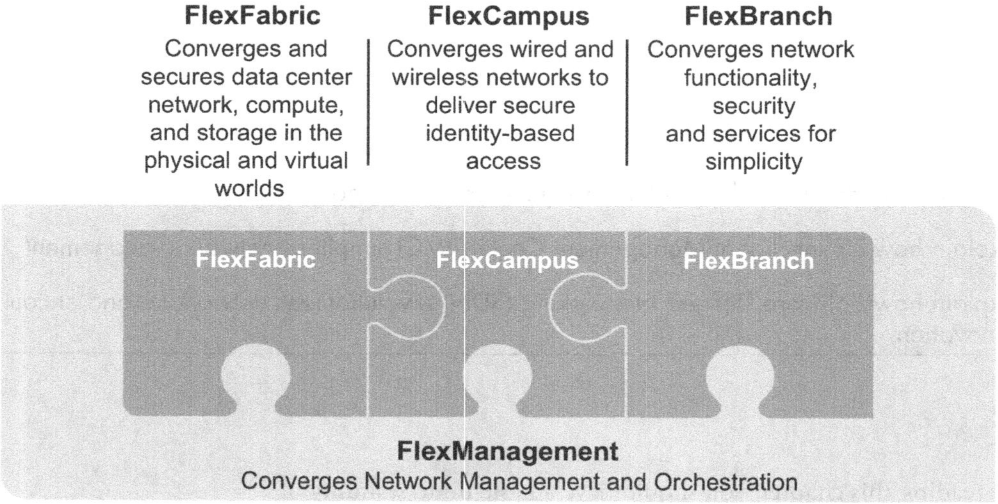


Figure 1-1: HP FlexNetwork architecture overview

## HP FlexFabric

HP FlexFabric solutions form the network foundation of an agile, automated, and secure data center. HP data center networking solutions connect servers to a virtualized, high-performance, low-latency network that consolidates multiple protocols into a single fabric to significantly lower network complexity and cost. This unique, wire-once approach will enable businesses to combine Ethernet and storage networks onto one converged fabric that can easily adapt to changing workloads.

## HP FlexCampus

HP FlexCampus solutions create an optimized, converged campus network that unifies wired and wireless access and supports multimedia applications. HP campus networking solutions are based on advanced two-tier (or sometimes three-tier) architectures that improve the performance of media-rich collaboration applications by reducing latency and accelerating network throughput as a whole. You will learn more about two-tier and three-tier architectures later in this module.

### HP FlexBranch

The branch office plays an important role in an organization's ability to attract, service, and retain customers and to increase revenue. However, branch users often struggle with limited or slow network services. HP FlexBranch solutions ensure that branch users receive the same, satisfying network experience as users at the main office.

### HP FlexManagement

You can manage the entire HP networking solution—from the data center to the main campus to each branch—from HP IMC.

IMC version is a centralized network management platform that allows you to manage both your physical and virtual networks. It uses SNMP among other protocols to discover, monitor, and manage devices. IMC integrates fault management, element configuration, and network monitoring from a central vantage point.

HP IMC supports both physical and virtual device management. It also supports both HP and third-party network devices. In fact, IMC supports thousands of network devices from dozens of vendors, including Cisco.

IMC has a modular architecture, which allows you to extend its capabilities in areas such as Bring-Your-Own Device (BYOD) management, wireless and wired user management, server health monitoring, and many others.

You can learn more about IMC in other HP training.

## HP data center solutions

As Figure 1-2 shows, HP offers data center solutions that meet organizations' requirements for modern data centers:

- High scalability and density
- High-performance computing
- Resiliency and high availability
- Large Layer 2 networks for virtualized environments
- Control over and visibility into virtual networks
- Network and storage convergence
- Network automation
- Comprehensive switching and routing features

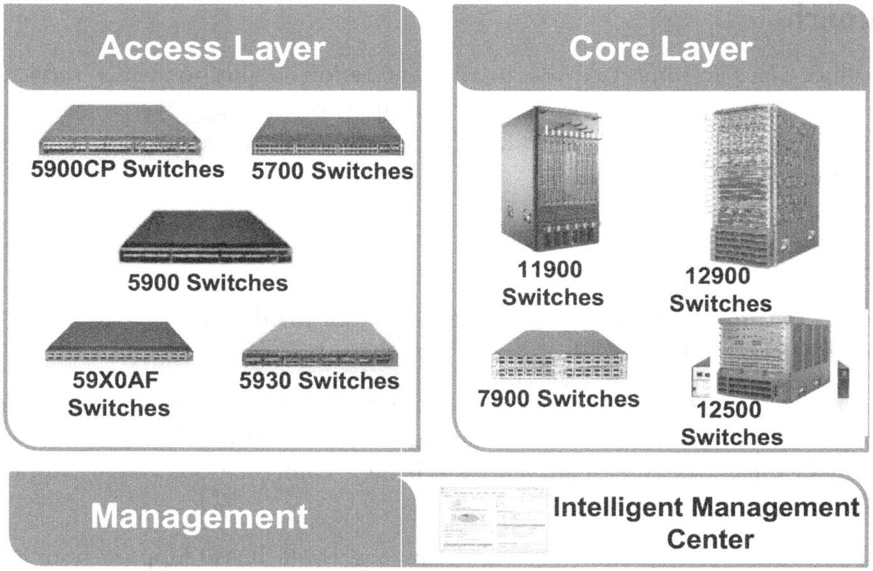


Figure 1-2: HP data center solutions

HP offers a range of data center switches with a high density of 10 GbE, 40 GbE, or even 100 GbE ports. Figure 1-2 shows some of the specific switches that HP offers for the access layer and core layer. (For a complete list of data center solutions, visit the HP networking website at www.hp.com/networking.)

All of the data center switches provide:

- Enhanced resiliency
- Visibility and control over connections to virtual machines (VMs)
- Converged data and storage
- Technologies for large Layer 2 networks and multipathing

In this study guide, you will learn how HP IRF can provide resiliency and other enhancements at any layer of the network. However, this course does not focus on managing connections to VMs, converged data, and storage, or the advanced networking technologies (such as Transparent Interconnection of Lots of Links [TRILL], DataCenter Bridging [DCB], and Shortest Path Bridging Mac-in-Mac Mode [SPBM]) that are used to implement data center solutions. If you are interested in learning more about these technologies, visit the ExpertOne website (www.hp.com/expertone).

## Access layer

HP 5700 Series switches are designed for companies that need converged data storage, technologies for large Layer 2 networks, multipathing, and enhanced resiliency but do not need complete

Layer 3 functionality. The 5700 Series switches provide light Layer 3 functionality and include options for 1 G and 10 G access connectivity.

HP 5900 Series switches provide full Layer 2 through 4 capabilities and support Fibre Channel (FC) over Ethernet (FCoE) for LAN/SAN convergence. You also have a choice between 1GbE switches and 10GbE switches.

The HP 5900AF Switch Series provides full Layer 2 through 4 capabilities and support FCoE for LAN/SAN convergence. The series includes various models, including 48 1 Gbps port versions and 48 10Gbps switch port versions, with 4 × 40 Gbps uplink ports. There is also a model with 48 1/10 Gbps ports, with 4 × 40 Gbps uplink connections. The 1/10 Gbps port version is especially convenient for data centers which are migrating servers from 1 to 10 Gbps interfaces.

The 5900CP Series includes top of rack (ToR) switches that support LAN/SAN convergence. The "CP" in the 5900CP model actually stands for Converged Ports: both FCoE and native FC are supported. With native FC ports you can connect FC storage directly to the same switch as servers. The module installed in each port determines whether that port functions as a 10 Gbps FCoE port, or as an 8 Gbps FC port.

The HP 5930 Series includes ToR switches with a high density of 40 GbE ports. These switches can be used to terminate 40 gigabit connections from blade server enclosures or can be deployed as a distribution or aggregation layer devices to concentrate a set of HP 5900 Switch Series. Each of the 40 Gbps ports can be split out as four 10 Gbps ports with a special cable. This means that the 32 40 Gbps ports could become 128 10 Gbps ports, available in a 1U device. These switches also support technologies for virtual overlay, helping customers connect VMs across the data center transparently.

### Corelayer

The HP 12900 Switch Series is a modular data center core switch designed to support virtualized data centers and the evolving needs of private and public cloud deployments.

The 12900 switch delivers high-performance, buffering, scalability, and availability with high-density 10 GbE and 40 GbE today, as well as a readiness to provide 100 GbE in the future. It has a nonblocking, lossless Clos architecture. Traditional switch architectures using a cross-bar fabric will not scale to the high densities of 40 GbE and 100 GbE connectivity. Switches with a Clos architecture provide a high-density on nonblocking 10 GbE connections and can scale to 40 GbE and 100 GbE environments.

Ready for SDN, the switch supports full Layer 2 and Layer 3 features, including advanced features such as TRILL and IRF, which provides the ability to build large, resilient switching fabrics. The HP 12900 Switch Series also supports fully redundant and hot-swappable components to complement its other enterprise-class capabilities.

- DCB and FCoE convergence
- Enhanced modularity with control and data plane separation

**Note**

The term "Clos" is not an acronym, but rather a reference to a Clos network, which is a type of circuit switching network that was first formalized by Charles Clos.

The HP 11900 Switch Series is a high-performance, data center aggregation switch that provides a high density of 10 GbE and 40 GbE ports. HP 11900 switches are the industry's first data center aggregation switches to support mature virtualization technology such as IRF, as well as standards-based protocols such as TRILL, SPB, and OpenFlow 1.3. (OpenFlow provides support for SDN.)

With latency as low as 3 µs, the HP 11900 Switch Series can scale to 384 1 GbE/10 GbE and 64 40 GbE ports. The switch delivers up to 7.7 Tb/s switching capacity and 5.8 Bpps forwarding throughput. A broad variety of interface options is available, including GbE, 10 GbE and 40 GbE.

Ready for SDN, the switch supports full Layer 2 and Layer 3 features, including advanced features such as TRILL and IRF, which provides the ability to build large, resilient switching fabrics. The 11900 Switch Series also supports fully redundant and hot-swappable components to complement its other enterprise-class capabilities.

- High-performance Clos-based switching architecture
- Large layer 2 scaling with TRILL and HP IRF
- Feature-rich routing with IPv4/IPv6, MPLS, and QoS
- Enhanced modularity with control and data plane separation
- High 10GbE and 40 GbE density; 100 GbE ready across 7.7 Tb/s switch fabric

The HP **12500 Switch Series** delivers high performance with a modern Clos/VOQ architecture (up to 13.3 Tbps). These switches are designed for customers with large networks, who want a future-proof, nonblocking solution. You also need to consider the switch fabric and connectivity modules, which need to provide orthogonal connections between ports and the fabric to take advantage of the Clos architecture.

- Advanced architecture: mid-plane, Clos
- 13.32 Tb switching capacity
- High-density 10GbE with 288 1:1, 576 4:1 ports
- 40/100 GbE ready
- Redundant switching fabric, power supply, fan tray
- Provide high density of 10 GbE/40 GbE ports

The HP 12500 Switch Series also support HP IRF, which provides unprecedented levels of performance and high availability. And Open Application Architecture (OAA) modules, which enable flexible deployment options for new services.

In addition, these switches also have energy-efficiency features that drive down operational expenses and are ideal for organizations contemplating large-scale data center consolidations, business continuity and disaster recovery sites, metropolitan area network deployments, and other applications requiring a robust, high-performance switching platform.

The **HP 7900 Switch Series**

- Is positioned for large-scale spine-and-leaf scale out for 40 GbE/100 GbE
- Provides high density 10 GbE, 40 GbE, and 100 GbE interfaces in a smaller footprint than a traditional modular chassis.
- Has a modern Clos/VOQ architecture for high capacity (up to 9.6 Tbps) and low latency
- Supports OpenFlow

## HP Converged Campus solutions

Figure 1-3 shows some of the switches HP offers for campus networks.

Now that you have reviewed the capabilities that are required in switches at the data center, turn your attention to campus networks. What capabilities do you think are needed at the campus? Make a list of the capabilities.

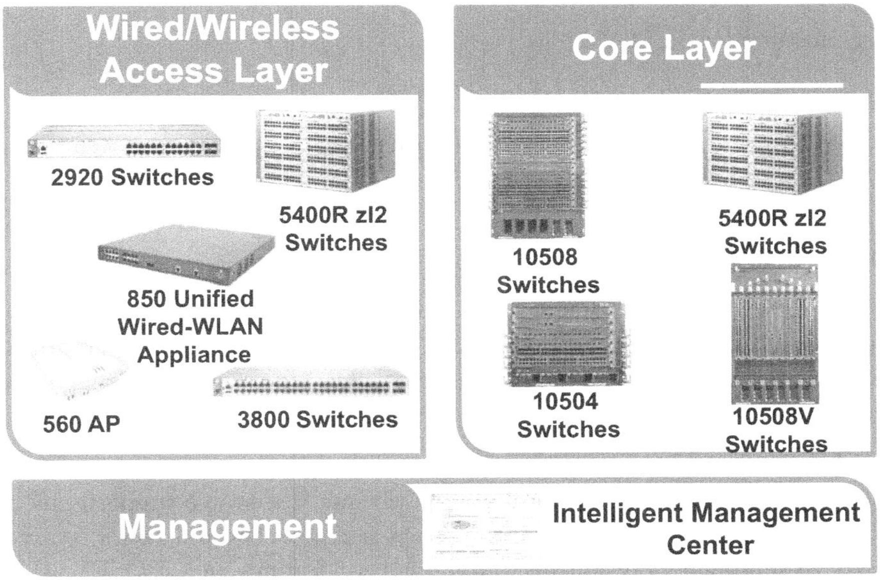


Figure 1-3: HP Converged Campus solutions

## Answers

Technologies and features that are useful in campus networking include many of the technologies covered in this study guide:

- VLANs
- STP
- Link aggregation
- Wireless technologies
- Dynamic IP routing protocols such as Open Shortest Path First (OSPF)
- Management technologies such as SNMP
- Switch virtualization technologies such as HP IRF

You might also have thought of technologies that are covered in other HP training such as:

- Port-based authentication and access control
- Quality of Service (QoS) for multimedia and unified communications (UC) traffic
- Voice VLANs and LLDP-MED
- Virtual Router Redundancy Protocol (VRRP)
- WAN technologies for the WAN edge
- Security technologies such as access control lists (ACLs)

These lists are not comprehensive. You might have listed other technologies as well.

## Access layer

The HP 5400R zl2 Switch Series, which can be deployed at the access or core, consists of advanced intelligent switches available in either a 6-slot or a 12-slot chassis. The foundation for the switch series is a purpose-built, programmable ProVision ASIC that allows the most demanding networking features, such as QoS and security, to be implemented in a scalable yet granular fashion.

It supports redundant management modules as well as redundant power supplies, providing high availability for environments that cannot tolerate down time. The switch supports any combination of 10/100, Gigabit Ethernet, and 10 Gigabit Ethernet interfaces and full Power over Ethernet Plus (PoE+) on all ports simultaneously. Further, it offers Layer 3 features such as OSPF, BGP, VRRP, and PIM, and support for emerging technologies such as SDN.

The HP 3800 Switch Series is a family of nine fully managed Gigabit Ethernet switches available in 24-port and 48-port models, with or without PoE+ and with either SFP+ or 10GBASE-T uplinks.

The HP 3800 Switch Series is built on the latest HP ProVision ASIC technology and advances in hardware engineering to deliver one of the most resilient and energy-efficient switches in the industry. Backplane stacking technology is implemented in the HP 3800 Switch Series to deliver chassis-like resiliency in a flexible, stackable form factor, providing:

- High-performance stacking—provides up to 336 Gbps of stacking throughput; each 4-port stacking module can support up to 42 Gbps in each direction per stacking port
- Ring, chain, and mesh topologies—supports up to a 10-member ring or chain and 5-member fully meshed stacks; meshed topologies offer increased resiliency vs. a standard ring
- Virtualized switching—when stacked, switches appear as a single chassis, providing simplified management (can eliminate the need for spanning tree in a switched network and reduce the number of routers in a routed network)
- Fully managed Layer 3 stackable switch series
- Low-latency, highly resilient architecture
- SFP+, 10GBASE-T, PoE+, modular stacking
- Highly resilient meshed stacking technology
- Industry-leading lifetime warranty

The HP 2920 Switch Series consists of four switches: the HP 2920-24G and 2920-24G-PoE+ Switches with 24 10/100/1000 ports and the HP 2920-48G and 2920-48G-PoE+ Switches with 48 10/100/1000 ports. Each switch has four dual-personality ports for 10/100/1000 or SFP connectivity.

In addition, the 2920 switch series supports up to four optional 10 Gigabit Ethernet (SFP+ and/or 10GBASE-T) ports, as well as a two-port stacking module. These options provide you with flexible and easy-to-deploy uplinks and stacking.

Together with static and RIP routing, robust security and management, enterprise-class features, free lifetime warranty, and free software updates, the 2920 switch series is a cost-effective, scalable solution for customers who are building high-performance networks. These switches can be deployed at the enterprise edge, in remote branch offices, and in converged networks.

- High-performance Gigabit Ethernet access switch
- Four optional 10 GbE (SFP+ and/or 10GBASE-T) ports
- Stacking capability with a total of four switches
- Layer 2 and Layer 3 plus static and RIP routing, PoE and PoE+ support
- Lifetime warranty, sFlow, ACLs, OpenFlow, and rate limiting

The HP 830 Unified Wired-WLAN Switch Series integrates both wireless controller and 1000 Mb/s Ethernet switch functions. The switch series provides 1000 Mb/s Ethernet ports, with each supporting a maximum power of up to 30 W PoE+ and IEEE 802.11a/b/g/n APs while delivering unified wired and wireless access control functions. The HP 830 24-Port PoE+ Unified Wired-WLAN Switch provides two 10GbE slots on the rear panel to relieve transmission bottleneck at the core of a WLAN network. The HP 830 Unified Wired-WLAN Switch Series are part of the HPN Mobility enterprise solution.

---

**Note:**

At the time this study guide was published, the HPE Aruba wireless solutions are not part of the HP0-Y52 exam. For more information about Aruba wireless solutions, please visit http://www.arubanetworks.com/. For more information about training, please visit http://community.arubanetworks.com/t5/Training-Certification-Career/bd-p/training-certification-career.

---

This product set provides edge-to-core unified access and consistent WLAN services to small/medium branch offices for enterprises that are deploying the HP 10500/7500 20G Unified Wired-WLAN Module at their central (or main) office.

- Unified wired and wireless services for branch offices
- System-wide approach to WLAN reliability through Wi-Fi Clear Connect
- PoE+ capability
- Built-in IEEE 802.1X and portal authentication servers
- 8-port and 24-port versions available

## Core layer

The HP 10500 Switch Series sets a new benchmark for performance, reliability, and scalability with a next-generation Clos architecture. Designed for enterprise campus core networks, the HP 10500 Switch Series enables a cloud-connected and media-rich capable infrastructure. The HP 10500 series provides industry-leading 10GbE/40 GbE port density, 3 microsecond latency and very low energy consumption. With HP IRF technology, the scalability and resiliency of the HP 10500 Switch Series can be extended and virtualized across up to four chassis with a single management interface, enabling flatter, more agile networks.

- High-speed fully distributed architecture—up to 11.52 Tbps switching capacity with current line cards and up to 13.72 Tbps switching fabric capacity with newer fabric modules provides non-blocking wire-speed 10GbE/40 GbE performance and future 100 GbE expansion capability
    - With four fabrics, the switch delivers up to 8.571 billion pps throughput
    - All switching and routing is performed in the I/O modules
    - Meets the demand of bandwidth-intensive applications today and in the future

- Scalable system design—backplane is designed for bandwidth increases; provides investment protection to support future technologies and higher-speed connectivity
- Flexible chassis selection—enables the tailoring of product selections to a budget with a choice of four chassis: 10504 switch (four open module slots), 10508 switch (eight open module slots), 10508-V switch (eight vertical open module slots), and 10512 switch (12 open module slots)
- Advanced next-generation Clos architecture
- Feature-rich, including IPv6 and MPLS functionality
- HP IRF technology virtualizes up to four chassis
- Ultra-high 1/10/40 GbE density; 100 GbE ready

HP also offers a range of APs that support 802.11n, 802.11ac, and 802.11a/b/g.

## HP branch solutions

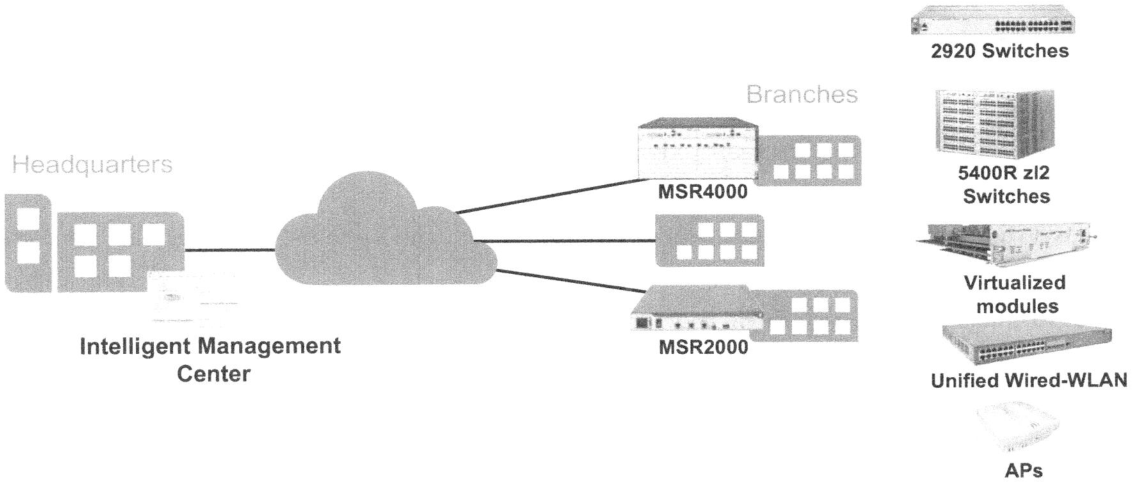


Figure 1-4: HP branch solutions

The branch office plays an important role in an organization's ability to attract and retain customers. Branch office employees—the face of the corporation—can significantly impact customer satisfaction and loyalty; particularly if supported by safe, dependable, and fast access to the corporate knowledge base (people, resources, and information); and to the productivity-enhancing tools in which the organization has invested. In addition, with today's increasing security vulnerabilities, branch office networks must be protected by comprehensive security policies and enforcement to help ensure business continuity.

Figure 1-4 shows some of the branch office solutions HP offers. The HP branch office solutions converge infrastructure and network applications to significantly improve performance, simplify deployments, centralize management, and reduce IT costs.

## WAN edge

HP offers a full range of routers—for the branch, campus, and data center. At the branch, companies typically deploy one of Multi-Service Router (MSR) Series: MSR4000, MSR3000, MSR2000, MSR1000, and MSR93RX. The course provides more details about the MSR4000 and MSR2000 Series.

The HP MSR4000 router series features a modular design that delivers unmatched application services for large branch offices, headquarters, and campus while reducing complexity, simplifying configuration, deployment, and management. The MSR4000 router series provides a full-featured, resilient routing platform, including IPv6 and MPLS, up to 20 Mpps forwarding capacity and 8 Gbps of IPsec VPN encrypted throughput. The MSR4000 router series leverages separated data and control planes, dual Main Processing Units (MPUs) and support for up to four power supplies ensuring the highest performance and reliability.

- Up to 20 Mpps forwarding performance; supports multiple concurrent services
- High reliability with separated hardware data and control planes and dual MPUs
- Open Application Platform for HP AllianceOne applications like WAN acceleration and Microsoft Lync
- Powerful aggregation capacity; integrated 10 GbE LAN, and up to 64 E1 or eight E3/T3 ports supported
- Zero-touch solution with single-pane-of-glass management

The HP MSR2000 router series features a modular design that delivers unmatched application services for small to medium sized branch offices while reducing complexity, simplifying configuration, deployment, and management. The MSR2000 router series provides a full-featured, resilient routing platform, including IPv6 and MPLS, up to 1 Mpps forwarding capacity and 400 Mbps of IPsec VPN encrypted throughput. The MSR2000 router multi-core processors ensures high performance and reliable routing, switching, SIP/voice, 3 G/4 G WAN and security services all in one device.

- Up to 1 Mpps forwarding; converged high-performance routing, switching, security, voice, mobility
- Embedded security features with hardware-based encryption, firewall, NAT, and VPNs
- Industry-leading breadth of LAN and WAN connectivity options
- No additional licensing complexity; no cost for advanced features
- Zero-touch solution, with single-pane-of-glass management

## Access layer

You have already learned about the 2920 and 5400R zl2 Switch Series and the Unified Wired-WLAN solutions and APs.

## Services modules

HP FlexBranch delivers a plug-and-play remote network that provides simplicity and convergence of functionality and services. The MSR4000 router supports OAP modules, and the 5400R switch supports zl2 service modules, which provide a platform for delivering services to the branch. HP AllianceONE partners offer applications on these modules, allowing you to integrate best-in-class applications in your router or switch.

# HP Intelligent Management Center (IMC)

As Figure 1-5 shows, IMC is a centralized network management platform that allows you to manage your physical networks (both wired and wireless) and your virtual networks. You can monitor and manage network traffic and devices from a single interface.

IMC allows you to manage a heterogeneous environment because it supports both HP and third-party network devices. In fact, IMC supports thousands of network devices from hundreds of vendors, including Cisco.

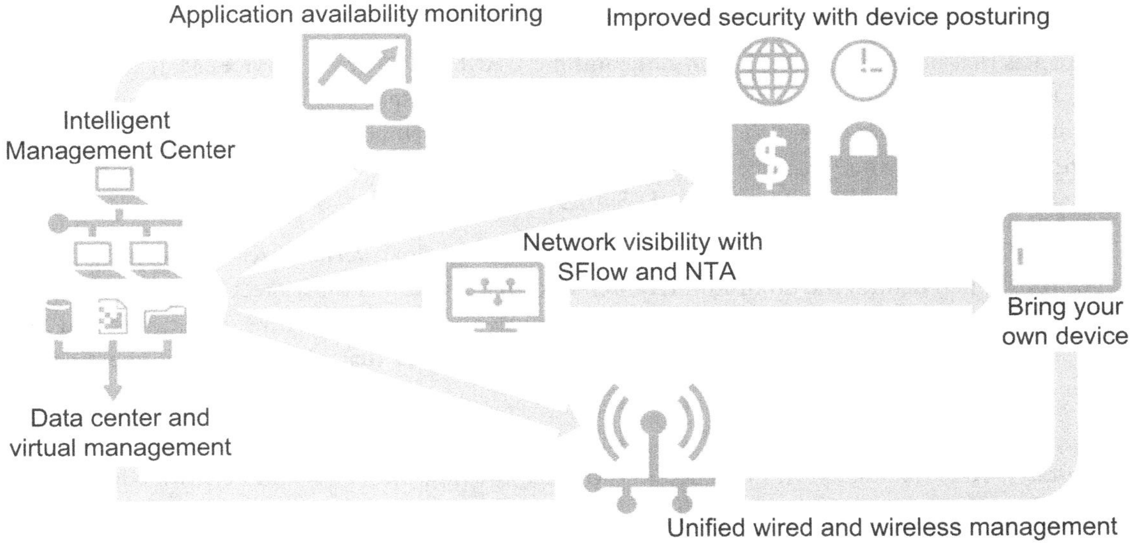


Figure 1-5: HP Intelligent Management Center (IMC)

With the IMC auto-discovery feature, you can use and locate all the devices on the network. IMC categorizes discovered devices into types of network devices (such as switches, routers, servers, APs, and desktops) and maps them on a network topology. You can view the network devices based on IP address or device type. You can also create custom views to make it easier to view and manage devices.

In addition, IMC allows you to establish baseline configurations and software images. You can compare configurations, track versions, and establish alerts if a configuration changes are made.

IMC has a modular architecture, allowing you to add management capabilities that your company needs. These modules are fully integrated into the IMC platform, allowing the modules to share information and functionality.

For example, Network Traffic Analyzer (NTA) is a graphical network-monitoring tool that provides real-time information about users and applications consuming network bandwidth. You can use NTA to plan, monitor, enhance, and troubleshoot networks, as well as identify bottlenecks and apply corrective measures for enhanced throughput.

The Wireless Services Manager unifies the management of wired and wireless networks on the IMC platform. The Wireless Services Manager adds wireless devices to the IMC network topology and allows you to configure and apply policies to these devices. You can configure WLANs and use radio frequency (RF) heat mapping to plan and adjust wireless coverage.

User Access Manager (UAM) allows you to translate business policies for access controls into network configurations. You can create service policies and then apply these service policies to user accounts. These policies can be enforced no matter where and how users connect, whether through local Ethernet connections, wirelessly through APs, or remotely through a VPN.

Companies that need a BYOD solution can combine UAM and Wireless Services Manager. The HP BYOD solution allows you to strike the right balance between users' convenience and your ability to control access to the network. The solution enables BYOD in three easy steps—onboarding, provisioning, and monitoring. You can learn more about setting up an HP BYOD solution in the *HP Unified Wired-Wireless Networks and BYOD course.*

Endpoint Admission Defense (EAD) helps protect networks from internal threats. Companies create a "posture" which defines the minimum requirements that endpoints must meet before accessing the network. For example, companies can dictate that devices and applications must be patched and running current anti-virus software before accessing the network so that hackers cannot exploit these vulnerabilities.

To help you streamline, automate, and orchestrate processes, HP provides several Virtual Application Networks (VAN) modules. VAN Connection Manager orchestrates VMs and network connections. It also allows you to apply policies with templates and manage VM migrations.

VAN Resource Manager provides drag-and-drop tools, which you can use to design network services. It helps you provision the network and accelerate the delivery of services and applications.

VAN Fabric Manager helps you manage large Layer 2 data centers, converged LAN/SAN, and connectivity across multiple data centers.

If you need more visibility into applications, servers, or VMs, you can add the Application Health Manager, Server Health Manager, and vMon.

These are just a portion of the management capabilities you can add through IMC modules. For a comprehensive list, visit the HP website at *www.hpe.com/networking*.

## HP simplifies network designs

Although this is not a network design class, you should understand at a basic level how HP networking solutions simplify network architectures. In Figure 1-6, you see the traditional three-tier architecture, which includes the access layer, distribution layer, and core layer. Designed for network environments that were common 10 years ago, this architecture no longer meets the needs of many companies. Because data centers are highly virtualized, traffic patterns have changed. In the past, a majority of network traffic was exchanged between the clients and the servers—a traffic flow called north-south. Now a lot of traffic is exchanged among VMs in data centers and campus networks—a traffic flow called east-west. The traditional three-tier network is not designed to handle this traffic.

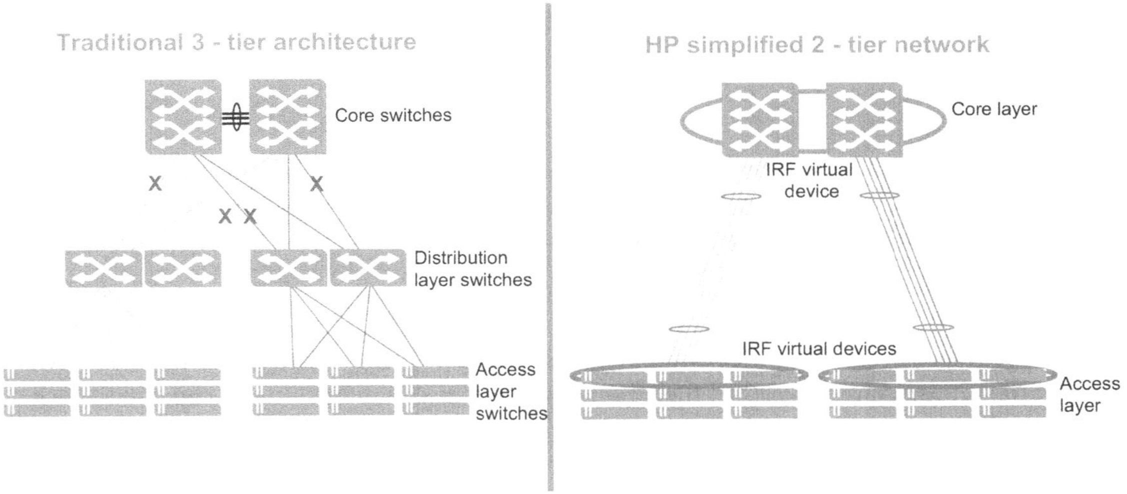


Figure 1-6: HP simplifies network designs

Further, users are demanding applications—such as multimedia, collaboration, and voice over IP (VoIP)—that require high-performance computing. The three-tier network introduces latency, disrupting these applications.

The three-tier network also relies on technologies such as Multiple Spanning Tree Protocol (MSTP) to help provide redundancy and reliability. Although MSTP is still widely in use, it fails over much too slowly to support today's networks.

HP IRF helps simplify networks by allowing you to combine multiple switches into a single, ultra-resilient virtual device. IRF virtual devices operate with incredible efficiency, providing redundancy and high availability, while eliminating the need to run outdated protocols that slow down the network and waste valuable network bandwidth. IRF virtual devices provide the high performance and rapid failover required for today's delay-sensitive, mission-critical applications.

With IRF, switches can be clustered across tiers to collapse and dramatically simplify the architecture. You can eliminate the distribution layer and unnecessary network hops. This two-tier architecture provides more direct, higher capacity connections between users and networks services.

At the core, IRF virtual devices eliminate a single point of failure while simplifying router redundancy using a single IP address rather than complex router redundancy protocols.

At the edge of the network, IRF virtual devices can interconnect hundreds of VMs, reducing latency, optimizing performance for high-volume server-to-server traffic flows and eliminating unnecessary network hops.

The two-tier topology is also connected with aggregated links—increasing bandwidth and providing redundancy. IRF can load balance traffic across all the links.

## HP leads in SDN

HP leads networking vendors in its commitment to innovative, open source-based technologies such as SDN. SDN has begun to revolutionize networking in much the way that standardized operating systems and virtualization revolutionized server environments.

In a traditional network, each network infrastructure device (such as a ProVision switch) has its own control plane and data plane. As Figure 1-7 shows, SDN, using open standard technologies such as OpenFlow, decouples the control plane from the data plane. Switches continue to forward traffic at the infrastructure layer, but they do it under the direction of an SDN controller (such as the HP VAN SDN Controller) at the control layer. The SDN controller communicates with a variety of applications at the application layer using open APIs. It submits information about the network to these applications and receives information from the applications that helps it to make forwarding decisions.

In short, the network infrastructure becomes programmable on the fly. SDN exposes the network's status and capabilities directly to applications, allowing business systems to request services from the network directly, rather than forcing network administrators to anticipate the needs and provision in advance.

HP's commitment to SDN opens up innovation:

- Network optimization
- Innovative network security
- Automated network provisioning
- Cloud solutions

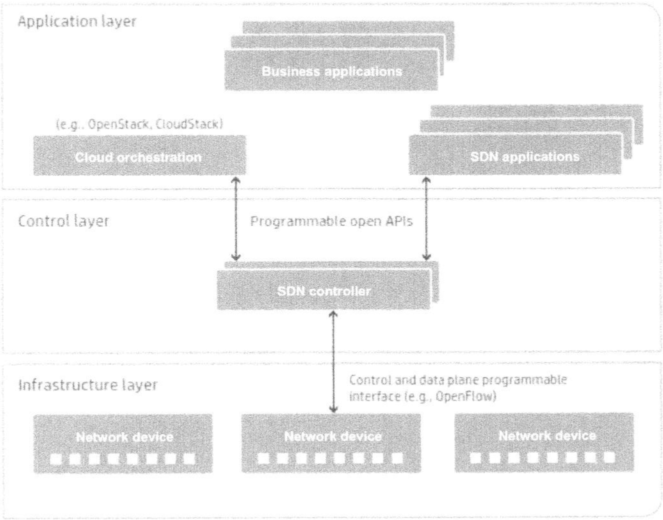


Figure 1-7: HP leads in SDN

SDN applications can help to optimize the network for particular applications, such as voice, video, and collaboration applications. They can revolutionize security by providing more information to central intrusion prevent systems (IPSs) and then drawing on the power of the complete network infrastructure to block threats. SDN applications can automate network provisioning by reprogramming the infrastructure on the fly, making SDN a key enabler for cloud solutions.

One of the most exciting features of SDN is that companies can unlock new innovations by developing new SDN apps for the application layer. HP encourages this innovation with its SDN Developer Kit, which helps SDN application developers create apps for HP SDN solutions, and its SDN App Store, which provides a marketplace for both HP SDN applications and other vendors' apps.

You can learn more about SDN in the *Creating HP Software-defined Networks* course.

## HP provides SDN solutions now

The media writes a lot about the "promise" of SDN. HP is delivering on that promise now, as Figure 1-8 shows. HP has the broadest infrastructure portfolio that's programmable for SDN. It starts with a wide array of SDN-ready switches and routers. These products span the entire network from the data center to the campus and the branch. HP also provides a consistent SDN implementation that goes from where the applications are run in the data center to where they are consumed by users in the campus and the branch.

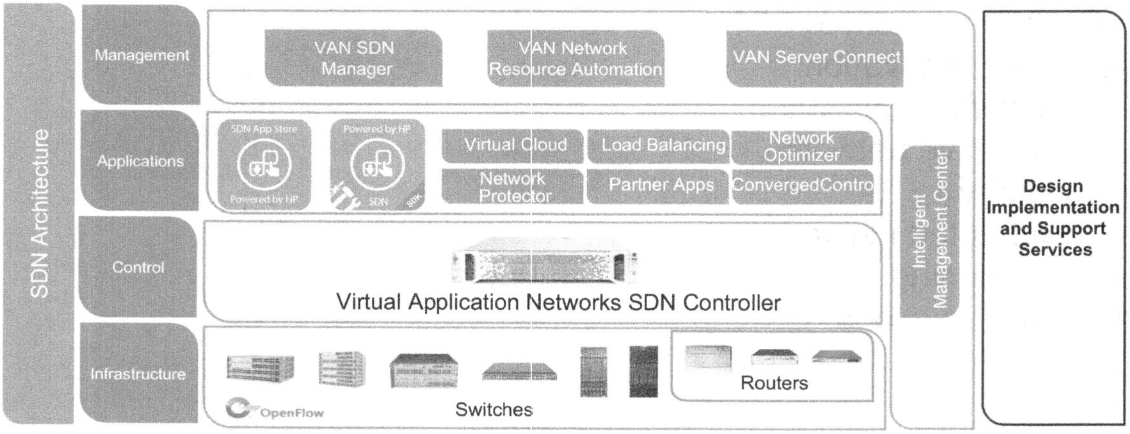


Figure 1-8: HP provides SDN solutions now

The HP VAN SDN Controller supports not only HP applications but also applications created by our alliance partners and independent software vendors. These new applications will be easily accessible to all customers through the HP SDN App Store. (To learn more about the HP SDN App Store, visit the HP networking website at www.hp.com/networking.)

## Summary

This module introduced you to some of the solutions HP provides for data center, Converged Campus, and Branch. You also learned about IMC, which allows you to manage physical (both wired and wireless) and virtual networks. Because IMC supports thousands of devices from hundreds of vendors, it allows you to manage a heterogeneous network.

You learned that HP simplifies network design, collapsing the traditional three-tier network architecture to two tiers. Finally, you learned that HP is embracing SDN to drive innovation across the network.

# 2 Basic Switch Setup

## EXAM OBJECTIVES

- ✓ Describe out-of-band management
- ✓ Navigate the HP ProVision and Comware command line interfaces (CLIs) and complete the initial setup on each
- ✓ Verify your configuration settings

## ASSUMED KNOWLEDGE

Before reading this chapter, you should have a basic understanding of:

- IP addressing
- Ethernet
- OSI model

## INTRODUCTION

This chapter describes out-of-band management and how to navigate and configure ProVision and Comware switches.

## Out-of-band management

When you initially configure a switch, you will typically use out-of-band management. For out-of-band management, you must have physical access to the switch. You connect your management station to the switch's console port using the serial cable that ships with the switch, as illustrated in Figure 2-1. This connection is dedicated to the management session. Out-of-band management does not require the switch to have network connectivity.

HP ProVision switches have one console port. Comware switches have an auxiliary (aux) port, which is used for console access. Comware routers use the console (con) port for their default console access, but most of the Comware routers also support an additional auxiliary port for a secondary CLI access. This could optionally be used for dial-in access.

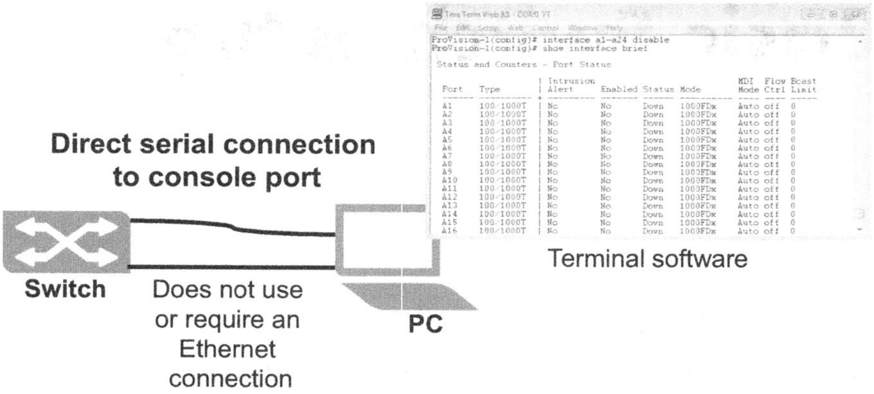


Figure 2-1: Out-of-band management

To open a management session with the switch and access the CLI, you use terminal software such as Tera Term or PuTTY, which are available as freeware, or Microsoft HyperTerminal, which is available free from Microsoft.

The switches use default settings for the terminal emulation software:

- 9600 bps
- 8 data bits
- No parity
- 1 stop bit
- No flow control

## ProVision CLI contexts

The ProVision switch CLI is organized into different contexts or levels. You can determine your current context from the switch prompt, as shown in Figure 2-2.

These contexts are described in more depth below.

```
ProVision>
```

The > symbol in the switch prompt indicates you are at the basic level. At this level you can view statistics and configuration information.

To move to the `enable` context, enter `enable`.

```
ProVision #
```

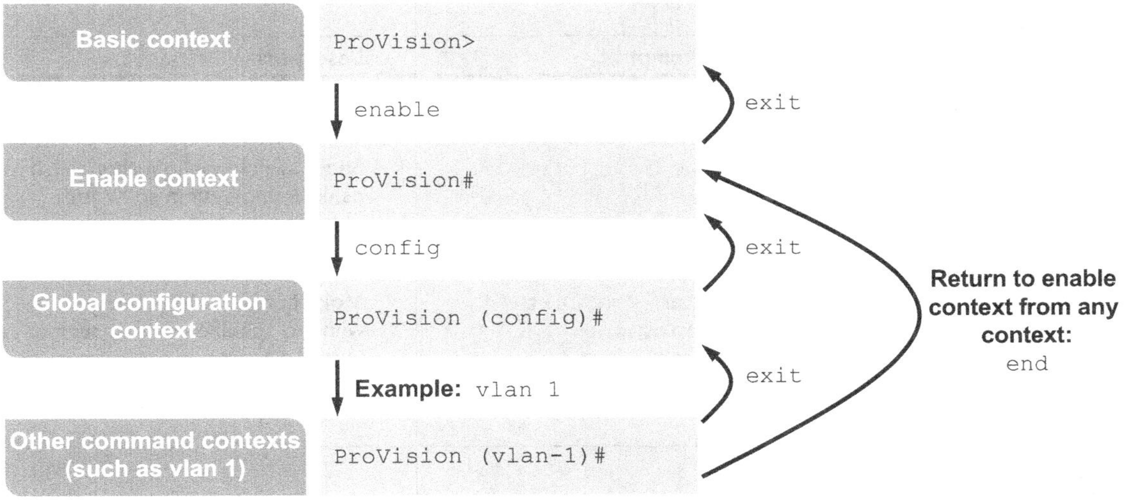


Figure 2-2: ProVision CLI contexts

The # symbol in the switch prompt appears at the enable context. From this context, you can view additional information about the switch and begin configuring the switch. For example, you can update the switch software.

To move to the global configuration context, enter `config`.

```
ProVision(config)#
```

The word "(config)" in the switch prompt indicates you are at the global configuration context. At this context, you can make configuration changes to the system's software features.

```
ProVision (<context>)
```

From the global configuration context, you can access other command contexts by entering the appropriate commands. In the switch prompt, <context> will be replaced with a context such as VLAN, a routing protocol such as Open Shortest Path First (OSPF), or port contexts:

```
Switch(vlan-1)#
```

```
Switch(ospf)#
```

You can then make configuration changes to that specific context.

Table 2-1 summarizes these contexts.

Note that at factory default settings, you have unrestricted access to the switch; no passwords are configured. To protect your network environment, you should control access, as you will learn in Chapter 3.

Table 2-1: Contexts in the ProVision CLI

| Context | CLI Prompt | Description |
|---|---|---|
| Basic | `Switch>` | View a limited number of statistics and configuration settings. |
| Enable | `Switch#` | Begin switch configuration (such as updating system software). |
| Global configuration | `Switch(config)#` | Make configuration changes to the system's software features. |
| Other command contexts | `Switch(<context>)#`<br>Examples:<br>`Switch(vlan-1)#`<br>`Switch(rip)#` | Make configuration changes within a specific context, such as to a VLAN, one or more ports, or routing protocols. |

**Note**

As you will see as you explore the ProVision and Comware CLI, each switch has its own commands. However, the ProVision CLI does accept certain Comware commands including a number of display commands, undo rather than no to negate commands, and quit.

## Comware CLI views (contexts)

The Comware switch CLI is divided into views, which are similar in concept to the ProVision contexts, as you can see in Figure 2-3. Each Comware view has a set of related commands. As with the ProVision CLI, the Comware switch prompt indicates your current view:

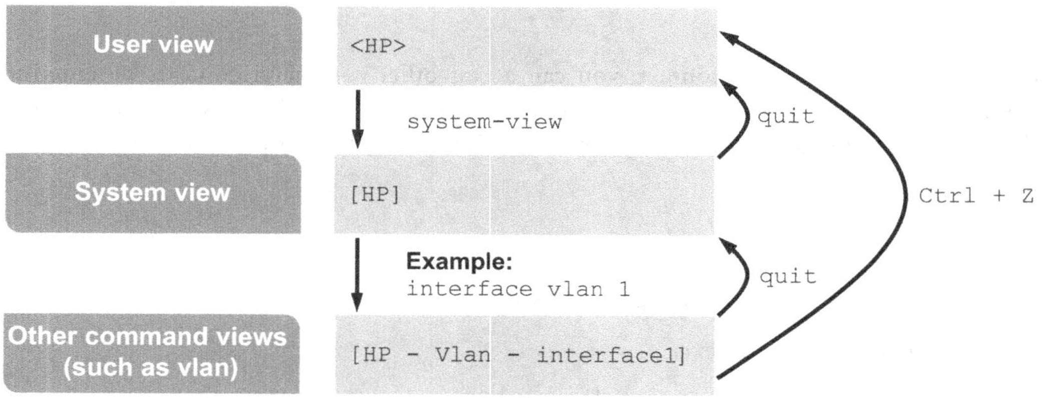


Figure 2-3: Comware CLI views (contexts)

```
<HP>
```

In the user view, the system name is enclosed in angle brackets (< >). In the user view, you can view settings, troubleshoot system problems, and manage files.

You can move to the system view by entering the command: `system-view`.

```
[HP]
```

In the system view, the system name is enclosed in square brackets ([ ]). In this view, you can make configuration changes to the switch' software. You can also access other command views.

You can return to the user view by entering `quit`.

```
[HP-<view>]
```

Other command views allow you to configure interfaces, VLANs, and switch features such as routing protocols, access control lists (ACLs), and so on.

To exit a specific view and return to the system view, enter `quit`.

After returning to the system view, you can return to the user view by pressing [Ctrl]+[Z].

Table 2-2 summarizes the main Comware views.

Table 2-2: Comware CLI views

| View name | Prompt | Description |
|---|---|---|
| User view | `<DeviceName>` | View settings, perform troubleshooting, and manage configurations and files. Move to the System View by entering `system-view`. |
| System view | `[DeviceName]` | Make changes to the switch's configuration. Also access other command views. Move to the User View by entering `quit`. |
| Other command views | `[DeviceName-<view>]` | Configure settings for specific physical or virtual interfaces, user interfaces, or other features. Move back a view by entering quit. Move to the User View by pressing [Crtl+Z]. |

## Help keys

Both ProVision and Comware CLIs provide help keys. You can use these help keys to find the appropriate command to complete a particular task or to help you complete a command. Table 2-3 provides a reference for help keys. Note that, if you enter a string and ? without a space, you see commands that start with those letters. If you enter a command and ? after a space, you see options for that command.

Table 2-3: Context-sensitive help

| CLI | Description |
|---|---|
| `? or help` | See a brief description for all available commands at your context or view. |
| `<string>?` | See commands that start with certain letters. |
| `<command> ?` | See options for the command and a brief description of each option. |
| `<string><Tab>` | Auto complete a command or a command option: Type as many characters as necessary to identify the command uniquely and press **<Tab>** |

**<Tab>** auto completes a command or a command option, which can be useful for verifying that you are entering the correct command. However, you do not need to enter the full command; the CLI will accept the command as long as you enter enough characters to identify that command uniquely.

If you have access to a ProVision or a Comware switch CLI, you should practice using these help keys.

## Accessing HP ProVision switch interfaces

On a fixed port switch, the interfaces are numbered from 1 to the maximum number of ports supported on that switch. On a modular switch, each module installed in the switch is assigned a letter, based on the slot in which it is installed. The ports in the module each have a unique number. You reference each port with the letter of the slot and the number of the port. For example, if the module is installed in the first slot and you want to configure settings on port 10, you would enter a10.

You can configure settings on a range of interfaces at the same time. To do so, you simply specify multiple IDs rather than the individual IDs in the command. Use a comma (,) to separate port IDs and a dash (-) to indicate a range. For example, enter:

```
ProVision(config)# interface 1,3-6
```

You then access a prompt that indicates the range of interfaces. For example, if you entered the command above, the prompt would reflect these interfaces:

```
ProVision(eth-1,3-6)#
```

You can then configure the settings for this range of interfaces.

Port references for ProVision switches are summarized in Figure 2-4.

- **Fixed switches:**
  - interface <1-24>
- **Modular switches:**
  - interface <slot_letter><1-24>
- **Examples:**
  - interface 19
  - interface a14
- **Configuring a range:**
  - interface 1,3-6

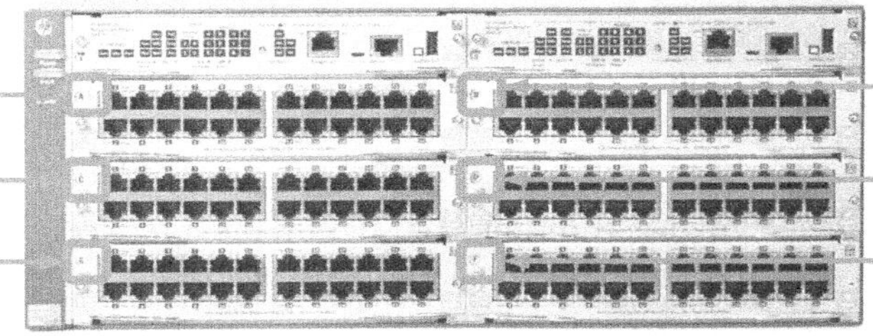


Figure 2-4: Accessing HP ProVision switch interfaces

## Accessing HP Comware switch interfaces

To access an interface context on Comware switches, you use the following command:

```
interface <type_slot_#>/<sub_slot>/<port_#>
```

The <Type> option refers to the Data Link Layer type, such as GigabitEthernet or Ten-GigabitEthernet. When you enter commands, you can use a shortcut for interface types. For example, GigabitEthernet can be represented as g, and Ten-GigabitEthernet can be represented as ten.

You replace <slot_#> with the slot in which the module or the card is installed. For both modular and fixed port switches, <slot_#> begins at 1 on the switch and <sub_slot> begins at 0. The ports are numbered from 1 to the maximum number of ports supported on that switch.

For example, interface G1/0/1 indicates the port is Gigabit Ethernet port, and it is the first port on module 1.

## Creating a range of interfaces on HP Comware switches

On Comware switches that are running the appropriate software version you can create an interface range.

> **Note**
> The software versions that support the interface range command depend on the switch model. Refer to the documentation and release notes for your switches.

When you define the interface range, you specify the range of interfaces that belong to it, and you can also give the interface range a name.

For example, you can enter:

```
[Comware] interface range name EdgePorts g1/0/1 g1/0/3 to g1/0/6
```

You can then configure settings on all of the interfaces at once. The interface range persists in the device configuration. You can access it again by specifying the range name. For example:

```
[Comware] interface range name EdgePorts
```

You can then configure additional settings on the interfaces that you previously defined in this range.

### Creating a port group on HP Comware switches

Comware switches support another way of defining settings on multiple physical interfaces at once: manual port groups. Manual port groups are available in older versions of software. You define a manual port group by name. You then add interfaces to the group. For example:

```
[Comware] port-group manual EdgePorts
[Comware-port-group-manual-EdgePorts] group-member g1/0/1 g1/0/3 to g1/0/6
```

You can then configure settings on all of the interfaces at once. The port group does not persist in the config. If you want to configure settings on a range of interfaces again, you must create the port group again.

## Completing basic configuration tasks

Now that you have a basic understanding of the ProVision and Comware CLI, you will learn more about how to use the CLI to complete the following common management tasks:

- Return HP Comware and HP ProVision switches to factory default settings
- Configure a hostname on ProVision switches and a sysname on Comware switches
- Disable and enable interfaces
- Use CLI help commands to navigate the CLI and enter commands

If possible, you should practice completing these tasks on both a Comware and a ProVision switch. You can download the HP Network Simulator, which allows you to practice entering commands in the Comware operating system. For more information about this simulator, go to the HP Network Simulator web page at http://h17007.www1.hp.com/lv/en/networking/products/simulator/index.aspx#.VgL9Vt9VhBc.

## Return HP switches to factory default settings

This section is organized into two parts: HP ProVision switches and HP Comware switches.

### HP ProVision switches

To determine whether or not a ProVision switch is at factory default settings, you first need to power on the switch and establish a connection to access the console. Then press **<Enter>** until a switch prompt is displayed.

At the manager context, enter the `show running-config` command. Remember, you do not need to memorize the syntax for this, or any other, command. However, you should be familiar with common commands and know what happens when you enter a particular command.

If the switch is at factory default settings, you should see output similar to the following:

```
Running configuration:

; J9575A Configuration Editor; Created on release #KA.15.10.0011

; Ver #03:03.1f.ef:f0

hostname "HP-3800-24G-2SFP+"

snmp-server community "public" unrestricted

oobm

   ip address dhcp-bootp

   exit

vlan 1

   name "DEFAULT_VLAN"

   ip address dhcp-bootp

   exit
```

If the output you see matches what is displayed above, the switch is already at factory default settings. You can tell this is the case by the small number of commands included in the output as well as the settings. For example, VLAN 1 still has the default name and no ip address is assigned to it. No other VLANs have been created on the switch.

If the output includes other commands, however, you can return the ProVision switch to factory default settings using the following command:

```
ProVision# erase startup-config
```

During the process of returning the switch to factory default settings, you will be prompted to confirm that you want to reboot the switch; press **y**. The switch will be rebooted.

## HP Comware switches

To determine if the Comware switch is at factory default settings, enter the `display current-configuration` command.

If the switch is at factory default settings, you should see output similar to the following:

```
<HP> display current-configuration

version 5.20, Release 1211P09

#

 sysname HP

#

 irf mac-address persistent timer

 irf auto-update enable

 undo irf link-delay

#

 domain default enable system

#

 telnet server enable

#

vlan 1

#

radius scheme system

 server-type extended

 primary authentication 127.0.0.1 1645

 primary accounting 127.0.0.1 1646

 user-name-format without-domain

#

domain system

 access-limit disable

 state active
```

```
 idle-cut disable
 self-service-url disable
#
user-group system
#
interface NULL0
#
interface GigabitEthernet1/0/1
<-output omitted->
interface Ten-GigabitEthernet1/0/28
#
 load xml-configuration
#
user-interface aux 0
user-interface vty 0 15
#
return
```

If the Comware switch is not running the default configuration, you can return it to factory default settings by entering:

```
<HP> reset saved-configuration
```

When the following prompt is displayed, press **y** and then <**Enter**>.

```
The saved configuration file will be erased. Are you sure? [Y/N]: Y
Configuration file in flash is being cleared.
Please wait ...
MainBoard:
Configuration file is cleared.
```

You then need to reboot the switch, using the `reboot` command as shown below. If a prompt is displayed, asking if you want to save the current configuration, press **n** and then press **<Enter>**. When prompted to continue the reboot, press **y** and then **<Enter>**.

```
<HP> reboot

Start to check configuration with next startup configuration file, please wait.........DONE!

 This command will reboot the device. Current configuration will be lost, save current configuration? [Y/N]:n

 This command will reboot the device. Continue? [Y/N]:y

#May  1 14:00:07:594 2000 Comware DEVM/1/REBOOT:

 Reboot device by command.

%May  1 14:00:07:694 2000 Comware DEVM/5/SYSTEM_REBOOT: System is rebooting now.

Starting......
```

Watch as the switch reboots. Press **<Ctrl+C>** when you see the following output:

```
System is starting...

Startup configuration file does not exist.

It will take a long time to get configuration file, please wait...

Press CTRL_C to break
```

## Configure a hostname or sysname

You can assign a custom name to your switch, which is called a *hostname* on ProVision switches and a *system name*, or *sysname*, on Comware switches. The hostname or sysname is displayed in the switch prompt so that you can easily identify the switch when you access the CLI.

### HP ProVision switches

To configure a hostname on a ProVision switch, simply access the global configuration mode by typing `configure` at the switch prompt. Then you can enter `hostname [name]` to make your changes. For example, if you wanted to assign a ProVision switch the hostname ProVision-1, you would enter:

```
HP-3800-24G-2SFP+ (config)# hostname ProVision-1

ProVision-1 (config)#
```

### HP Comware switches

Custom names for Comware switches are called system names, or sysname. To assign a system name, first access the system view. From there, enter the `sysname [name]` command. If you wanted to assign the switch a sysname of Comware-1, you would enter:

```
[HP] sysname Comware-1

[Comware-1]
```

## Disable and enable interfaces

By default, all interfaces on ProVision and Comware switches are enabled. You may want to disable all *unused* ports. Disabling ports can strengthen security, preventing anyone from connecting unauthorized devices to the network. It can also help prevent Layer 2 loops.

### HP ProVision switches

To disable ports—for example, ports 2 to 18 and ports 20 to 26—on a ProVision switch, enter the following command:

```
ProVision(config)# interface 2-18,20-26 disable
```

If you enter this command on a switch that has a total of 26 ports, only interfaces 1 and 19 will be enabled.

To enable an interface, enter the following command:

```
ProVision(config)# interface 2 enable
```

To check which interfaces are disabled or enabled on a Provision switch, enter the following command:

```
ProVision(config)# show interfaces brief
```

The output you see will look similar to the following:

```
 Status and Counters - Port Status

                | Intrusion                              MDI  Flow Bcast
Port  Type      | Alert     Enabled  Status Mode     Mode Ctrl Limit
----- --------- + --------- -------- ------ -------- ---- ---- -----
   1  100/1000T | No        Yes      Down   1000FDx  Auto off  0
   2  100/1000T | No        No       Down   1000FDx  Auto off  0
<-output omitted->
   19 100/1000T | No        Yes      Up     1000FDx  Auto off  0
 <-output omitted->
```

As you can see, ports 1 and 19 are enabled, but port 2 is disabled.

You may also want to add descriptions to interfaces so that you can remember how they are used. For example, you can enter a description that indicates the device to which the interface is connected. In the example that follows, the description indicates interface 1 is connected to a server and interface 19 is connected to a switch with the sysname Comware-1:

```
ProVision(config)# interface 1 name Server

ProVision(config)# interface 19 name Comware-1
```

## HP Comware switches

To disable all unused ports on a switch, you may want to first create a port group and add all unused ports into the group. Once all the unused ports are in the group, you can shut down the ports and return to the system-view. For example, if you wanted to disable all 1 GB and 10 GB ports to the group, called myports, you would enter:

```
[Comware] port-group manual myports

[Comware-port-group-manual-myports] group-member g1/0/1 to ten1/0/28

[Comware-port-group-manual-myports] shutdown

[Comware-port-group-manual-myports] quit
```

If you wanted to disable just one port, you would enter:

```
[Comware] interface g1/0/19

[Comware-GigabitEthernet1/0/19] shutdown
```

To check which interfaces are disabled or enabled on a Comware switch, enter the following command:

```
[Comware] display interface brief
```

The output should be similar to the following:

```
The brief information of interface(s) under route mode:

Link: ADM - administratively down; Stby - standby

<-output omitted->

Speed or Duplex: (a)/A - auto; H - half; F - full

Type: A - access; T - trunk; H - hybrid

Interface           Link Speed  Duplex Type PVID Description

GE1/0/1             ADM  auto   A      A    1

<-output omitted->
```

```
GE1/0/19          UP    1G(a)  F(a)    A    1    Comware-2
GE1/0/20          ADM   auto   A       A    1
GE1/0/21          UP    1G(a)  F(a)    A    1    ProVision-1
<-output omitted->
```

Ports that have been disabled should be administratively down (ADM or ADM DOWN).

To enable ports, enter the `undo shutdown` command at the appropriate interface context.

```
[Comware] interface g1/0/19
[Comware-GigabitEthernet1/0/19] undo shutdown
```

When you enable ports, you may want to add descriptions to help you remember how the port is being used. For example, if you wanted to enable port g1/0/19, you might enter the following description:

```
[Comware-GigabitEthernet1/0/19] description Connection to Comware-2
[Comware-GigabitEthernet1/0/19] quit
```

## Practice using CLI help commands

Help commands can provide you with a list of:

- Commands available at a certain context in the CLI
- Commands that begin with a certain character string
- Keywords associated with a particular command.

### HP ProVision switches

To see which commands are available at the enable context `(ProVision#)` in the ProVision CLI, enter the ? command:

```
ProVisicn# ?
```

Some important commands available at the enable context include:

- `show`, which enables you to examine current configuration parameters
- `copy`, which enables you to back up the switch configuration
- `ping` and `traceroute`, which are connectivity test tools

To list the parameters available for the `show` command, enter:

```
ProVision# show ?
```

You can scroll through the output or abort it by pressing <Ctrl+C>.

To see what options are available for the `show cpu` command, enter:

```
ProVision# show cpu ?
slot          Display module CPU statistics.
<1-300>    Time (seconds) over which to average CPU utilization.
<cr>
```

Notice the `<cr>` at the end. This means that you can execute the command without supplying any further parameters.

### HP Comware switches

Use the "?" command to display the commands available at the user view (`<Comware>`) in the Comware CLI:

```
<Comware> ?
```

Some important commands available at this level include:

- `display`, which enables you to examine current configuration parameters
- `save`, which enables you to back up a switch configuration
- `ping` and `traceroute`, which are connectivity test tools

To list the parameters available for the `display` command, enter:

```
<Comware> display ?
```

To list the `display` command parameters that begin with the letter "a" enter:

```
[Comware] display a?
```

## Save configuration settings

When you make configuration changes to the switch, you should save the changes to the startup-config file. If you do not save these changes, they will be lost if the switch is rebooted.

### HP ProVision switches

To save the configuration on a ProVision switch, enter the following command:

```
ProVision(config)# write memory
```

### HP Comware switches

To save the configuration on a Comware switch, enter:

```
[Comware] save
```

When prompted if you are sure you want to save the config, press **y**. When prompted to input a filename, accept the default and press **<Enter>**.

```
The current configuration will be written to the device. Are you sure?
[Y/N]: y

Please input they file name(*.cfg)[flash:/config.cfg]

(To leave the existing filename unchanged, press the enter key): <Enter>

Validating file. Please wait....

Saved the current configuration to mainboard device successfully
```

## View commands previously executed on the switch

You can view the commands that have been entered on both the ProVision and Comware CLI. On a ProVision switch, enter:

```
ProVision-1# show history
```

You will see output similar to the following:

```
6       configure

5       hostname ProVision-1

4       show interfaces brief

3       interface 2-18,20-26 disable

2       interface 1 name Server

1       interface 19 name Comware-1
```

As the example output shows, the administrator moved to the global configuration mode, configured the hostname, view the interfaces, disabled interfaces 2–18 and 20–26, and configured descriptions for interfaces 1 and 19.

On a Comware switch, enter:

```
[Comware-1] display history
```

You will see output such as:

```
system-view

  sysname Comware-1

  port-group manual myports
```

```
group-member g1/0/1 to ten1/0/28
shutdown
quit
```

As this example output shows, the administrator moved to the System view, configured a sysname, created a port group, added interfaces to the port group, shut down the interfaces assigned to the port group, and moved back to the System view.

## Turn off messages that display in the HP Comware CLI

As you practice using the Comware CLI, you might notice that the switch displays messages to the CLI terminal as events occur. Although these messages are helpful in understanding switch operations, you may want to stop these messages from displaying while you are configuring the switch. You can turn off these messages by entering the `undo terminal monitor` command at the User view:

```
[Comware] quit
<Comware> undo terminal monitor
<Comware> system-view
```

## Access other management interfaces on HP switches

The CLI is the primary management interface for both ProVision and Comware switches. This study guide focuses on the CLI.

ProVision switches also support a Menu interface and a web browser interface. Some Comware switches support a web browser interface. The Menu interface can be helpful for new administrators who need to configure basic settings for a new switch.

To access the Menu interface, enter the menu command in the ProVision CLI:

```
ProVision# menu
```

The Menu interface is displayed as shown below:

```
ProVision                                          9-Oct-2014  14:29:21
===================- CONSOLE - MANAGER MODE -====================
                         Main Menu
   1. Status and Counters...
   2. Switch Configuration...
   3. Console Passwords...
   4. Event Log
```

```
   5. Command Line (CLI)

   6. Reboot Switch

   7. Download OS

   8. Run Setup

   9. Stacking...

   0. Logout

Provides the menu to display configuration, status, and counters.

To select menu item, press item number, or highlight item and press <Enter>.
```

When using the menu, you can use the <Up> and <Down> arrow keys to make a selection or just type the number of the selection.

For example, you might want to select **1. Status and Counters,** and then **1. General System Information**.

You can then see what operating system version of code is the switch is running and determine the CPU utilization of the switch. To exit the Menu interface, you simply enter the option for Back until you reach the main screen. Then enter the option for returning to the Main Menu (0) and logging out (0).

## Learning check 1

You will now review what you have learned about the ProVision and Comware CLIs. As you prepare to take the HP0-Y52 exam, you should use this and other learning checks in this book to ensure that you have a thorough understanding of the concepts being tested.

1. In the Comware CLI, which feature did you use to help you disable multiple interfaces more easily?

   How might you use this feature to configure other settings on Comware switches?

2. Which type of switch uses sysname: ProVision or Comware?

   What benefit is there to using hostnames and sysnames on switches?

3. How can you determine if a port is enabled in the ProVision CLI?

4. How can you determine if a port is disabled in the Comware CLI?

5. Which help commands can you use on both the Comware CLI and the ProVision CLI?

6. How do you return a ProVision switch back to factory default settings?

## Answers to learning check 1

1. In the Comware CLI, which feature did you use to help you disable multiple interfaces more easily?

    You used a manual port group, which lets you assign interfaces to a group and then configure settings on all interfaces in that group. The port group is not added to the configuration, so you need to create a new one whenever you want to configure multiple settings on a switch at once.

    How might you use this feature to configure other settings on Comware switches?

    You could use port groups to configure any settings that are common to many interfaces. For example, you might configure many edge ports in the same VLAN and with other edge port settings.

2. Which type of switch uses sysname: ProVision or Comware?

    Comware

    What benefit is there to using hostnames and sysnames on switches?

    To better organize the network and make it easier to identify and configure switches and connections.

3. How can you determine if a port is enabled in the ProVision CLI?

    You can use the `show interfaces brief` command to view the port status output. In the status column, there will be a Yes if the port is enabled.

4. How can you determine if a port is disabled in the Comware CLI?

    You can enter the `display interface brief` command. A table of interface information will appear. If a port's description in the Link column is ADM or DOWN, it is disabled.

5. Which help commands can you use on both the Comware CLI and the ProVision CLI?

    You can use ? and <Tab> for help in both CLIs.

6. How do you return a ProVision switch back to factory default settings?

    Enter `erase startup-config`. When you are prompted to confirm that you want to reboot the switch, press **y**. The switch will be rebooted.

## Link Layer Discovery Protocol (LLDP)

In a heterogeneous network, devices from different vendors need to be able to discover one another and exchange configuration information. To enable this exchange of information, the Internet Engineering Task Force (IETF) defined LLDP in IEEE 802.1AB. The protocol operates at the Data Link layer, enabling directly connected devices to exchange information.

With LLDP, devices exchange local device information such as its major functions, management IP address, device ID, and port ID. Each device sends this information as type, length, and value (TLV) in LLDP data units (LLDPDUs) to directly connected devices. At the same time, the device receives LLDPDUs from neighbors that support LLDP. The local device saves the information it receives in a standard management information base (MIB). Simple Network Management Protocol (SNMP) programs, such as HP Intelligent Management Center (IMC), can use the LLDP information stored in MIBs to quickly detect Layer 2 network topology changes and identify each change. This information is summarized in Figure 2-5.

All current HP ProVision switches, Comware switches, Comware routers, and access points support LLDP.

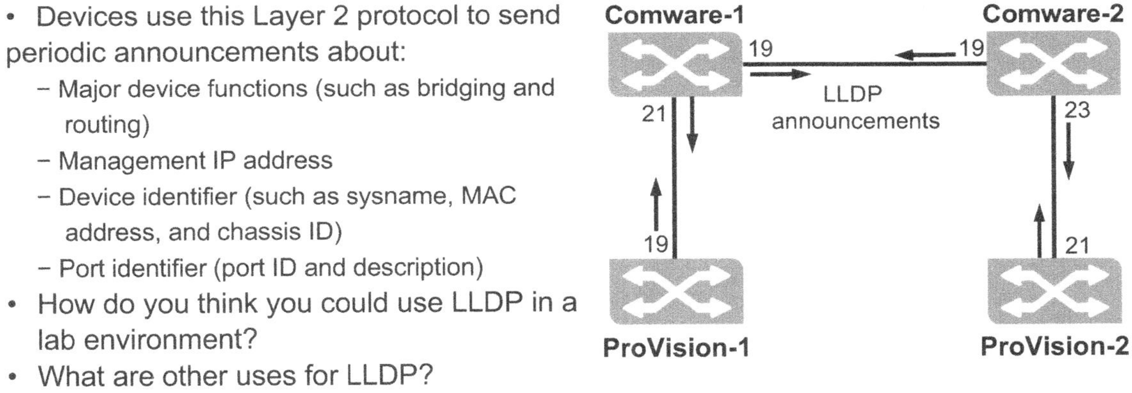


Figure 2-5: Link Layer Discovery Protocol (LLDP)

### Questions

1. How do you think you could use LLDP in a lab environment?

2. What are other uses for LLDP?

## Answers

1. How do you think you could use LLDP in a lab environment?

   You can use LLDP information to determine how devices are connected in the lab topology, which is particularly important when you do not have physical access to the switches. You can verify connectivity and check the port connections and other basic configuration settings (such as IP address).

2. What are other uses for LLDP?

   LLDP can be used for troubleshooting and to locate devices on the network and gather information about those devices. SNMP programs and other applications can use LLDP to discover devices on the network and create a Layer 2 topology of the environment.

## Assign IP addresses to switches

On HP switches, you typically configure IP addresses on a virtual LAN (VLAN) basis. The switches have a default VLAN, VLAN 1, which is often used to manage the switches.

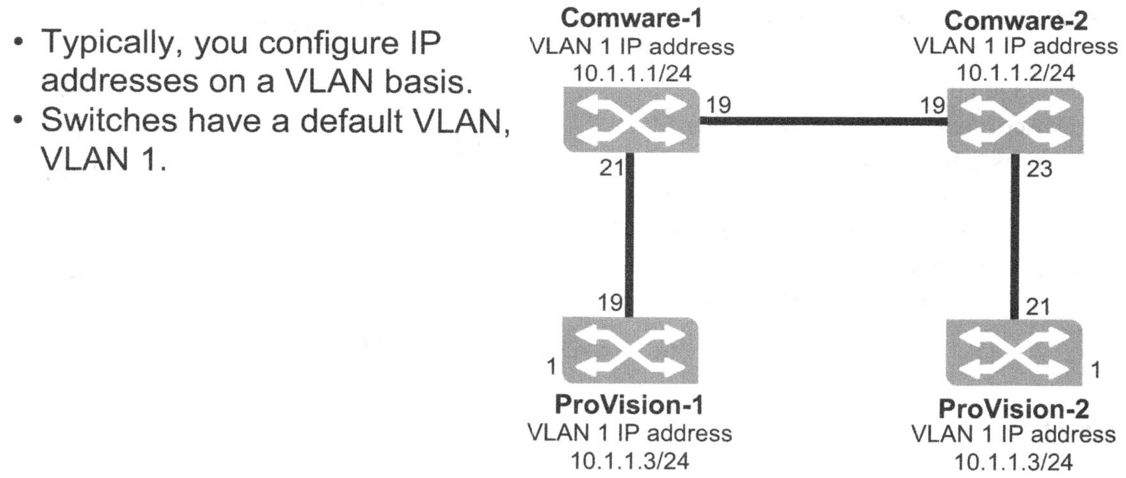


Figure 2-6: Assigning IP addresses to switches

HP ExpertOne If you need to know more about IP addressing, access the HP Aurasma application on your smartphone and use the application to focus on Figure 2-6. This action will launch a video about IP addressing. (Note that this study guide uses IPv4 addresses.)

You will learn more about using VLANs later in this study guide.

## Configuring IP addresses on switches

### HP ProVision switches

To manually configure an IP address on a ProVision switch, first access the terminal session. From there, you can access the global configuration mode context, move to the desired VLAN context, and assign that VLAN an IP address. The following commands, for example, assign VLAN 1 the IP address 10.1.1.3 255.255.255.0.

```
ProVision# configure
ProVision(config)# vlan 1 ip address 10.1.1.3/24
ProVision(config)# exit
```

To verify the configuration, you enter:

```
ProVision# show ip
```

The output will look similar to the following:

```
Internet (IP) Service
 IP Routing : Disabled
 Default Gateway :
 Default TTL     : 64
 Arp Age         : 20
 Domain Suffix   :
 DNS server      :
             |                                           Proxy ARP
VLAN         | IP Config  IP Address  Subnet Mask     Std  Local
------------ + --------- ---------- -------------   ----------
DEFAULT_VLAN | Manual     10.1.1.3    255.255.255.0   No   No
```

#### Using DHCP to assign IP address on ProVision switches

Rather than manually assigning the VLAN interface on a ProVision switch, you can use a Dynamic Host Configuration Protocol (DHCP) server to assign the interface an IP address. By default, ProVision switches are configured to accept an IP address from a DHCP server for VLAN 1, which you might have noticed when reviewing the default configuration:

```
ProVision# show running-config
Running configuration:
; J9575A Configuration Editor; Created on release #KA.15.10.0011
; Ver #03:03.1f.ef:f0
hostname "HP-3800-24G-2SFP+"
snmp-server community "public" unrestricted
oobm
   ip address dhcp-bootp
   exit
vlan 1
   name "DEFAULT_VLAN"
   ip address dhcp-bootp
   exit
```

When a ProVision switch that still has this default setting for VLAN 1 is booted, it immediately begins sending DHCP request packets on the network. You can configure your DHCP server to send switches an IP address (using the DHCP offer packet). This process is illustrated in Figure 2-7.

You should configure the DHCP server to send the ProVision switches a particular IP address so that you know what that IP address is and the switch always receives the same IP address.

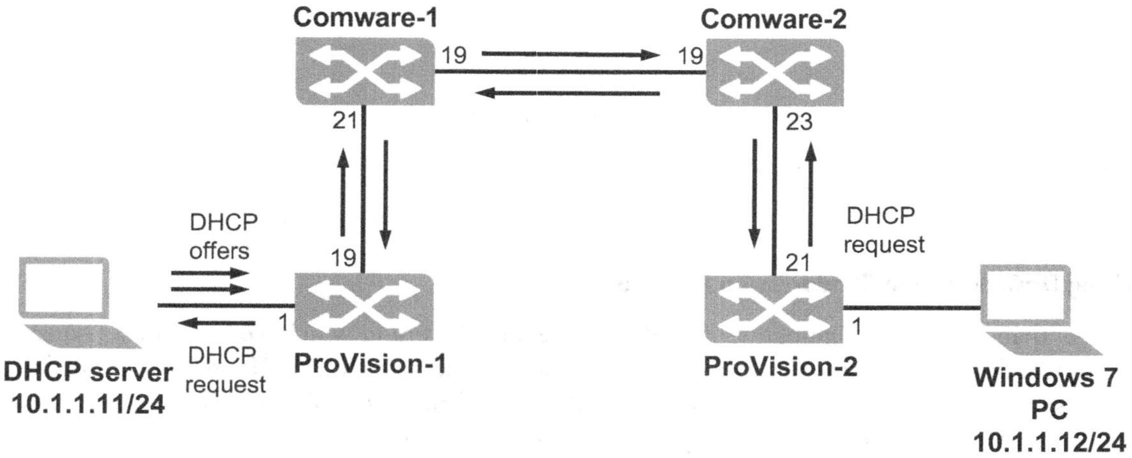


Figure 2-7: Assign IP addresses to switches

HP ExpertOne You can learn more about IP addressing by accessing the HP Aurasma application on your smartphone and focusing on Figure 2-7. This action will launch a video that covers more advanced IP addressing concepts.

Because the ProVision switches support DHCP assignment of IP addresses on VLAN 1, this gives you another option for initially accessing and configuring the switch. You can use out-of-band management, which you will learn more about in the next chapter.

If the switch does not receive a reply to its DHCP requests, it continues to periodically send request packets, but with decreasing frequency.

### HP Comware switches

To configure an IP address on a Comware switch, you move to the system view, then to the view for the appropriate VLAN interface, and enter the `IP address` command. In the following example, VLAN 1 is assigned IP address 10.1.1.1 with a 24-bit mask.

```
<Comware> system-view

System View: return to User View with Ctrl+Z.

[Comware] interface vlan 1

[Comware-Vlan-interface1] ip address 10.1.1.1 24
```

To verify the configuration, enter the `display this` command. You can use this command at any CLI context to display the commands that apply to that particular context.

```
[Comware-Vlan-interface1] display this
```

The output should look similar to the following:

```
interface Vlan-interface1

 ip address 10.1.1.1 255.255.255.0

#

return
```

You can also verify the VLAN interface is up, using the following command:

```
 [Comware] display interface brief

The brief information of interface(s) under route mode:

Interface     Link     Protocol-link  Protocol type    Main IP

NULL0         UP       UP(spoofing)   NULL             --

Vlan1         UP       UP             ETHERNET         10.1.1.1

<-output omitted->
```

# Verify connectivity

This section explains how to verify connectivity to other network devices on both ProVision and Comware devices.

## HP ProVision switches

To verify Layer-2 connectivity on a switch, you can examine its LLDP table. On a ProVision switch, you can list the LLDP neighbors by entering:

```
ProVision(config)# show lldp info remote-device
```

You should see output similar to the following:

```
LLDP Remote Devices Information

LocalPort | ChassisId        PortId PortDescr SysName

--------- + ---------------- ------ --------- -----------------

          | d07e28-ced31e    Gig... Gigabi... Comware-2
```

You can also view more detailed information about the switch connected to a specific port, in this case port 21, by entering:

```
ProVision# show lldp info remote-device 21

LLDP Remote Device Information Detail

Local Port     : 21

ChassisType    : mac-address

ChassisId      : d07e28-ced31e

PortType       : inte...

PortId         : GigabitEthernet1/0/23

SysName        : Comware-2

System Descr : HP Comware Platform Software, Software Version 5.20, Rele...

PortDescr      : GigabitEthernet1/0/23 Interface

Pvid           : 1

System Capabilities Supported  : bridge, router

System Capabilities Enabled    : bridge, router
```

```
Remote Management Address
   Type    : ipv4
   Address : 10.1.1.2
<-output omitted->
```

To verify that a ProVision switch can ping other devices, use the `ping` command. For example, you might want to see if the ProVision switch can reach another switch that has the IP address 10.1.1.2.

```
ProVision# ping 10.1.1.2
```

If you want to examine the ARP table on the ProVision switch, use the `show arp` command. As the example output below shows, the ARP table displays the device, its IP address, and its MAC address.

```
ProVision# show arp
IP ARP table
IP Address       MAC Address        Type    Port
---------------  ----------------- ------- ----
15.255.128.1     00000c-07ac00     dynamic B1
15.255.131.19    00a0c9-b1503d     dynamic B1
15.255.133.150   000bcd-3cbeec     dynamic B1
```

You can also examine the MAC address table on ProVision switch by using the `show mac-address` command. The table displays the device, its MAC address, and its port.

## HP Comware switches

To use the LLDP table to verify Layer-2 connectivity on Comware switches, enter:

```
<Comware> display lldp neighbor-information list
```

In the example output below, you see ProVision-1 and Comware-2 as LLDP neighbors.

```
System Name  Local Interface Chassis ID     Port ID
Comware-2    GE1/0/19        d07e-28ce-d31e GigabitEthernet1/0/19
ProVision-1  GE1/0/21        d4c9-ef85-1680 19
```

To view detailed information about a switch connected to a particular port (g1/0/21 in this case), enter:

```
<Comware> display lldp neighbor-information interface g1/0/21
```

```
LLDP neighbor-information of port 21[GigabitEthernet1/0/21]:
  Neighbor index    : 1
  Update time       : 0 days,2 hours,37 minutes,0 seconds
  Chassis type      : MAC address
  Chassis ID        : d4c9-ef85-1680
  Port ID type      : Locally assigned
  Port ID           : 19
  Port description : 19
  System name           : ProVision-1
  System description : HP J9575A 3800-24G-2SFP+ Switch, revision KA.15.10.0011, ROM KA.15.09 (/ws/swbuildm/KA_rel_irvine_qaoff/code/build/tam(swbuildm_KA_rel_irvine_qaoff_rel_irvine))
  System capabilities supported : Bridge,Router
  System capabilities enabled   : Bridge
  Management address type           : ipv4
  Management address                : 10.1.1.3
  Management address interface type : IfIndex
  Management address interface ID   : Unknown
  Management address OID            : 0
  Port VLAN ID(PVID): 1
<-output omitted->
```

Notice that in the detailed information, you can see the hostname and management IP address.

Return to the Comware CLI and verify it can ping its neighbors.

```
<Comware> ping 10.1.1.2
```

To examine the ARP table on a Comware switch, use the `display arp` command. You will see output similar to the following:

```
[Switch] display arp
Type: S-Static D-Dynamic O-Openflow M-Multiport I-Invalid
10
```

```
IP address MAC address VLAN Interface Aging Type
192.168.1.1 00e0-fc01-0000 10 N/A N/A M
```

To examine the MAC address table, use the `display mac-address` command.

## Summary

Take a minute to think about the configuration settings you learned about in this chapter and write them down. You do not need to remember the exact command syntax because you know how to use help commands to discover the syntax of the commands you want to enter. Instead, focus on the task you completed. For example, you might remember that you disabled interfaces and configured IP addresses.

## Learning check 2

The following questions will help you measure your understanding of the material presented in this chapter.

1. Which commands can you use to help verify connectivity?
2. What is an ARP table?
3. How can you use the ARP table to troubleshoot?
4. On a ProVision switch, what tasks can you complete at the basic context? At the enable context?
5. On a Comware switch, what tasks can you complete at the User view? At the System view?
6. What does the port-group feature allow you to do?
7. Of the following information, which cannot be seen in the LLDP table of a switch for neighboring devices?
    a. Major device functions
    b. Current users logged in
    c. Management IP address
    d. Device identifier
    e. Local port identifier

## Answers to learning check 2

1. Which commands can you use to help verify connectivity?

   You can troubleshoot with show and display commands such as:

   show interface brief

   display interface brief

   show lldp info remote

   display lldp neighbor list

2. What is an ARP table?

   An ARP table shows the MAC address that is associated with an IP address. A switch learns the MAC addressees associated with IP addresses that it needs to process. For example, if you ping a device in the same subnet, the switch will use ARP to resolve the device's IP address. The ARP table might also show the forwarding interface for reaching the MAC address, which the switch learns from its MAC forwarding table.

3. How can you use the ARP table to troubleshoot?

   You can see whether the switch is able to resolve IP addresses. You can see whether the resolved MAC address is the device that you think owns that IP address. You can see which forwarding interface the switch has learned for a particular IP address and MAC address.

4. On a ProVision switch, what tasks can you complete at the basic context? At the enable context?

   At the basic context, you can view statistics and configure information. At the enable context, you can begin configuring the switch. You can also access the global configuration context, where you configure more settings and access other contexts.

5. On a Comware switch, what tasks can you complete at the User view? At the System view?

   The User view allows you to view settings, troubleshoot issues, and manage configurations and files. The System view allows you to make configuration changes to the switch and access other command views.

6. What does the port-group feature allow you to do?

   It allows you to apply a command to multiple Comware interfaces simultaneously.

7. Of the following information, which cannot be seen in the LLDP table of a switch for neighboring devices?

   ☑ **B**. Current users logged in is the correct answer.

   ☒ **A**, **C**, **D**, and **E** are incorrect.

# 3 Protecting Management Access

## EXAM OBJECTIVES

✓ Protect in-band and out-of-band management access to HP switches using local authentication

✓ Control the privilege level allowed to managers

## ASSUMED KNOWLEDGE

Before reading this chapter, you should be able to:

- Access the HP ProVision and Comware command line interfaces (CLIs) out of band
- Navigate the ProVision and Comware CLIs
- Complete basic management tasks on ProVision and Comware switches, including returning switches to factory default settings, disabling or enabling switch interfaces, and configuring an IP address on VLAN 1
- Verify connectivity between switches

## INTRODUCTION

This chapter explains how to control basic management access to HP switches, focusing on local authentication. You will learn how to implement simple password protection for various forms of CLI access.

## In-band management

In the previous chapter, you learned how to access a switch's CLI out-of-band, using a serial connection to the switch's console port. You will now learn about an alternative way to reach the CLI: in-band. In-band management is illustrated in Figure 3-1.

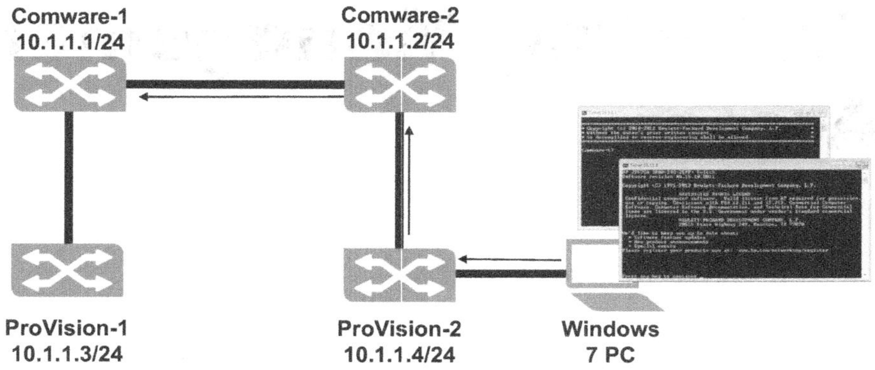


Figure 3-1: In-band management

With in-band management, your management communications run over network connections. You require IP connectivity to the networking device through a direct or an indirect Ethernet connection. To open an in-band management session to access the network device's CLI, you must use terminal emulation software, such as Putty or Tera Term. This software communicates with the switch using a network protocol:

- Telnet—Carries the terminal session in plaintext
- Secure Shell (SSH)—Establishes a secure tunnel for the session, using authentication and encryption

This chapter focuses on using Telnet to access a switch CLI, but SSH is the more secure option in a real-world environment.

To manage a switch through in-band (networked) access, you must configure the switch with an IP address and subnet mask that is reachable on your network. You might also need to configure other settings, which are described in the following pages.

## Comparing in-band and out-of-band management

Think about what you have learned about in-band management and out-of-band management. What are the advantages of using in-band management? What are the disadvantages of using in-band management?

In-band management is convenient. You can access any network device from any location that has IP connectivity to that device. You can open multiple sessions and manage multiple devices at the same time. Sometimes you need to access devices remotely so that you can respond to problems and troubleshoot in a more timely fashion, decreasing network downtime and providing better availability.

On the other hand, in-band management is less secure. Out-of-band management is inherently more secure because you have to have physical access to the device to establish a connection. As a result, you can control in-band management access by physically securing the device. In addition, the connection is dedicated to the management session so that other users cannot eavesdrop on the traffic exchanged between the device and the management station.

With in-band management, any user who can reach a network infrastructure device's IP address can attempt to log in to the device. A hacker who obtains access to infrastructure devices can collect a great deal of valuable information, cause denial of service (DoS) attacks, redirect traffic, eavesdrop, and so on. Therefore, you must be careful to secure in-band management. This is why SSH is recommended for real-world environments.

## Control access to HP ProVision switches

At factory default settings, no passwords are configured on the switch. If a user knows an IP address on the switch, that user can access the switch. You should always control who can access the switch.

### Operator and manager roles

When you protect management access to a ProVision switch, you also control the role that is assigned to a user who logs in. The user can have one of these roles:

- **Operator**: An operator has access to basic mode, which provides limited `show` commands. The operator can view only statistics and configuration information.
- **Manager**: A manager can access the enable mode and from the enable mode, move to the global configuration mode. The manager has complete read-write access to the switch. As a manager you can make configuration changes as well as view information.

If you have not set up any protections on the switch, all users log in as managers.

### Access an HP ProVision switch through a Telnet session

The Telnet server is enabled on ProVision switches by default. You do not have to enter any commands to establish a Telnet session.

### Configure local passwords

You can protect management access to the switch's CLI by configuring a manager username and password, an operator username and password, or both. When you enter the commands for setting the password, you are not required to specify a username. If you do not, the switch keeps the default user names, which are **manager** for manager and **operator** for operator. If you prefer, you can configure your own usernames.

To configure an operator password, access the ProVision switch CLI using either in-band or out-of-band management and move to the global configuration mode.

```
ProVision-1# configure
```

To configure an operator password, you use the `password operator` command. When you enter this command, you will be prompted to type the password. The password you enter is case sensitive. In the following example, the password will be set to "hp."

```
ProVision-1(config)# password operator

New password for operator: hp

Please retype new password for operator: hp

DHCP-based image file download from a TFTP server is disabled when an operator or a manager password is set.

DHCP-based config file download from a TFTP server is disabled when an operator or a manager password is set.

Autorun is disabled as operator/manager password is configured.
```

When users access the switch CLI, they will be prompted for a username. As you recall, if you do not set a username, the default username—in this case, operator—is used. When the users are prompted for a username, they can enter `operator`.

```
Username: operator
```

Users will then be prompted for a password. For the example used above, users would enter `hp`. Note that when entering the password, users will not see the letters that they type.

```
Password: hp
```

If users enter the correct password for the operator, they will be at the basic mode. From here, users can try to move to the enable mode by entering:

```
ProVision-1> enable
```

If a manager username and password have not been configured, users will move directly to the enable mode. If a manager username and password have been configured, users will be prompted to enter those credentials.

To configure a manager password, you enter the `password manager` command from the global configuration mode. As before, you will be prompted to enter the password.

```
ProVision-1(config)# password manager

New password for manager: hp

Please retype new password for manager: hp
```

When you configure an operator username and password or a manager username and password, users can no longer log directly in to the switch using a console session or a Telnet session. Instead users are prompted to enter a username and password.

- If they enter the correct credentials for the operator role, they enter the operator context. From this context, they can enter limited `show` commands.
- If users enter the correct credentials for the manager role, they enter the enable mode and receive complete read-write access to the ProVision switch.

When you configure these passwords, you can set just *one* username and password for the manager role and *one* username and password for the operator role. The passwords apply to all in-band and out-of-band management access methods.

If you want to change the manager username, you can add the `user-name` option to the `password manager` command and then specify another username. The command below changes the manager username to admin.

```
ProVision-1(config)# password manager user-name admin

New password for manager: hp

Please retype new password for manager: hp
```

When users try to log in as the manager, they will need to enter the new username you configured—in the example, admin:

```
Username: admin

Password: hp
```

If users try to enter `manager` for the username, the login will fail because the new credentials have replaced the old ones.

You can configure a different username for the operator, using the same `user-name` option:

```
ProVision-1(config)# password operator user-name miriam

New password for operator: hp

Please retype new password for operator: hp
```

## Remove the manager or operator password

You can remove the password for the operator account by entering:

```
ProVision-1(config)# no password operator

Password protection for operator will be deleted, continue [y/n]?  y
```

If the switch has a manager password but no operator password, users are still required to log in to the switch. They must enter the manager credentials. In this configuration, the switch does not provide limited operator privileges.

To remove the manager password, enter:

```
ProVision(config)# no password manager
```

```
Password protection for manager will be deleted, continue [y/n]?  y
```

When you make configuration changes to the switch, you should save the changes by entering:

```
ProVision(config)# write memory
```

## Understand where local credentials are saved

By default, the credentials are not included in the ProVision configuration. The same holds true for other credentials listed in Table 3-1. Because the credentials are not stored in the configuration, they are not included if you save the configuration to a file and download that file from the switch. In addition, if you reset the configuration to factory default settings, the passwords are still applied. You must manually remove the passwords by using the `no` form of the `password` command.

Table 3-1: Credentials not included in the configuration by default

| Credentials | Included in configuration by default? |
|---|---|
| Local manager and operator usernames and passwords | No |
| Simple Network Management Protocol (SNMP) security credentials | No |
| Usernames and passwords used for 802.1X credentials | No |
| TACACS+ encryption keys | No |
| RADIUS shared secret | No |
| SSH public keys | No |

If you want to store the credentials in the internal flash and view them in the config, you should enter the `include-credentials` command.

```
ProVision(config)# include-credentials
```

After you enter this command, the currently configured manager and operator usernames and passwords, RADIUS shared secret keys, SNMP and 802.1X authenticator (port-access) security credentials, and SSH client public-keys are stored in the running configuration. To view the currently configured security settings in the running configuration, enter:

```
ProVision# show running-config
```

You can also configure ProVision switches to authenticate management users to a RADIUS or TACACS+ server, but this setup is beyond the scope of this study guide.

## Controlling access to HP Comware switches

You will now learn about controlling management access to Comware switches.

### Privileges

As you learned, ProVision switches permit users one of the two forms of access to the CLI: manager or operator. Comware switches, on the other hand, permit users one of the *four* different privilege levels. The privilege level determines which commands the user is allowed to enter. At the default settings, the four privilege levels are the following:

- **Level 0 (Visitor)** includes commands for network diagnosis and commands for accessing an external device. Commands at this level include ping, tracert, telnet, ssh2, and certain display commands. These are read-only type commands, which cannot be saved and do not survive a device restart. Upon device restart, the commands at this level will be restored to the default settings.
- **Level 1 (Monitor)** includes commands for system maintenance and service fault diagnosis. Commands at this level include debugging, terminal, refresh, reset, send, and certain display commands. Again, command configurations cannot be saved and do not survive a device restart.
- **Level 2 (System)** includes service configuration commands, including interface, VLAN, and routing configuration commands, as well as commands for configuring services at different network levels. In fact, this level permits all configuration commands except those at the manage level.
- **Level 3 (Manage)** includes commands that influence the basic operation of the system and that configure system support modules. These commands include ones for managing the file system; configuring FTP, TFTP, and Xmodem download; managing users and privileges; and configuring system-local parameter settings (settings that are not defined by any protocols or RFCs).

A user at a certain privilege level can enter any commands at that level or a level below. For example, a user with privilege level 3 (manage) can enter all the manage, system, monitor, and visitor commands.

You can change which commands are associated with various privilege levels. The command for making those changes is a level 3 (manage) command by default.

## User interfaces

To control management access to Comware switches, you configure settings on user interfaces. Each form of access to a Comware device is associated with a user interface, as illustrated in Figure 3-2.

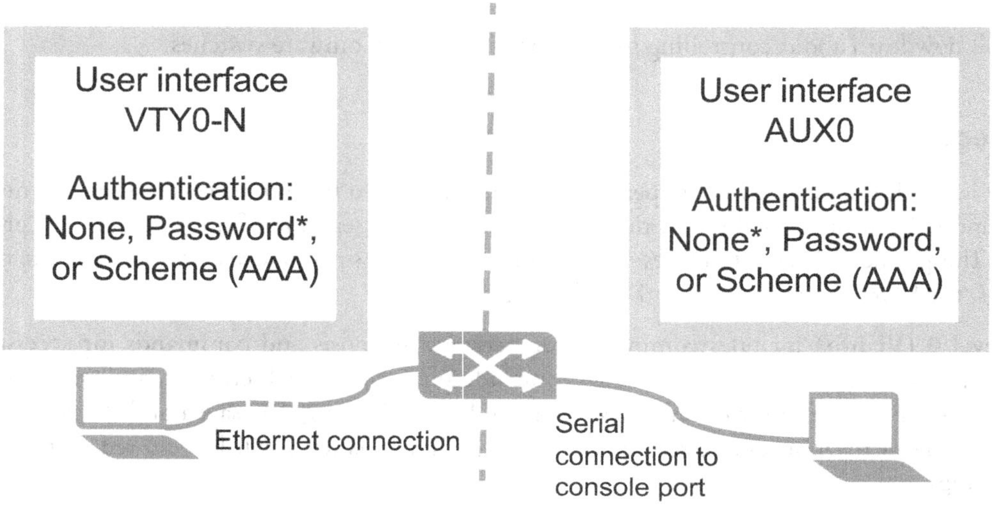


Figure 3-2: Comware: User interfaces

- **AUX user interface**—Manages and monitors users who log in through the console port (out-of-band management) on Comware switches. The type of the console port is EIA/TIA-232 DTE. The interface ID is aux0.
- **CON user interface**—Manages and monitors users who log in through the console port (out-of-band management) on Comware routers. The interface ID is con0.
- **VTY user interface**—Manages and monitors users who log in through Telnet or SSH (in-band management). A Comware switch can support more than one such session, so it has multiple VTY interfaces. The interface IDs are vty 0, vty1, and so on, to vty n (in which n is the number of sessions supported minus one). To find out the number of sessions supported on a particular switch, refer to the switch documentation to. You can also enter the `user-interface vty 0 ?` command in the Comware CLI to find the range for VTY interface IDs. The user interface (also called a *line*) allows you to manage and monitor terminal sessions between a management station and the switch. The Comware CLI has a view associated with each interface, which you access with this command: `user-interface <interfaceID>`.

You can configure various parameters for controlling the session and how users log in to it. These settings will apply whenever any user attempts to log in through the corresponding interface. For example, the settings for the aux0 user interface apply to any user attempting to access the switch CLI through the console port. The settings for the vty 0 user interface apply to any user attempting to access the switch CLI through Telnet or SSH if no other session is active. The vty 1 user interface settings apply to any user attempting Telnet or SSH access if another session is active, and so on.

Each user interface has a default privilege level:

- On the routers, con 0 has a default privilege level of 3
- On the switches, aux 0 has a default privilege level of 3
- All other user interfaces have a default privilege level of 0

You can configure multiple VTY interfaces at once by entering a range of interface IDs for the `user-interface` command. In this way, you can set up consistent settings for managing Telnet (or SSH) access, regardless of how many sessions are currently active.

Telnet access is disabled by default on Comware switches. To allow Telnet access, you must first enable the switch's Telnet server.

An important setting for user interfaces is the authentication method, which controls whether a user needs to authenticate and, if so, how. Comware switches support three authentication methods:

- **None**—The user does not need to log in. The CLI prompt appears immediately after the session connects. None is the default setting for the AUX user interface.
- **Password**—The user is prompted for a password when logging in. The switch checks the password against the password set on the interface. Password is the default setting for VTY interfaces.
- **Scheme**—The user is prompted for a username and a password when logging in. The switch checks user credentials according to global Authentication, Authorization, and Accounting (AAA) settings. These settings might direct the switch to authenticate the user locally or against a remote RADIUS or TACACS+ server.

Because Comware switches have different settings for each user interface, you can set up different methods and credentials on different interfaces. For example, you could require passwords on VTY interfaces for Telnet access but leave console access open by using the none authentication method on the aux 0 interface.

You will learn more about configuring the password and scheme methods on the next pages.

## Restricting access with passwords

You will first learn how to configure Comware switches to enforce password authentication on the VTY interfaces. As Figure 3-3 shows, when you configure this form of authentication, a user is prompted for just a password (no username). You set the password in the user interface. The user interface also controls the user's privilege level.

When you set the password for VTY interfaces, you can use the `simple` keyword, which stores the password in the config in cleartext, or the `cipher` keyword, which stores the password in an encrypted format. For real-world environments, the cipher option is more secure.

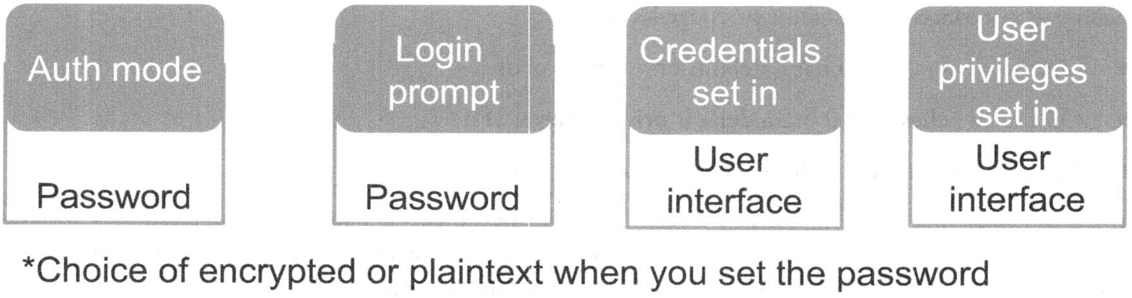


Figure 3-3: Comware: Restricting access with passwords

When you specify the password (or none) authentication method, the user interface controls the privilege level for users who successfully log in. The default privilege level for the user interface depends on the type:

- On routers, con 0 has a default privilege level of 3.
- On switches, aux 0 has a default privilege level of 3.
- All other user interfaces have a default privilege level of 0.

You can change the privilege level.

### Set up password authentication for Telnet access on Comware switches

To set up password authentication on a Comware switch, access a console session with the switch and navigate to the system view.

```
<Comware-1> system-view
```

From there, you enable the Telnet server on the switch:

```
[Comware-1] telnet server enable

% Start Telnet server
```

If you try to establish a Telnet session at this point, you will be able to make a connection, but the connection will quickly close because a password has not been configured for the VTY interfaces.

```
****************************************************************************

* Copyright (c) 2004-2009 Hangzhou H3C Tech. Co., Ltd. All rights reserved. *

* Without the owner's prior written consent,                               *

* no decompiling or reverse-engineering shall be allowed.                  *

****************************************************************************
```

```
Login password has not been set!

Connection to host lost.
```

You can configure a password for one VTY interface or a range of VTY interfaces at the same time. To access the entire range of VTY interfaces, enter:

```
[Comware-1] user-interface vty 0 15
```

The interfaces already use password authentication, but the command for setting that method is `authentication-mode password`. The example below sets the password to "hp."

```
[Comware-1-ui-vty0-15] authentication-mode password

[Comware-1-ui-vty0-15] set authentication password simple hp

[Comware-1-ui-vty0-15] quit
```

However, this configuration still does not give the user complete privileges on the switch; that is, the user cannot execute certain commands. To determine the rights that a user who is currently logged in to the switch has, you can enter the `display users` command. As you can see in the output below, VTY interface (Telnet) users have privilege level 0 after these commands are entered.

```
[Comware-1] display users

The user application information of the user interface(s):

  Idx UI      Delay    Type Userlevel

F 3   AUX 0   00:00:00      3

  29  VTY 0   00:00:08 TEL  0

Following are more details.

VTY 0    :

        Location: 10.1.1.12

 +     : Current operation user.

 F     : Current operation user work in async mode.
```

You can change the privilege level for users who log in through Telnet by configuring the privilege level for the VTY user interfaces.

For example, to set the privilege level to 2, enter:

```
[Comware-1] user-interface vty 0 15

[Comware-1-ui-vty0-15] user privilege level 2
```

This change will take affect when the user logs out of the switch and then logs in again. You can verify the change by entering the `display users` command when the user logs in to the switch again.

```
<Comware-1> display users
The user application information of the user interface(s):
  Idx UI        Delay     Type Userlevel
  3   AUX 0     00:03:02       3
 +29  VTY 0     00:00:00 TEL  2
```

When you make configuration changes to the switch, enter the following command to save the configuration: `save [force]`

## Learning check 1

Answer these questions to assess what you have learned in this chapter so far.

1. Assume that you have set only a manager password on a ProVision switch. Can users receive operator access without logging in? Explain how users log in and receive privileges with this configuration.

2. Assume that you have set only an operator password on a ProVision switch. Can users receive manager level access? Explain how users log in and receive privileges in this scenario.

3. How many manager usernames and passwords does a ProVision switch support? How many operator usernames and passwords does a ProVision switch support?

4. You logged in to a Comware switch with Telnet but could not enter some commands. Why? What privilege level did you have by default?

5. What sets the Comware privilege level for the Telnet user? How can you change this level?

6. ProVision switches do not include the manager and the operator passwords in the config—nor other credentials, passwords, and keys—unless you enter the `include-credentials` command. What is a use case for using the `include-credentials` command to add these credentials to the configuration?

## Answers to learning check 1

1. Assume that you have set only a manager password on a ProVision switch. Can users receive operator access without logging in? Explain how users log in and receive privileges in this scenario.

   No, users do not receive operator access without logging in. As soon as you set one password, all users are forced to log in. When a user successfully logs in, the user is placed in enable mode and has manager privileges. In other words, in this scenario, users are either denied access or have manager-level access.

2. Assume that you have set only an operator password on a ProVision switch. Can users receive manager level access? Explain how users log in and receive privileges in this scenario.

   Yes, users can receive manager access. In this scenario, users must log in with the operator credentials. They are then placed in operator mode. But they can receive manager-level access by entering `enable`. Because no manager password is set, the user moves to enable mode and receives full privileges without entering credentials. In other words, in this scenario, users are again either denied access or can easily obtain manager-level access.

3. How many manager usernames and passwords does a ProVision switch support? How many operator usernames and passwords does a ProVision switch support?

   A ProVision switch supports a single manager account (username and password) and a single operator account (username and password). New credentials write over old credentials.

   Note that you can use local-user accounts and authorization groups to add more accounts, but this configuration is not covered in this study guide.

4. You log in to a Comware switch with Telnet, but you could not enter some commands. Why? What privilege level did you have by default?

   You could not enter the commands because you were not authorized to enter commands of that level. Your privilege level was 0, which permits only simple diagnostic commands.

5. What sets the privilege level for the Telnet user? How do you change this level?

   The VTY interface settings set the privilege level. Privilege level 0 is the default for this type of interface. You changed the level from the VTY interface.

6. ProVision switches do not include the manager and the operator passwords in the config—nor other credentials, passwords, and keys—unless you enter the `include-credentials` command. What is a use case for using the `include-credentials` command to add these credentials to the configuration?

   You might want to use the `include-credentials` command when you are first setting up a base configuration for your switches:

   - You can more easily see the credentials, verify them, and test them.
   - You can save the credentials in the switch's startup configuration and then upload the configuration to a TFTP server. You can then use that configuration as a baseline configuration for other switches. Because the credentials are included, you do not have to reconfigure the credentials on all of your switches separately.
   - You can store different credentials in different config files. (ProVision switches running the latest software support three config files.) You can test new credentials by loading a new config and revert to the old credentials by loading the previous config.

   After you test the configuration and deploy it to all your switches, you might want to disable the `include-credentials` command again for security.

## HP Comware: Change the privilege level

Earlier in this chapter, you learned how to change the privileges Telnet users receive by configuring the privilege level in the VTY interface settings. However, this configuration changes the level for all users. You might want to give Telnet users level 0 privileges by default, but then allow certain users higher privileges. One way to do so is to use the `super` command and `super passwords` to permit users to change their privilege level.

The `super <level>` command permits a user to move to a different privilege level. If the user attempts to move to a higher privilege level, the switch prompts the user for a password or denies the command if no password is set for the specified level. By default, no super passwords are set. To permit users to raise their privileges to a specific level, simply set a super password for that level:

```
[Comware] super password [level-number] <cipher | plaintext> <string>
```

If you do not specify a level number in the command, you set the super password for level 3.

After you set the `super password`, users who know the correct password can switch to that level temporarily without logging out and terminating the current connection. The users can continue to configure the switch, but the commands that they can execute have changed to the new level. After the user logs out and that user or a different user logs in again, the user privilege is restored to the original level, maintaining security.

Note that you can also use the `super` command to move to a lower level than your current level. You can always move to a lower level; you are *not* prompted for a password. While you are managing a switch, you do not typically need to move to a lower level. (You can always use commands at a lower privilege level than your own without changing levels.) But you might want to change to a lower level before turning a session over to a different manager.

## Super passwords versus other credentials

By default, the ability to raise the privilege level with a `super` command is controlled by local passwords, as described above. However, you can alternatively use the scheme authentication to control access. (The `super authentication mode` command configures this setting.) Scheme authentication uses usernames and passwords, which might be stored locally or on a centralized authentication server. Setting up these credentials requires additional configuration settings, which are beyond the scope of this study guide.

You can also use both scheme authentication and local passwords. In this configuration, local passwords act as a backup if the centralized authentication server is not available.

## Use super passwords to permit managers to raise their privilege level on the Comware switches

In this section, you will consider an example scenario for using super passwords. In this example, Telnet users receive limited privileges, but the company wants to enable managers to raise their privilege level, provided that they know the correct password. For this purpose, you can set super passwords. The following example shows how to create a super password of "hp" for privilege levels 1–3. Remember that passwords are case-sensitive.

```
[Comware-1] super password level 1 simple hp
[Comware-1] super password level 2 simple hp
[Comware-1] super password level 3 simple hp
```

---

**Note**

In a real-world environment, you would, of course, configure different passwords for each level, set more complex and secure passwords, and store the passwords as encrypted.

---

To see how the super user works, look at the output below, which shows the available commands if you log in as a user with the privilege level 0.

```
<Comware-1> ?
User view commands:
  cluster   Run cluster command
  display   Display current system information
  ping      Ping function
  quit      Exit from current command view
  ssh2      Establish a secure shell client connection
```

```
  super     Set the current user priority level
  telnet    Establish one TELNET connection
  tracert   Trace route function
```

The output below shows the `display` commands that you can enter if you have privilege level 0.

```
<Comware-1> display ?
  gvrp  VLAN Protocol
  stp   Spanning tree protocol
```

Notice how restrictive the list of commands are.

If the `super` commands listed earlier were entered on the switch, you could use a super password—hp, in this example—to move to a different privilege level.

```
<Comware-1> super 1
 Password: hp
User privilege level is 1, and only those commands can be used
whose level is equal or less than this.
Privilege note: 0-VISIT, 1-MONITOR, 2-SYSTEM, 3-MANAGE
```

You could then enter the `display users` command and see that your privilege level had been changed.

```
<Comware-1> display users
The user application information of the user interface(s):
  Idx UI      Delay    Type Userlevel
+ 29  VTY 0   00:00:00 TEL  1
<-output omitted->
```

You could also enter the ? to list the commands you could execute:

```
<Comware-1> ?
User view commands:
  cluster        Run cluster command
  debugging      Enable system debugging functions
  display        Display current system information
  ipc            Interprocess communication
  ping           Ping function
```

```
  quit            Exit from current command view
  refresh         Do soft reset
  reset           Reset operation
  screen-length Specify the lines displayed on one screen
  send            Send information to other user terminal interface
  ssh2            Establish a secure shell client connection
  super           Set the current user priority level
  telnet          Establish one TELNET connection
  terminal        Set the terminal line characteristics
  tracert         Trace route function
  undo            Cancel current setting
```

Likewise, you could enter the following command to list the `display` commands you could execute:

```
<Comware-1> display ?
  acfp            Specify ACFP configuration information
  acl             Specify acl configuration information
  acsei           Specify ACSEI (a method for exchanging information between
                  ACFP client and ACFP server) configuration information
  archive         Display archive information
  arp             Display ARP information
  bfd             BFD status and configuration information
  bgp             BGP information
  bootp           BOOTstrap  Protocol
<-output omitted->
```

If you tried to move to system view, however, you would see the following message:

```
<Comware-1> system-view
           ^
 % Unrecognized command found at '^' position.
```

You cannot reach the system view and use configuration commands at this level.

In this example, you could use the `super password`—hp in the example—to switch to privilege level 2:

```
<Comware-1> super 2

 Password: hp

User privilege level is 2, and only those commands can be used

whose level is equal or less than this.

Privilege note: 0-VISIT, 1-MONITOR, 2-SYSTEM, 3-MANAGE
```

With this privilege level, you could move to the system view.

```
<Comware-1> system-view

[Comware-1] system-view
```

As you see, you can now access all the system view commands. If you returned to the user view and attempted to view the file directory, however, you would find that you do not have rights to enter this command.

```
[Comware-1] quit

<Comware-1> dir

           ^

% Unrecognized command found at '^' position.
```

You would also lack access to some other user view commands related to file, system, and user management. You could use "?" to see the commands available to you.

```
<Comware-1> ?

User view commands:

  backup          Backup next startup-configuration file to TFTP server

  boot-loader     Set boot loader

  bootrom         Update/read/backup/restore bootrom

  cluster         Run cluster command

  debugging       Enable system debugging functions

<-output omitted->
```

In this example, you could then switch to level 3, using the super password, hp.

```
<Comware-1> super 3

 Password: hp

User privilege level is 3, and only those commands can be used whose
level is equal or less than this.

Privilege note: 0-VISIT, 1-MONITOR, 2-SYSTEM, 3-MANAGE
```

With this privilege level, you could view the directory, using the following command:

```
<Comware-1> dir
```

If you entered "?" you would see that you now have access to more user view commands.

```
<Comware-1> ?

User view commands:

  archive          Specify archive settings

  backup           Backup next startup-configuration file to TFTP server

  boot-loader      Set boot loader

  bootrom          Update/read/backup/restore bootrom

  cd               Change current directory

  clock            Specify the system clock

  cluster          Run cluster command

  copy             Copy from one file to another

  debugging        Enable system debugging functions

  delete           Delete a file

  dir              List files on a file system

<-output omitted->
```

If you moved to system view, you would see that you have access to those commands as well.

```
<Comware-1> system-view
```

The next time you logged in using a Telnet session, you would receive the privilege level 0. You could test this by completing the steps outlined above and then logging out of the switch.

```
[Comware-1] quit

<Comware-1> quit
```

In this example, you were at privilege level 3 when you logged out. But what would your default Telnet privilege be when you logged back in? You could check this by initiating a new Telnet session and logging back in. At the same time, you should launch a console session and, from that session, view users.

```
<Comware-1>display users
The user application information of the user interface(s):
  Idx UI      Delay    Type Userlevel
F 3   AUX 0   00:00:00      3
  29  VTY 0   00:00:05 TEL  0
```

As the above output shows, your Telnet privilege level is 0. If you knew the super passwords, you could change your privilege level as you did before.

In this example, the passwords were set on the VTY interface, but these settings do not control access to the console. As you recall, the default privilege for the console session is 3.

## Learning check 2

1. How can you use super passwords as part of a secure management solution?

### Answer to learning check 2

1. How can you use super passwords as part of a secure management solution?

   You can set a super password for each privilege level 1–3. You can then configure a lower privilege level on user interfaces so that users, by default, have limited privileges when they log in. You can then inform managers who should have higher privileges of the super password for their level.

## Summary

In this chapter, you have learned how to secure management access on HP switches using local passwords. You learned that ProVision switches offer operator and manager-level access. You learned about the four privilege levels on Comware switches and the way to configure the level assigned to users who access the switch CLI in-band or out-of-band. You also learned how to use super passwords to permit users to raise their privilege level.

## Learning check 3

The following questions will help you measure your understanding of the material presented in this chapter.

1. Which is *not* a supported authentication method for a Comware user interface?
   a. Asymmetric keys
   b. None
   c. Password
   d. Scheme

2. Which are the user types for ProVision switches? (Select two.)
   a. Monitor
   b. Manager
   c. Operator
   d. System
   e. Visitor

3. On Comware switches, what is the number for the system privilege level?

4. Enter the Comware command that changes a user's privilege level:

5. Which is true when HP switches are at their factory default settings?
   a. Telnet is disabled on both Comware and ProVision switches.
   b. Telnet is enabled on Comware switches and disabled on ProVision switches.
   c. Telnet is enabled on ProVision switches and disabled on Comware switches.
   d. Telnet is enabled on both Comware and ProVision switches.

## Answers to learning check 3

1. Which is *not* a supported authentication method for a Comware user interface?

   ☑ **A**. Asymmetric keys is correct.

   ☒ **B**, **C**, and **D** are incorrect.

2. Which are the user types for ProVision switches? (Select two.)

   ☑ **B** and **C**. Manager and Operator are correct.

   ☒ **A**, **D**, and **E** are incorrect.

3. On Comware switches, what is the number for the system privilege level?

   2

4. Enter the Comware command that changes a user's privilege level:

   super [level]

5. Which is true when HP switches are at their factory default settings?

   ☑ **C**. Telnet is enabled on ProVision switches and disabled on Comware switches is correct.

   ☒ **A**, **B**, and **D** are incorrect.

# 4 Managing Software and Configurations

## EXAM OBJECTIVES

✓ Manage software and upgrade the operating system on HP switches

✓ Manage configuration files on HP switches

## ASSUMED KNOWLEDGE

Before reading this chapter, you should be able to:

- Access and navigate the HP ProVision and Comware CLIs
- Control management access to ProVision and Comware CLIs

## INTRODUCTION

This chapter explains how to manage software and configuration files on ProVision and Comware switches.

## HP ProVision: Software file management

ProVision switches feature two flash memory locations for storing switch software image files, which are illustrated in Figure 4-1.

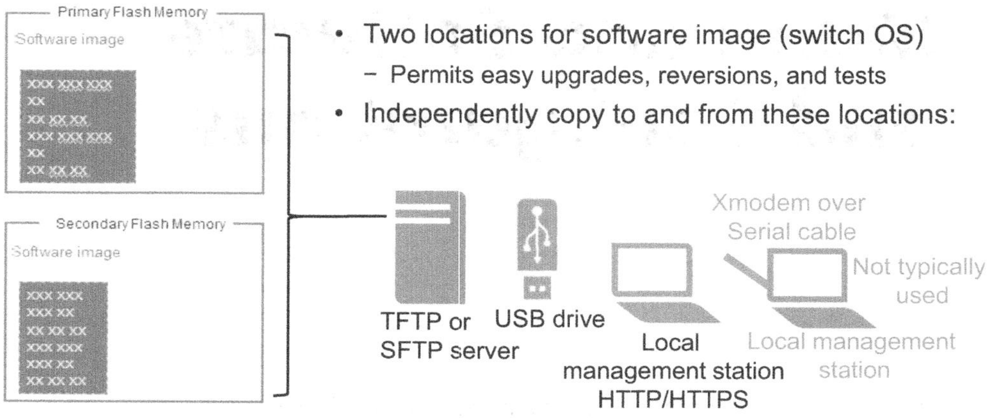


Figure 4-1: ProVision: Software file management

- **Primary flash**—Default storage for a switch software image.
- **Secondary flash**—Additional storage for either a redundant or an alternate switch software image.

The switch boots from and uses only one image at a time. You can choose the location for the software from which the switch boots. This location is not shown in the switch configuration but is rather stored in a separate file in the flash. After you change the location in the configuration, the switch must reboot in order for the switch to boot the software in the new location. An administrator with console access to the switch can also choose the software to boot during the boot process, overriding the default location selected in the configuration.

Having two flash memory locations for the software image helps you to manage the switch more easily. You can test a new image in your system without having to replace the previous image; if necessary, you can revert to the previous image easily. You can also use the two image locations for troubleshooting. For example, you can copy a problem image into secondary flash for later analysis and place another, proven image in primary flash to run your system.

**Note**

Before you load new software, you should make sure that a switch has backup software in flash in case the new software has problems. You can then easily boot from the old software without having to reinstall it.

As Figure 4-1 shows, you can copy a new software image to either the primary or the secondary flash location in several ways:

- From a TFTP server or an SFTP server
- From a storage device attached to the USB port, if the switch has such a port

- From a management station, which is logged in to the switch's Web browser interface
- From a management station, using Xmodem over a console connection

Using Xmodem over the console connection slows down the update process because the fastest clock rate of the console connection is 115K. You would have to wait a very long time for a large software image to copy. (Xmodem over Telnet would have a higher speed.)

You should use Xmodem only when absolutely necessary: when the switch cannot reach a TFTP or an SFTP server and you do not have physical access to the USB slot. (Usually, if you have console access to a switch, you also have physical access. It is possible to receive a remote connection to the console port through a Telnet-to-serial server.)

It is best practice to copy new software to a different location from the software that is currently in use. The switch runs a check when it copies the software, verifying that the software image is valid for this switch.

You can also copy software images in the primary or the secondary flash to any of the locations listed above. In this way, you can backup and archive images.

The job aids at the end of this chapter provide steps for upgrading ProVision software.

## HP Comware: Software file management

As is shown in Figure 4-2, Comware devices have an open flash drive. You can save any file, such as software images, configuration files, and others, to this flash drive. In fact, you can save two, three, or more software images to the switch flash. As a result, you can easily upgrade software, revert to a previous version, store a software version that you want to test, and so on.

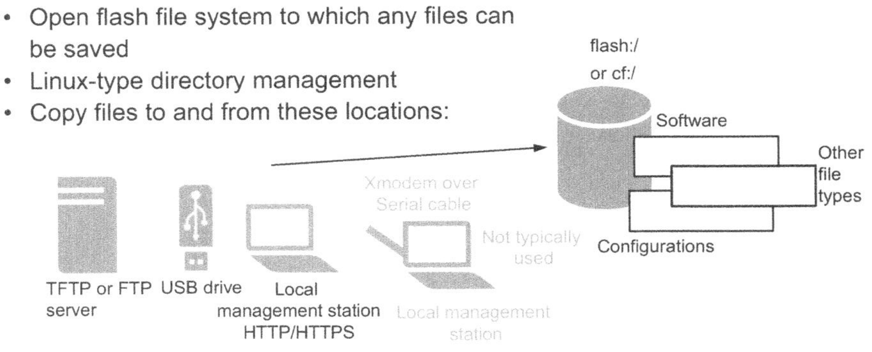


Figure 4-2: Comware: Software file management

Most Comware switches have their flash memory on the system board, and the drive is named *flash:*. In addition, most Comware routers and switches have a compact flash card, which is named *cf:*. If the device has more than one compact flash slot, they are referred to as *cfa:,cfb:,* and so on.

Comware devices also support USB port or ports, which can support USB drives that offer additional flash memory locations. To reference these locations, use *uf:*.

You can manage files and directories in the flash memory in the same way that you might manage a Linux system. You can create subdirectories, move files to new locations, copy files, delete files, read the contents of files, and so on. Note that the `dir` command displays the contents of the flash file system on the system board by default. A job aid at the end of this chapter provides commands for managing the directory.

You have similar options for copying new software images to the flash on Comware switches as on ProVision switches:

- From a TFTP server or an FTP server
- From a storage device attached to the switch USB port
- From a management station, which is logged into the switch's Web browser interface
- From a management station, using Xmodem over a console connection

Again, TFTP servers, FTP servers, and USB devices are the typical choices for a CLI software upload. An Xmodem transfer is relatively slow. You can use the Web browser interface as well.

The job aids at the end of this chapter provide steps for upgrading Comware software using the CLI.

## View software images on HP switches

In this section, you will learn how to view the software images stored on an HP switch and determine which image that switch is using.

### View software images on HP ProVision switches

If you want to know which software version a ProVision switch is running, access the switch CLI and then enter the `show version` command. Examine the output to determine which boot image the switch used when it booted last:

```
ProVision-1# show version

Image stamp:

/ws/swbuildm/KA_rel_irvine_qaoff/code/build/tam(swbuildm_KA_rel_
irvine_qaoff_rel_irvine)

                Mar 13 2013 11:26:17

                KA.15.10.0011

                542

Boot Image:     Primary
```

In the example output, the switch booted with KA.15.10.0011, which is the primary boot image.

If you want to examine the boot files or boot ROM version in the switch's flash, use the `show flash` command:

```
ProVision-1# show flash
Image                 Size (bytes) Date     Version
----------------- ------------ -------- --------------
Primary Image     :     14881496 03/13/13 KA.15.10.0011
Secondary Image   :     14881496 03/13/13 KA.15.10.0011

Boot ROM Version : KA.15.09
Default Boot     : Primary
```

Also note the boot ROM version. Sometimes you can upgrade a switch's software without upgrading the boot ROM, but sometimes you must upgrade the boot ROM. The release notes for the new software version will indicate the required boot ROM version, so you should keep track of which version your switch is using.

## HP Comware switches

When you manage a Comware switch, you will use `display` commands instead of `show` commands.

You can use the `display version` command to view the software version and the boot ROM version the switch is using. This information is highlighted in the example output below. Remember that some software upgrades might require an upgrade to the boot ROM. The release notes for each software version will note if a boot ROM upgrade is required.

```
<Comware-1> display version
HP Comware Platform Software
Comware Software, Version 5.20, Release 1211P09
Copyright (c) 2010-2012 Hewlett-Packard Development Company, L.P.
HP A5800-24G Switch uptime is 0 week, 0 day, 3 hours, 47 minutes
HP A5800-24G Switch with 2 Processors
1024M   bytes SDRAM
4M      bytes Nor Flash Memory
512M    bytes Nand Flash Memory
```

```
Config Register points to Nand Flash
Hardware Version is Ver.B
CPLD Version is 003
BootRom Version is 215
[SubSlot 0] 24GE+4SFP Plus Hardware Version is Ver.B
[SubSlot 1] No Module
```

You can also use the `display boot-loader` command to determine the default image (switch OS) used when booting the switch.

```
<Comware-1> display boot-loader
Slot 1
The current boot app is:  flash:/a5800_5820x-cmw520-r1211p09.bin
The main boot app is:     flash:/a5800_5820x-cmw520-r1211p09.bin
The backup boot app is:   flash:/
```

To view the files in the flash file system of the switch, enter.

```
<Comware-1> dir
Directory of flash:/
   0     -rw-   7032287  Apr 26 2000 12:01:32   logfile.log
   1     -rw-       151  Apr 26 2000 14:17:59   system.xml
   2     -rw-      1983  Apr 26 2000 14:18:01   config.cfg
   3     -rw-      2406  Apr 26 2000 12:30:46   iccrunning.cfg
<-output omitted>
   28    drw-         -  Apr 28 2000 03:05:32   ams
   29    drw-         -  Apr 28 2000 03:05:33   apj
   30    drw-         -  Apr 28 2000 03:05:34   emea
   31    -rw-  21624756  Dec 02 2009 13:50:12   a5800_5820x-cmw520-
r1211p09.bin
```

In the above output, notice the configuration file, config.cfg (File 2). Files 28, 29, and 30 are directories. File 31 is an operating system. Table 4-1 lists the meaning of the letters in the second column.

Table 4-1: Directory legend

| Letter | Meaning |
|---|---|
| D | directory (The item is a directory) |
| R | read (You have read access to this file or directory.) |
| W | write (You have write access to this file or directory.) |

# Boot from different software images on HP switches

You should know how to configure HP switches to boot from different software images.

## Boot from different images on HP ProVision switches

To reconfigure the switch to boot from the secondary software location, you should use the `boot system flash secondary` command. This command reboots the switch immediately. You will be prompted to confirm the reboot. You may also be prompted to save the current configuration.

```
ProVision-1# boot system flash secondary

System will be rebooted from secondary image. Do you want to continue
[y/n]?  y

Do you want to save current configuration [y/n/^C]?  y
```

After you enter this command, notice that as the switch reboots, its default boot profile is "secondary."

```
Boot Profiles:

0. Monitor ROM Console

1. Primary Software Image

2. Secondary Software Image

Select profile (secondary):
```

After the switch reboots, you can use the `show flash` command to see which flash image was used on the last reboot. In the example output below, the secondary image was used.

```
ProVision-1# show version

Image stamp:

/ws/swbuildm/KA_rel_irvine_qaoff/code/build/tam(swbuildm_KA_rel_
irvine_qaoff_rel_irvine)
```

```
                Mar 13 2013 11:26:17
                KA.15.10.0011
                542
Boot Image:     Secondary
```

If you want to set the default flash location to primary again, you use the `boot set-default flash primary` command. Confirm the change by entering `y` when prompted.

```
ProVision-1# boot set-default flash primary

This command changes the location of the default boot. This command will
change the default flash image to boot from primary image.

Hereafter, 'reload' and 'boot' commands will boot from primary image.

Do you want to continue [y/n]?  y
```

Note that the `boot set-default flash primary` command does not initiate an immediate reboot. If you view the flash, it shows the software selected for *subsequent* reboots.

```
ProVision-1# show flash

Image                 Size (bytes) Date     Version
----------------- ------------ -------- --------------
Primary Image     :     14881496 03/13/13 KA.15.10.0011
Secondary Image   :     14881496 03/13/13 KA.15.10.0011
Boot ROM Version : KA.15.09
Default Boot      : Primary
```

However, if you view the version information, you will see that the switch is still using the secondary image, from which it booted before.

```
ProVision-1# show version

Image stamp:

/ws/swbuildm/KA_rel_irvine_qaoff/code/build/tam(swbuildm_KA_rel_
irvine_qaoff_rel_irvine)

                Mar 13 2013 11:26:17
                KA.15.10.0011
                542
Boot Image:     Secondary
```

If you want the switch to use the primary image, you can initiate a reboot. Because the switch has been configured to use the primary image on subsequent reboots, it will boot with this image.

```
ProVision-1# reload

System will be rebooted from primary image. Do you want to continue
[y/n]?  y
```

### Delete a file from an HP ProVision switch

You can delete a file by entering:

```
ProVision-1# delete <filename>
```

## Boot from different images on HP Comware switches

In this section, you will first learn how to make a backup copy of the current software image and then boot the switch from this image. If you decide to practice completing these steps, you should ensure that there is enough free space available to accommodate the backup file. The switch may already be storing multiple software images or configuration files.

Use the `dir` command to determine the free space available.

```
<Comware-1> dir

Directory of flash:/

   0     -rw-      4327  May 02 2000 07:39:24   config.cfg

<-output omitted->

   17     -rw-   27443204   May 01 2000 15:37:28    a5800_5820x-cmw520-
r1211p09.bin

<-output omitted->

515712 KB total (458474 KB free)
```

From this output, you can determine the total space free. You can then compare that to the size of the file that you are copying. Or if you are upgrading to a new software image, you can compare the total space free to the size of the new software image.

The free space is indicated in KB, but the file size of software images stored on the switch is indicated in bytes. If you divide the file size by 1000, you can quickly determine if you have enough space for the copied software image.

If you do not have enough space, you will need to delete one or more files. See the "Delete a file from HP Comware switches" on page 88.

If there is enough free space, you can use the `copy` command to make the backup copy of the software image. In the example command below:

- The first filename is the name of the software image that will be copied. You can begin typing the filename and then use <Tab> to complete it. But make sure that the name exactly matches the file that you want to copy; your switch might have multiple files with similar names.
- The second filename is "backupapp.bin." (If the latter file already exists, you will be prompted to overwrite it.)
- Remember to confirm the copy.

```
<Comware-1> copy a5800_5820x-cmw520-r1211p09.bin backupapp.bin

Copy flash:/a5800_5820x-cmw520-r1211p09.bin to flash:/backupapp.bin?[Y/N]: y

.....................................................................

%Copy file flash:/a5800_5820x-cmw520-r1211p09.bin to flash:/backupapp.bin...
Done
```

To change the software image that the switch uses to boot, you use the `boot-loader` command. The following example command makes the backupapp.bin file the new main OS from which the switch boots.

```
<Comware-1> boot-loader file backupapp.bin slot all main

  This command will set the boot file of the specified board. Continue?
[Y/N]: y

  The specified file will be used as the main boot file at the next reboot
on slot 1!
```

To verify that the software image for the next boot (the "main boot app") has been changed, you use the `display boot-loader` command.

```
<Comware-1> display boot-loader

 Slot 1

The current boot app is:   flash:/a5800_5820x-cmw520-r1211p09.bin

The main boot app is:      flash:/backupapp.bin

The backup boot app is:    flash:/
```

You can then reboot the switch to use the new software image. If you made any configuration changes without saving them, you will be prompted to save them.

```
<Comware-1> reboot
 Start to check configuration with next startup configuration file, please wait.
........DONE!
 This command will reboot the device. Continue? [Y/N]: y
#Apr 26 13:18:28:855 2000 Comware-1 DEV/1/REBOOT:
 Reboot device by command.
%Apr 26 13:18:28:956 2000 Comware-1 DEV/4/SYSTEM REBOOT:
 System is rebooting now.
```

After the switch has rebooted, you can log back in and verify the boot loader status.

```
<Comware-1> display boot-loader
 Slot 1
The current boot app is:  flash:/backupapp.bin
The main boot app is:     flash:/backupapp.bin
The backup boot app is:   flash:/
```

To reset the boot loader file back to the original file name, enter:

```
<Comware-1> boot-loader file a5800_5820x-cmw520-r1211p09.bin slot all main
  This command will set the boot file of the specified board. Continue? [Y/N]: y
  The specified file will be used as the main boot file at the next reboot on slot 1!
```

If you want to set the same software file as the backup boot loader file, you use this command:

```
<Comware-1> boot-loader file a5800_5820x-cmw520-r1211p09.bin slot all backup
  This command will set the boot file of the specified board. Continue? [Y/N]: y
  The specified file will be used as the backup boot file at the next reboot on slot 1!
```

You can verify your changes by entering:

```
<Comware-1> display boot-loader
Slot 1
The current boot app is:  flash:/backupapp.bin
The main boot app is:     flash:/a5800_5820x-cmw520-r1211p09.bin
The backup boot app is:   flash:/ a5800_5820x-cmw520-r1211p09.bin
```

### Delete a file from HP Comware switches

To remove a file from the flash on a Comware switch, simply enter:

```
<Comware-1> delete <filename>
```

The `delete` command places the file in the recycle bin in the Comware flash file system. The file still exists and can be recovered. To empty the recycle bin, enter:

```
<Comware-1> reset recycle-bin
```

## Learning check 1

You will now answer several questions, based on what you have learned about managing software images on HP switches.

1. What are the similarities and differences between these ProVision commands:
   - `show version`
   - `show flash`
2. What are the similarities and differences between these ProVision commands:
   - `boot system flash <primary | secondary>`
   - `boot set-default flash <primary | secondary>`
3. How many software images can you store on a Comware switch? How do you choose which image is used when the switch boots?

## Answers to learning check 1

1. What are the similarities and differences between these ProVision commands:

   - `show version`
   - `show flash`

   Both commands give information about the software images on the switch. The `show version` command gives details about the software from which the switch last booted. The `show flash` command displays information about *both* software images on the switch. It indicates which software image location will be used for *subsequent* boots. This location might be different from the location in the `show version` command if you have changed the default and not yet rebooted the switch.

2. What are the similarities and differences between these ProVision commands:

   - `boot system flash <primary | secondary>`
   - `boot set-default flash <primary | secondary>`

   Both commands configure which software image location the switch uses for subsequent boots.

   The first command initiates an immediate, full reboot.

   The second command does not initiate a reboot. To make the new software take effect, you must enter `boot` (full reboot) or `reload` (slightly faster reboot with fewer subsystem tests) command. If the switch has redundant management modules, this command could also cause the switch to failover from the current management module to the standby one.

3. How many software images can you store on a Comware switch? How do you choose which image is used when the switch boots?

   The Comware switch can store as many images as fit in its flash. You choose the image used for subsequent boots with the `boot-loader file <filename> main` command.

## Configuration file management on HP switches

The remainder of this chapter focuses on managing configuration files on HP switches.

Network infrastructure devices have a running-config and a startup-config:

- **Running-config**—The current configurations that are running on the device and are stored in RAM (volatile memory)
- **Startup-config**—A configuration file that is stored in flash memory and loaded when the switch boots

The running-config file exists in volatile memory and controls switch operation. If no configuration changes have been made in the CLI since the switch was last booted, the running-config file is identical to the startup-config file. The startup-config file exists in flash (nonvolatile) memory and is used to preserve the most recently saved configuration as the permanent configuration. Booting the switch replaces the current running-config file with a new running-config file that is an exact copy of the current startup-config file.

## HP ProVision: Configuration file management

ProVision switches have three config file slots in their flash memory, as illustrated in Figure 4-3. When you save the running-config with the `write memory` command, the switch, by default, saves the config to a file in slot 1, which is named config1. This file is the active file, and it is used as the startup-config file by both the primary and secondary software.

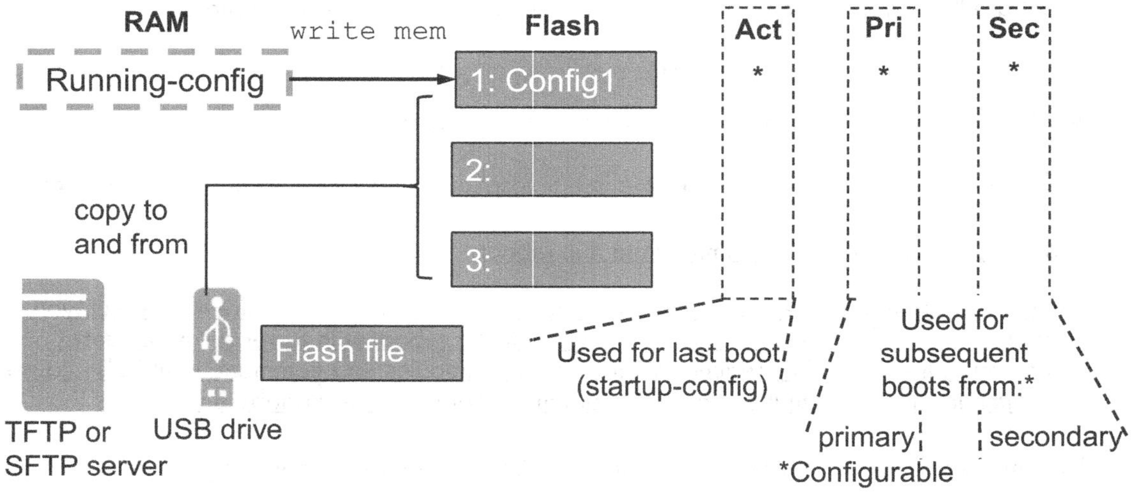


Figure 4-3: ProVision: Configuration file management

You can copy config files to any slot in a number of ways:

- From another config file
- From a TFTP server or an SFTP server
- From a USB device

Figure 4-3 shows the options available with the CLI `copy` command. You can alternatively use the switch's Web browser interface to copy files to and from your local management station.

You can use Xmodem as well, but this method is not commonly used.

You can configure the switch to use a new config file as the startup-config file for subsequent system boots. You can choose the startup-config file separately for the primary and for the secondary software. For example, you might have a config file with new features that is associated with new software; a config file without those features is associated with older software. If you need to revert to the secondary software image for any reason, the switch will use the config file associates with the software image.

To view the config files in the ProVision switch's flash, you can use the `show config files` command.

```
ProVision-1# show config files

Configuration files:

 id | act pri sec | name

 ---+-------------+--------------------------------------------------

  1 |  *   *   *  | config1

  2 |             |

  3 |             |
```

In the output for the `show config files` command, you will see three columns:

- **Act**—An asterisk (*) in this column indicates that the switch last booted with this config file. Entering the `write memory` command saves the startup-config to this file.
- **Pri**—An asterisk (*) in this column indicates that the switch is configured to boot with this config file when it boots the primary software.
- **Sec**—An asterisk (*) in this column indicates that the switch is configured to boot with this config file when it boots the secondary software.

If you are using the primary software, and the active file is different from the primary file, the `boot` command boots the primary file, and the `reload` command boots the active file. The same holds true if you are using the secondary software.

The job aids at the end of this chapter provide commands for managing config files.

## HP Comware: Configuration file management

With Comware devices, you have more options in storing and managing multiple configuration files. As you learned, Comware devices have an open flash directory that stores both software images and configuration files as well as other types of files. You can store as many configuration files to the flash as the memory allows. Configuration files always need the .cfg extension.

As is shown in Figure 4-4, the `save` command copies the running-config to a config file. When you enter the command without a filename, you are prompted to enter a filename or retain the existing startup-config filename. In either case, this file will contain your saved configurations and will become the main startup-config. The switch executes this file at bootup.

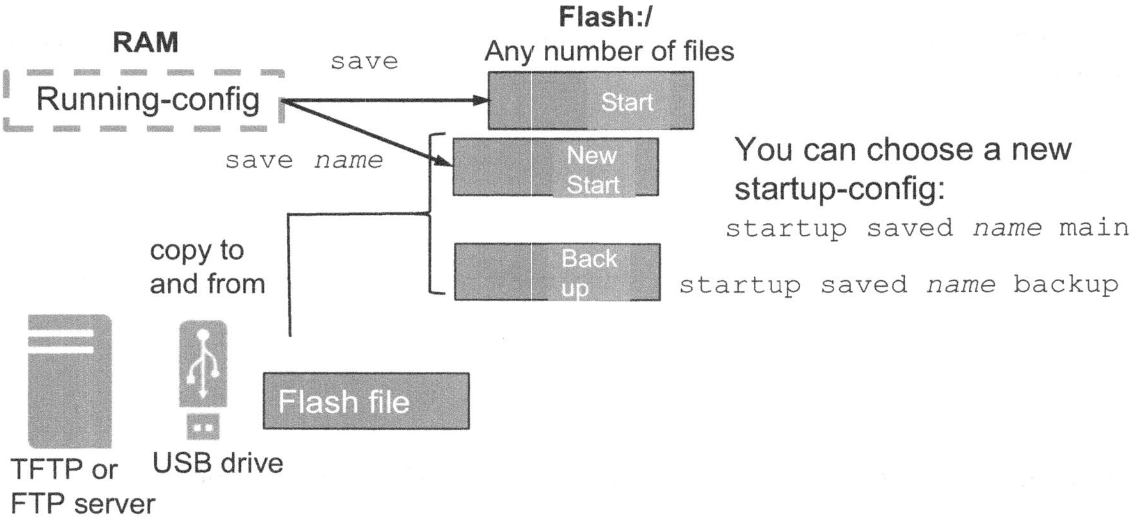


Figure 4-4: Comware: Configuration file management

When you enter the `save` command with a filename (for example, `save myconfig.cfg`), the running-config is saved as a named file in flash. However, the file does not become the startup-config.

You can also copy config files to and from a number of locations:

- A TFTP server or an FTP server
- A USB device
- Another file in flash

Figure 4-4 shows the options available with the `copy` command. You can alternatively use the switch's Web browser interface to copy files to and from your local management station.

You can choose a new startup-config from any of these config files using the `startup saved-configuration` command. The command on its own or with the `main` option sets the specified config file as the new startup-config. A switch can only have one main startup-config. Any file that used to be the main startup-config is retained in flash, but it is no longer the startup-config.

You can optionally use the `backup` option with the `startup saved-configuration` command to select one file as the backup startup-config. The switch executes this file if the main startup-config cannot successfully boot for some reason.

The job aids at the end of this chapter provide commands for managing config files.

## Change the configuration file that HP switches use to boot

In this section, you will learn how to change the configuration file an HP switch uses to boot.

### Change the configuration file that HP ProVision switches use to boot

To help you understand how to change the configuration file that a ProVision switch uses to boot, this section outlines some steps that you can complete on a standalone switch or a switch on a test network. If you cannot actually complete these steps, you can read through the steps and view the example output.

You will first create another configuration file and store it in slot 2 on the ProVision switch. To do so, you change the hostname on the existing configuration file and then copy the configuration to slot 2 on the ProVision switch.

From the global configuration view, change the switch's hostname to "ProVision-X" and save the configuration.

```
ProVision(config)# hostname ProVision-X

ProVision-X(config)# write memory
```

Copy the config1 file in the first slot to a file called config2 and verify the copy.

```
ProVision-X# copy config config1 config config2

ProVision-X# show config files

Configuration files:

 id | act pri sec | name

 ---+-------------+------------------------------------------------

  1 |  *   *   *  | config1

  2 |             | config2

  3 |             |
```

Change the current configuration's hostname to "ProVision-1" and save this configuration change.

```
ProVision-X(config)# hostname ProVision-1

ProVision-1(config)# write memory
```

At this point, the config2 file has a hostname of "ProVision-X" and the config1 file, the default configuration boot file, has a hostname of "ProVision-1."

ProVision switches give you two ways to boot from a config file:

- Set the config file as the default file used with the primary or secondary software (`startup-default` command)
- Boot from the primary or secondary software and use the config file just for this first reboot (`boot system` command)

This example uses the second option, using the `boot system` command to reboot the switch from its current software, primary, and from config file, config2. (If you complete this configuration and are prompted to save the configuration, do not.)

```
ProVision-1(config)# boot system flash primary config config2

System will be rebooted from primary image. Do you want to continue
[y/n]? y
```

After the switch reboots, the prompt should show that the switch booted from config 2.

```
ProVision-X#
```

View the configuration files. Notice that:

- config2 is the active file
- But config2 is not a primary or a secondary configuration file. The `boot system` command specified a one-time boot of this file.

```
ProVision-X# show config files

Configuration files:

 id | act pri sec | name
 ---+-------------+------------------------------------------
  1 |       *   *  | config1
  2 |  *           | config2
  3 |             |
```

With this configuration, if you make a change to the running-config and save it, the change is saved to config2. You can test this by making and saving a simple change and then viewing config2.

```
ProVision-X# config

ProVision-X(config)# hostname ProVision-Y

ProVision-Y(config)# write memory
```

```
ProVision-Y(config)# show config config2

Startup configuration:

; J9575A Configuration Editor; Created on release #KA.15.10.0011

; Ver #03:03.1f.ef:f0

hostname "ProVision-Y"

<output omitted>
```

If you reboot the switch, the primary config file, config1, will boot.

```
ProVision-1(config)# boot

System will be rebooted from primary image. Do you want to continue
[y/n]?  y
```

After the switch has rebooted, verify that the prompt has returned to the config1 prompt.

```
ProVision-1#
```

Verify that config1 is listed as the active configuration file.

```
ProVision-1# show config files

Configuration files:

 id | act pri sec | name
 ---+-------------+------------------------------------------------
  1 |  *   *   *  | config1
  2 |             | config2
  3 |             |
```

## Change the configuration file that HP Comware switches use to boot

To help you understand how to change the configuration file that a Comware switch uses to boot, this section outlines some steps that you can complete on a standalone switch, a switch on a test network, or the HP Comware simulator. If you cannot actually complete these steps, you can read through the steps and view the example output.

First, view which file the Comware switch used as the startup configuration file when it booted:

```
<Comware-1> display startup
```

```
MainBoard:
 Current startup saved-configuration file: flash:/config.cfg
 Next main startup saved-configuration file: flash:/config.cfg
 Next backup startup saved-configuration file:
```

In the example output, the switch used config.cfg.

Change the switch's hostname to "Comware-1" and save the configuration.

```
<Comware> system
[Comware] sysname Comware-1
[Comware-1] save
The current configuration will be written to the device. Are you sure?
[Y/N]: y
Please input the file name(*.cfg)[flash:/config.cfg]
(To leave the existing filename unchanged, press the enter key): <Enter>
 Validating file. Please wait....
 Saved the current configuration to mainboard device successfully
```

This will change the hostname and save the configuration to the config.cfg file.

Now change the switch hostname to "Comware-X."

```
<Comware> system
[Comware] sysname Comware-X
[Comware-X] quit
```

Save the modified configuration to a named configuration file, backup.cfg. If a file with this name is already stored on the switch, you will be prompted to save this configuration over an existing file. You may want to use a unique name for the file so you do not alter an existing configuration file. In this example, however, the response will be y to overwrite the file.

```
<Comware-X> save backup.cfg
The current configuration will be saved to flash:/backup.cfg. Continue?
[Y/N]:y
flash:/backup.cfg exists, overwrite? [Y/N]:y
Now saving current configuration to the device.
```

At this point, you have at least two configuration files in flash. The default file has a system name of "Comware-1," and the backup file has a system name of "Comware-X." Change the main startup-config file to backup.cfg by entering:

```
<Comware-X> startup saved-configuration backup.cfg main

Please wait ...

Setting the master board ...

... Done!
```

Reboot the switch. If you are prompted to save the configuration, DO NOT.

```
<Comware-X> reboot

 Start to check configuration with next startup configuration file, please wait.

........DONE!

 This command will reboot the device. Current configuration will be lost
in next startup if you continue. Continue? [Y/N]: y
```

Notice the prompt when you log in to the switch:

```
<Comware-X>
```

Verify that the backup.cfg configuration file was used to boot the switch.

```
<Comware-X> display startup

MainBoard:

 Current startup saved-configuration file: flash:/backup.cfg

 Next main startup saved-configuration file: flash:/backup.cfg

 Next backup startup saved-configuration file:
```

Now, reset the main startup-config file to "config.cfg."

```
<Comware-X> startup saved-configuration config.cfg main

Please wait ...

Setting the master board ...

... Done!
```

Reboot the switch so that it is once again using the config.cfg startup-config. If prompted to save the configuration, DO NOT.

```
<Comware-1> reboot

 Start to check configuration with next startup configuration file, please wait.

........DONE!

Current configuration will be lost, save current configuration? [Y/N] n

 This command will reboot the device. Current configuration will be lost
in next startup if you continue. Continue? [Y/N]: y
```

## Delete configuration files on HP Comware switches

If you want to delete a configuration file from flash, enter:

```
<Comware-1> delete backup.cfg

Delete flash:/backup.cfg?[Y/N]: y

%Delete file flash:/backup.cfg...Done.
```

To verify that the file has been deleted from flash, use the `dir` command.

```
<Comware-1> dir

Directory of flash:/

   1     -rw-      1983  Apr 26 2000 13:02:12   config.cfg

   2     -rw-       151  Apr 26 2000 13:02:10   system.xml

   3     -rw-   7032287  Apr 26 2000 12:01:32   logfile.log

   4     -rw-      2406  Apr 26 2000 12:30:46   iccrunning.cfg

<-output omitted->
```

Because the `delete` command places the file in the recycle bin in the Comware flash file system, the file still exists and can be recovered. Verify by viewing all files. File names surrounded by the square brackets indicate a file in the recycle bin.

```
<Comware-1> dir /all

Directory of flash:/

   0     -rw-      1983  Apr 26 2000 13:02:12   config.cfg

   1     -rw-       151  Apr 26 2000 13:02:10   system.xml

<-output omitted->
```

```
   35    -rw-   21624756   Dec 02 2009 13:50:12    s5800_5820x-cmw520-
r1109p01.bin

   36    -rw-       1983   Apr 26 2000 13:02:28    [backup.cfg]
```

You can use the `undelete` command to restore the file:

```
<Comware-1> undelete backup.cfg
```

Alternatively, you can empty the recycle bin. You will be prompted to permanently delete each file in the recycle bin.

```
<Comware-1> reset recycle-bin

Clear flash:/~/backup.cfg ?[Y/N]: y

Clearing files from flash may take a long time. Please wait...

%Cleared file flash:/~/backup.cfg.
```

## Plan a software upgrade

Although the steps you must take to upgrade software on a ProVision or a Comware switch are a fairly straightforward process, a software upgrade is not a trivial matter. You should carefully plan any software upgrade. Job Aid 4-1 outlines some of the factors that you should consider. For example, you must schedule downtime while the software is being updated. You should also carefully read the release notes, which will contain information about the update and the new software version. For example, the release notes will note if the boot ROM must be updated. The release notes will also list the new features available with the new version of the software.

In addition, you should plan what you will do if the upgrade does not go according to plan. If possible, you should first upgrade a switch in a test network first and then ensure the process goes smoothly.

You should also save the current running-config and back up the current configuration. You should also ensure you have a copy of the current software, in case you need to revert to that software version.

You should also talk to other people in your organization and identify other preparation steps you should take for your particular environment.

## Summary

In this chapter, you have learned how to manage software and configuration files on HP switches. You learned about the ProVision and Comware file systems and focused on practicing simple management tasks such as copying files, selecting the software image used for booting the switch, and selecting the startup-config file.

## Learning check 2

The following questions will help you measure your understanding of the material presented in this chapter.

1. How many software images can you install in flash on a ProVision switch?
   a. One
   b. Two
   c. Three
   d. Unlimited

2. To which config file are changes saved when you enter write memory on a ProVision switch?

3. You enter `save` on a Comware switch. When prompted for the file name, you press [Enter]. To which file are changes saved?

4. You enter `save` on a Comware switch. When prompted for the file name, you enter a new name. To which file are changes saved? Is this file used as the startup-config file?

5. You enter `save newConfig.cfg` on a Comware switch. Is this file used as the startup-config file?

6. How do you configure a Comware switch to use the newConfig.cfg file as its startup-config on subsequent reboots?

7. How do you configure a ProVision switch to boot from a config named newConfig once but keep the existing config file for other reboots?

8. How do you configure a ProVision switch to always boot from a config named newConfig when using the primary software?

9. You want to test out a new configuration on a switch. You first need to back up the existing configuration locally. You will then make the changes, which you want to be able to save. After you try out the changes, you want to revert to the backup configuration. Describe how you would complete this task on a ProVision switch and on a Comware switch.

10. Why would you enter the `reset recycle-bin` command on a Comware switch?

## Answers to learning check 2

1. How many software images can you install in flash on a ProVision switch?

   ☑ **B**. Two is correct

   ☒ **A**, **C**, and **D** are incorrect

2. To which config file are changes saved when you enter write memory on a ProVision switch?

   The config file that is marked as active (the last configuration with which the switch booted).

3. You enter `save` on a Comware switch. When prompted for the file name, you press [Enter]. To which file are changes saved?

   The changes are saved to the existing startup-config file.

4. You enter `save` on a Comware switch. When prompted for the file name, you enter a new name. To which file are changes saved? Is this file used as the startup-config file?

   When you specify a different file name from the current startup-config, the changes are saved to that file and that file becomes the startup-config.

5. You enter `save newConfig.cfg` on a Comware switch. Is this file used as the startup-config file?

   No. The changes are saved to the new file in flash called newConfig.cfg, but the file is not necessarily used as the startup-config file.

6. How do you configure a Comware switch to use the newConfig.cfg file as its startup-config on subsequent reboots?

   You use the `startup saved-configuration newConfig.cfg main` command.

7. How do you configure a ProVision switch to boot from a config named newConfig once but keep the existing config file for other reboots?

   You use this command: `boot system flash [primary | secondary] config newConfig.`

8. How do you configure a ProVision switch to always boot from a config named newConfig when using the primary software?

   You use this command: `startup-default primary config newConfig.`

9. You want to test out a new configuration on a switch. You first need to back up the existing configuration locally. You will then make the changes, which you want to be able to save.

After you try out the changes, you want to revert to the backup configuration. Describe how you would complete this task on a ProVision switch and on a Comware switch.

On a ProVision switch, you should follow a procedure like this:

- Save the current configuration (`write memory`).
- View the config files and find the name of the current startup-config (`show config files`), often config1. You should also check that a slot is free for the backup config. If no slot is free, you must delete a file. You might want to back up that file using a TFTP or a USB first.
- Copy the currently active config to a file named backup (`copy config config1 config` backup).
- Make the configuration changes, save them (`write memory`), and test them.
- You have a few choices for reverting to the backup config.
    - You could configure the switch to use backup as its default primary, secondary, or both software config. (`startup-default [primary | secondary] config backup`). You would then reboot the switch so that the backup config takes effect (`boot`).
    - You could archive the test configuration using another config slot or TFTP or USB (`copy config config1 <destination>`). You could then execute a one-time boot from the backup config (`boot system flash [primary | secondary] config backup`). You could then copy the backup config back over config1, which is still being used for primary and secondary boots (`copy config backup config config1`). Next time you reboot the switch, it will use the config1 file but have the configurations from the backup.

On a Comware switch, you should follow a procedure like this:

- Save the current configuration to a named file like backup.cfg (`save backup.cfg`).
- Make the configuration changes, save them (save with the existing or new file name), and test them.
- You can archive the test configuration changes to a new named file like test.cfg (`save test.cfg`).
- You have a few choices for reverting to the backup config:
    - You can copy backup.cfg to the current startup-config file name (`copy backup.cfg <config name>`).
    - You can configure the switch to use backup.cfg as its main startup-config (`startup saved-configuration backup.cfg`).
- In any case, end by rebooting the switch so that the backup config takes effect (`reboot`).

10. Why would you enter the `reset recycle-bin` command on a Comware switch?

    To permanently delete files that are in the recycle bin

# Software and configuration job aids

## Job Aid 4-1: Prework before a software update

| | Step | ProVision command | Comware command |
|---|---|---|---|
| ☐ | Schedule down time. | | |
| ☐ | Check release notes to determine whether a boot ROM update is required. | | |
| ☐ | Check release notes for known issues or gotchas like documented changes in system defaults. | | |
| ☐ | Have a recovery plan in case the update does not go according to plan. | | |
| ☐ | Consider testing the upgrade in a lab network or less critical segment. | | |
| ☐ | Back up the current configuration on all switches, making sure to save the running-config. | write memory<br>copy start tftp <IP address> <filename> | save force<br>tftp <IP address> put <startup-config name> |
| ☐ | Make sure that you have a copy or a backup of the current software image. | copy flash primary tftp <IP address> <filename> | tftp <IP address> put <filename> |

## Job Aid 4-2: Upgrade HP ProVision software (USB)

| | Step | Command | Notes |
|---|---|---|---|
| 1 | Complete the prework checklist. | | See Job Aid 4-1 |
| 2 | Copy the new software image to the USB device. | | |
| 3 | Insert the USB device in the switch's USB port. | | |
| 4 | Log in to the switch CLI. | | Make sure that you are in enable mode. |
| 5 | Check the location of the current software image (primary or secondary). | show version | Look for Primary or Secondary next to **Boot Image:** |

*Continued*

Job Aid 4-2: Continued

| | | | |
|---|---|---|---|
| 6 | View files on the USB device and note the image name. | dir | Depending on your terminal client, you can copy the image name. |
| 7 | Back up the current configuration and associate this file with the current software. | write memory<br>copy config <startup-config filename> config <backup filename><br>startup-default <primary \| secondary> config <backup filename> | Configurations with the new software might not be backward compatible with the current software, so this step will make it easier for you to revert to the current software if necessary.<br>If you saw Primary, in step 5, enter primary.<br>If you saw Secondary, in step 5, enter secondary. |
| 8 | Copy the new image to the software location that is *not* currently in use. | copy usb flash <image name> <primary \| secondary> | Depending on your terminal client, you can paste the image name.<br>If you saw Primary, in step 5, enter secondary.<br>If you saw Secondary, in step 5, enter primary. |
| 9 | You will be prompted: The Primary \| Secondary OS Image will be deleted, continue [y/n] | y | |
| 10 | Wait as the software is validated and copied. | | The process might take a few minutes. When the process is complete, you will see this message: |
| 11 | Choose whether you want to:<br>■ Immediately run a full reboot with all subsystem tests<br>■ Reboot later (full reboot or faster reboot) | | Your choice determines whether you complete step 12a or 12b. |
| 12a | Immediately run a full reboot | boot system flash <primary \| secondary> | The command changes the software from which the switch boots and initiates an immediate boot. |
| | You will be prompted to confirm. | y | |

*Continued*

Job Aid 4-2: Continued

| | | | |
|---|---|---|---|
| 12b | Reboot later | | |
| | Set the image for the next reboot. | `boot set-default flash <primary \| secondary>` | |
| | You will be prompted to confirm. | `y` | |
| | When you are ready, initiate or schedule a reboot. | Either:<br>■ `boot`<br>■ `reload [after <[dd:]hh:]mm> \| at <hh:mm [:ss]> [<mm/dd[/[yy]yy]>]]` | boot initiates a full reboot.<br>reload initiates a reboot.<br>reload after sets a delay of a certain length.<br>reload at schedules the reload for an absolute time. |
| | You will be prompted to confirm. | `y` | |
| 13 | Verify the new software version after the reboot. | `show version` | |
| 14 | Run acceptance tests and verify system operations. Check the configuration as some defaults might have changed. Backup and revert, as necessary | | |

Job Aid 4-3: Upgrade HP ProVision software (TFTP)

| | Step | Command | Notes |
|---|---|---|---|
| 1 | Complete the prework checklist. | | See Job Aid 4-1 |
| 2 | Copy the new software image to the TFTP server. | | Note the IP address of the TFTP server. You can use a hostname instead of an IP address if the switch has a DNS server configured.<br>Also note the image name. |
| 3 | Log in to the switch CLI. | | Make sure that you are in enable mode. |
| 4 | Make sure that the switch has connectivity to the TFTP server. | `ping <server IP address>` | If the switch cannot reach the server, you should check its IP address and default route. |
| 5 | Check the location of the current software image (primary or secondary). | `show version` | Look for Primary or Secondary next to **Boot Image:** |

*Continued*

Job Aid 4-3: Continued

| | | | |
|---|---|---|---|
| 6 | Back up the current configuration and associate this file with the current software. | `write memory`<br>`copy config <startup-config filename> config <backup filename>`<br>`startup-default <primary \| secondary> config <backup filename>` | Configurations with the new software might not be backward compatible with the current software, so this step will make it easier for you to revert to the current software if necessary.<br>If you saw Primary, in step 5, enter `primary`.<br>If you saw Secondary, in step 5, enter `secondary`. |
| 7 | Copy the new image to the software location that is *not* currently in use. | `copy tftp flash <server IP address> <image name> <primary \| secondary>` | If you saw Primary in step 5, enter `secondary`.<br>If you saw Secondary in step 5, enter `primary`. |
| 8 | You will be prompted: The Primary \| Secondary OS Image will be deleted, continue [y/n] | `y` | |
| 9 | Wait as the software is validated and copied. | | The process might take a few minutes. When the process is complete, you will see this message: |
| 10 | Choose whether you want to:<br>Immediately run a full reboot with all subsystem tests<br>Reboot later (full reboot or faster reboot) | | Your choice determines whether you complete step 11a or 11b. |
| 11a | Immediately run a full reboot | `boot system flash <primary \| secondary>` | The command changes the software from which the switch boots and initiates an immediate boot. |
| | You will be prompted to confirm. | `y` | |
| 11b | Reboot later | | |
| | Set the image for the next reboot. | `boot set-default flash <primary \| secondary>` | |
| | You will be prompted to confirm. | `y` | |
| | When you are ready, initiate or schedule a reboot. | Either:<br>■ `boot`<br>■ `reload [after <[dd:]hh:]mm> \| at <hh:mm[:ss]> [<mm/dd[/[yy]yy]>]]` | `boot` initiates a full reboot.<br>`reload` initiates a reboot.<br>`reload after` sets a delay of a certain length.<br>`reload at` schedules the reload for an absolute time. |
| | You will be prompted to confirm. | `y` | |

*Continued*

Job Aid 4-3: Continued

| | | | |
|---|---|---|---|
| 12 | Verify the new software version after the reboot. | `show version` | |
| 13 | Run acceptance tests and verify system operations. Check the configuration as some defaults might have changed. Backup and revert, as necessary | | |

Job Aid 4-4: Upgrade HP ProVision software (SFTP)

| | Step | Command | Notes |
|---|---|---|---|
| 1 | Complete the prework checklist. | | See Job Aid 4-1 |
| 2 | Copy the new software image to the SFTP server. | | Note the IP address of the SFTP server. You can use a hostname instead of an IP address if the switch has a DNS server.<br>Also note the image name. |
| 3 | Log in to the switch CLI. | | Make sure that you are in enable mode. |
| 4 | Make sure that the switch has connectivity to the SFTP server. | `ping <IP address \| hostname>` | If the switch cannot reach the server, you should check its IP address and default route. |
| 5 | If the switch does not already support SSH, set up SSH. | `crypto key generate ssh <rsa \| dsa> bits <bits>`<br>`ip ssh` | The `ip ssh` command might not be necessary, as SSH is enabled by default. |
| 6 | Check the location of the current software image (primary or secondary). | `show version` | Look for Primary or Secondary next to `Boot Image:` |
| 7 | Back up the current configuration and associate this file with the current software. | `write memory`<br>`copy config <startup-config filename> config <backup filename>`<br>`startup-default <primary \| secondary> config <backup filename>` | Configurations with the new software might not be backward compatible with the current software, so this step will make it easier for you to revert to the current software if necessary.<br>If you saw Primary, in step 6, enter `primary`.<br>If you saw Secondary, in step 6, enter `secondary`. |

*Continued*

Job Aid 4-4: Continued

<table>
<tr><td>8</td><td>Copy the new image to the software location that is not currently in use.</td><td><code>copy sftp flash {user <name> <IP address | hostname> | <username>@<IP address | hostname> | <IP address | hostname>} <image name> <primary | secondary></code></td><td>If you saw Primary in step 6, enter <code>secondary</code>.<br>If you saw Secondary in step 6, enter <code>primary</code>.</td></tr>
<tr><td>9</td><td>If you are prompted for a password, enter it.</td><td></td><td></td></tr>
<tr><td>10</td><td>You will be prompted: The Primary | Secondary OS Image will be deleted, continue [y/n]</td><td><code>y</code></td><td></td></tr>
<tr><td>11</td><td>Wait as the software is validated and copied.</td><td></td><td>The process might take a few minutes. When the process is complete, you will see this message:</td></tr>
<tr><td>12</td><td>Choose whether you want to:<br>Immediately run a full reboot with all subsystem tests<br>Reboot later (full reboot or faster reboot)</td><td></td><td>Your choice determines whether you complete step 13a or 13b.</td></tr>
<tr><td rowspan="2">13a</td><td>Immediately run a full reboot</td><td><code>boot system flash <primary | secondary></code></td><td rowspan="2">The command changes the software from which the switch boots and initiates an immediate boot.</td></tr>
<tr><td>You will be prompted to confirm.</td><td><code>y</code></td></tr>
<tr><td rowspan="5">13b</td><td>Reboot later</td><td></td><td></td></tr>
<tr><td>Set the image for the next reboot.</td><td><code>boot set-default flash <primary | secondary></code></td><td></td></tr>
<tr><td>You will be prompted to confirm.</td><td><code>y</code></td><td></td></tr>
<tr><td>When you are ready, initiate or schedule a reboot.</td><td>Either:<br>■ <code>boot</code><br>■ <code>reload [after <[dd:]hh:]mm> | at <hh:mm[:ss]> [<mm/dd[/[yy]yy]>]]</code></td><td><code>boot</code> initiates a full reboot.<br><code>reload</code> initiates a reboot.<br><code>reload after</code> sets a delay of a certain length.<br><code>reload at</code> schedules the reload for an absolute time.</td></tr>
<tr><td>You will be prompted to confirm.</td><td><code>y</code></td><td></td></tr>
<tr><td>14</td><td>Verify the new software version after the reboot.</td><td><code>show version</code></td><td></td></tr>
</table>

*Continued*

Job Aid 4-4: Continued

| | |
|---|---|
| 15 | Run acceptance tests and verify system operations. Check the configuration as some defaults might have changed. Backup and revert, as necessary |

Job Aid 4-5: Copy HP ProVision files

<table>
<tr><th>Copy from:</th><th>To:</th><th>Command</th></tr>
<tr><td>Config file in flash</td><td>Config file in flash</td><td>copy config <source filename> config <dest filename><br>*Copies to the active startup-config are not allowed.</td></tr>
<tr><td></td><td>TFTP server</td><td>copy config <source filename> tftp <IP address | hostname> <dest filename></td></tr>
<tr><td></td><td>SFTP server</td><td>copy config <source filename> sftp {user <name> <IP address | hostname> | <username>@<IP address | hostname> | <IP address | hostname>} <dest filename></td></tr>
<tr><td></td><td>USB</td><td>copy config <source filename> usb <dest filename></td></tr>
<tr><td>TFTP server</td><td>Config file in flash</td><td>copy tftp config <dest filename> <IP address | hostname> <source filename></td></tr>
<tr><td></td><td>Startup-config</td><td>copy tftp startup-config <IP address | hostname> <source filename></td></tr>
<tr><td></td><td>Software image</td><td>copy tftp flash <IP address | hostname> <source filename> <primary | secondary></td></tr>
<tr><td>SFTP server</td><td>Config file in flash</td><td>copy sftp config <dest filename> {user <name> <IP address | hostname> | <username>@<IP address | hostname> | <IP address | hostname>} <source filename></td></tr>
<tr><td></td><td>Startup-config</td><td>copy sftp startup-config {user <name> <IP address | hostname> | <username>@<IP address | hostname> | <IP address | hostname>} <source filename></td></tr>
<tr><td></td><td>Software image</td><td>copy sftp flash {user <name> <IP address | hostname> | <username>@<IP address | hostname> | <IP address | hostname>} <source filename> <primary | secondary></td></tr>
</table>

*Continued*

## Job Aid 4-5: Continued

| | | |
|---|---|---|
| USB | Config file in flash | `copy usb config <source filename> <dest filename>` |
| | Startup-config | `copy usb startup-config <source filename>` |
| | Software image | `copy usb flash <source filename> <primary \| secondary>` |
| Software | Primary software (Secondary software copies to primary) | `copy flash flash primary` |
| | Secondary software (Primary software copies to secondary) | `copy flash flash secondary` |
| | TFTP server | `copy flash tftp <IP address> <dest filename> <primary \| secondary>` |
| | SFTP server | `copy flash sftp {user <name> <IP address \| hostname> \| <username>@<IP address \| hostname> \| <IP address \| hostname>} <dest filename> <primary \| secondary>` |
| | USB | `copy flash usb <dest filename> <primary \| secondary>` |

## Job Aid 4-6: Manage HP ProVision config files

| Description | Command |
|---|---|
| Delete a config file. | `delete <filename>` |
| Rename a config file. | `rename config <filename> <new filename>` |
| Copy a config file. | `copy config <filename> <destination>`<br>*See Job Aid 4-5 for more details on copying to various destinations. |
| View the names of config files; view which files are active and used as startup-config with various software. | `show config files` |
| View the contents of a config file. | `show config <filename>` |
| Choose the config file used when the switch boots from primary software, secondary software, or both (no keyword). | `startup-default [primary \| secondary] config <filename>`<br>*You must use the `boot` command for the change to take effect. |

Job Aid 4-7: Upgrade HP Comware software (USB)

<table>
<tr><th></th><th>Step</th><th>Command</th><th>Notes</th></tr>
<tr><td>1</td><td>Complete the prework checklist.</td><td></td><td>See Job Aid 4-1</td></tr>
<tr><td>2</td><td>Copy the new software image to the USB device.</td><td></td><td></td></tr>
<tr><td>3</td><td>Insert the USB device into the switch's USB port.</td><td></td><td></td></tr>
<tr><td>4</td><td>Log in to the switch CLI.</td><td></td><td>Make sure that you are in the user view.<br>You also need level 3 privileges. (The display users command shows your privileges.)</td></tr>
<tr><td>5</td><td>Back up the current configuration.</td><td>save <backup filename></td><td>Configurations with the new software might not be backward compatible with the current software, so this step will make it easier for you to revert to the current software if necessary.</td></tr>
<tr><td>6</td><td>View files on the USB device and note the image name.</td><td>dir uf:/</td><td>If your devices has more than one USB slot, replace uf:/ with ufa:/ or ufb:/ and so on.<br>Depending on your terminal client, you can copy the image name.</td></tr>
<tr><td>7</td><td>Copy the new image to the switch flash.</td><td>copy uf:/<image name> flash:/<image name></td><td>If you are updating a router, replace flash:/ with cf:/ (cfa:/).</td></tr>
<tr><td>8</td><td>Configure the switch to boot from the new software image.</td><td>boot-loader file <image name> [slot <ID | all>] main</td><td></td></tr>
<tr><td>9</td><td>If desired, specify a different software image as the backup image. You can also set your backup configuration as a backup config.</td><td>boot-loader file <image name 2> [slot <ID | all>] backup<br>startup saved-configuration <backup filename> backup</td><td>For example, you might want to use the previous software image as a backup.</td></tr>
<tr><td>10</td><td>Reboot the switch.</td><td>reboot</td><td>If desired, you can complete this step after a delay.</td></tr>
</table>

*Continued*

Job Aid 4-7: Continued

| | | | |
|---|---|---|---|
| 11 | Verify the new software version. | display version | |
| 12 | Run acceptance tests and verify system operations. Check the configuration as some defaults might have changed. Backup and revert, as necessary | | |

Job Aid 4-8: Upgrade HP Comware software (TFTP)

<table>
<tr><th></th><th>Step</th><th>Command</th><th>Notes</th></tr>
<tr><td>1</td><td>Complete the prework checklist.</td><td></td><td>See Job Aid 4-1.</td></tr>
<tr><td>2</td><td>Copy the new software image to the TFTP server.</td><td></td><td>Note the IP address of the TFTP server. You can use a hostname instead of an IP address if the switch has a DNS server.<br>Also note the image name.</td></tr>
<tr><td>3</td><td>Log in to the switch CLI.</td><td></td><td>Make sure that you are in the user view.<br>You also need level 3 privileges. (The display users command shows your privileges.)</td></tr>
<tr><td>4</td><td>Back up the current configuration.</td><td>save <backup filename></td><td>Configurations with the new software might not be backward compatible with the current software, so this step will make it easier for you to revert to the current software if necessary.</td></tr>
<tr><td>5</td><td>Make sure that the switch has connectivity to the TFTP server.</td><td>ping <IP address | hostname></td><td>If the switch cannot reach the server, you should check its IP address and default route.</td></tr>
<tr><td>6</td><td>Copy the new image to the switch flash.</td><td>tftp <IP address | hostname> get <image name> [<dest image name>]</td><td>If you do not specify a second image name, the image is stored with the current filename in the switch flash.</td></tr>
<tr><td>9</td><td>Configure the switch to boot from the new software image.</td><td>boot-loader file <image name> [slot <ID | all>] main</td><td></td></tr>
<tr><td>8</td><td>If desired, specify a different software image as the backup image. You can also set your backup configuration as a backup config.</td><td>boot-loader file <image name 2> [slot <ID | all>] backup<br>startup saved-configuration <backup filename> backup</td><td>For example, you might want to use the previous software image as a backup.</td></tr>
</table>

*Continued*

Job Aid 4-8: Continued

| | | | |
|---|---|---|---|
| 9 | Reboot the switch. | `reboot` | If desired, you can complete this step after a delay, of course. |
| 10 | Verify the new software version. | `display version` | |
| 11 | Run acceptance tests and verify system operations. Check the configuration as some defaults might have changed. Backup and revert, as necessary | | |

Job Aid 4-9: Upgrade HP Comware software (FTP)

| | Step | Command | Notes |
|---|---|---|---|
| 1 | Complete the prework checklist. | | See Job Aid 4-1. |
| 2 | Copy the new software image to the FTP server. | | Note the IP address of the FTP server. You can use a hostname instead of an IP address if the switch has a DNS server configured.<br>Also note the image name. |
| 3 | Log in to the switch CLI. | | Make sure that you are in the user view.<br>You also need level 3 privileges. (The `display users` command shows your privileges.) |
| 4 | Back up the current configuration. | `save <backup filename>` | Configurations with the new software might not be backward compatible with the current software, so this step will make it easier for you to revert to the current software if necessary. |
| 5 | Make sure that the switch has connectivity to the FTP server. | `ping <IP address \| hostname>` | If the switch cannot reach the server, you should check its IP address and default route. |
| 6 | Initiate an FTP session to the server. | `ftp <IP address \| hostname>` | |
| 7 | If prompted for credentials, enter them. | | |

Continued

Job Aid 4-9: Continued

| | | | |
|---|---|---|---|
| 8 | Use binary mode. | binary | |
| 9 | Copy the new image to the switch flash. | get <image name> | |
| 10 | Configure the switch to boot from the new software image. | boot-loader file <image name> [slot <ID \| all>] main | |
| 11 | If desired, specify a different software image as the backup image. You can also set your backed configuration as a backup config. | boot-loader file <image name 2> [slot <ID \| all>] backup<br>startup saved-configuration <backup filename> backup | For example, you might want to use the previous software image as a backup. |
| 12 | Reboot the switch. | reboot | If you want, you can complete this step after a delay, of course. |
| 13 | Verify the new software version. | display version | |
| 14 | Run acceptance tests and verify system operations. Check the configuration as some defaults might have changed. Backup and revert, as necessary | | |

Job Aid 4-10: Manage the HP Comware flash file system

| Description | Command |
|---|---|
| Display current directory information. The /all parameter shows all files, including ones marked for deletion. | dir [/all] [<file-url>] |
| Display the current working directory | pwd |
| Change the current working directory | cd <directory> |
| Create a directory | mkdir <directory> |
| Remove a directory | rmdir <directory> |
| Restore a file from the recycle bin | undelete <file-url> |
| Display the contents of a file (currently only a text file can be displayed) | more <file-url> |
| Rename a file | rename <fileurl-source> <fileurl-dest> |
| Copy a file | copy <fileurl-source> <fileurl-dest> |
| Move a file | move <fileurl-source> <fileurl-dest> |
| Delete a file (place it in the recycle bin). | delete <file-url> |

Continued

## Job Aid 4-10: Continued

| | |
|---|---|
| Restore a file from the recycle bin | undelete <file-url> |
| Permanently delete a file | delete /unreserved <file-url> |
| Empty the recycle bin | reset recycle-bin |

## Job Aid 4-11: Save the HP Comware running-config

| Save | To: | Set as startup-config? | Command |
|---|---|---|---|
| Running-config | Named file set as the main startup-config | Yes | save force<br>or<br>save<br>■ Press [Enter] when prompted for filename. |
| Running-config | Named file different from the current main startup-config | Yes | save<br>■ Provide a name when prompted for filename. |
| Running-config | Named file | No | save <filename> |
| *To set a new file as the startup-config: startup saved-configuration <filename> [main] | | | |

## Job Aid 4-12: Copy HP Comware files

| Copy from: | To: | Command |
|---|---|---|
| File in flash | File in flash | copy <source filename> <dest filename> |
| | TFTP server | tftp <IP address \| hostname> put <filename> [new dest filename] |
| | FTP server | ftp <IP address \| hostname><br>*Enter credentials when prompted<br>binary<br>put <filename> [new dest filename] |
| | USB | copy [flash:/]<filename> uf:/<filename> |
| | Running-config | system-view<br>config replace file <filename> |
| Startup-config | TFTP server | backup startup-configuration to <IP address \| hostname> [new dest filename] |
| TFTP server | File in flash | tftp <IP address \| hostname> get <filename> [new dest filename] |
| | Startup-config | restore startup-configuration from <IP address \| hostname> <filename on server><br>*Filename becomes the new startup-config filename. The startup-config is used on the next boot. |
| FTP server | File in flash | ftp <IP address \| hostname><br>*Enter credentials when prompted<br>binary<br>get <filename on server> [new dest filename] |
| USB | File in flash | copy uf:/<filename> [flash:/]<filename> |

# 5 Virtual Local Area Networks (VLANs)

## EXAM OBJECTIVES

✓ Explain use cases for Virtual LANs (VLANs)

✓ Configure port-based VLANs on HP switches, using the appropriate settings and port types

✓ Implement basic IP routing between directly connected VLANs or links

## ASSUMED KNOWLEDGE

- IP addressing
- Open Systems Interconnection (OSI) model
- Local area networks (LANs)

## Introduction

VLANs help you to control endpoints and their traffic, allowing you to place them in isolated groups according to considerations such as device type, user identity, and location. In this chapter, you learn how to set up VLANs on HP switches.

## Use VLANs to isolate communications

You can define multiple VLANs on the same switch to isolate communications between devices. For example, Figure 5-1 shows a switch that has two VLANs—VLAN 11 and VLAN 21. VLAN 11 includes the devices used by the Financial Services department, and VLAN 21 contains devices used by the Marketing department.

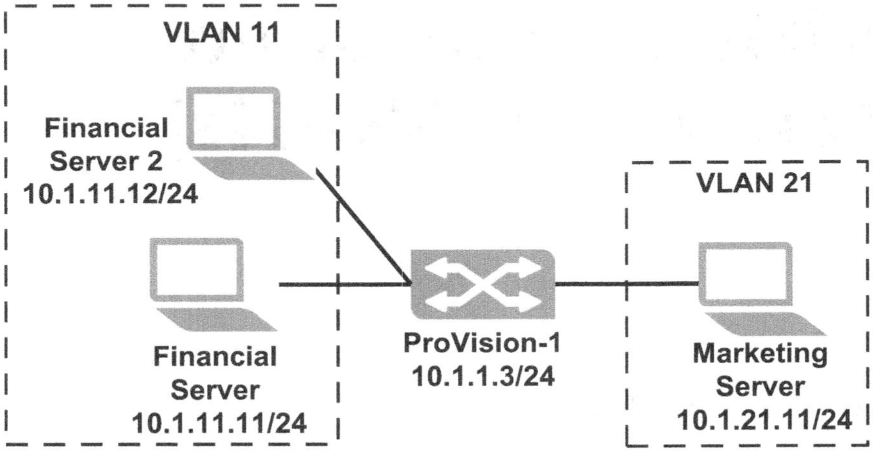


Figure 5-1: Use VLANs to isolate communications

A VLAN is a Layer 2 concept; it defines a broadcast domain. A switch will only switch frames that arrive on VLAN 11 on other ports that belong to VLAN 11. VLAN 11 broadcasts are only flooded out VLAN 11 ports, and so on.

A subnet is a Layer 3 concept. Devices in the same subnet can communicate without the need for routing. Nonetheless, a VLAN and a subnet have a close correspondence. Typically, a VLAN is associated with one subnet, and a subnet is associated with one VLAN, as VLAN 11 is associated with 10.1.11.0/24 in the example shown in Figure 5-1.

A switch can support a VLAN without having an IP address on that VLAN. However, if the switch does have an IP address on a VLAN, the subnet for that address becomes linked to that VLAN. For example, you cannot *assign* IP address 10.1.1.1/24 on VLAN 1 and IP address 10.1.1.2/24 on VLAN 2. The 10.1.1.0/24 subnet belongs to the VLAN 1 interface uniquely. (On the other hand, you can associate more than subnet with a VLAN through multinetting; however, multinetting is used less typically and is not covered in this study guide.)

In this chapter, you will explore the implications of moving devices to another VLAN and subnet.

## Assign an endpoint to a VLAN

When you assign an endpoint to a VLAN, you are actually assigning that endpoint's *traffic* to a particular VLAN. There are several ways to make this assignment. The most common one—and the one that you will use for this activity—is a port-based VLAN assignment.

On a ProVision switch, you make the port that connects to the endpoint an untagged member of the desired VLAN (see Figure 5-2). Then all traffic that arrives on that port is assigned to that VLAN. Actually, all *untagged* traffic is assigned to this VLAN, and tagged traffic is dropped.

You will learn about tagging in a moment. For now, just understand that Ethernet frames are untagged unless a device takes specific steps otherwise. Endpoints generally send untagged traffic.

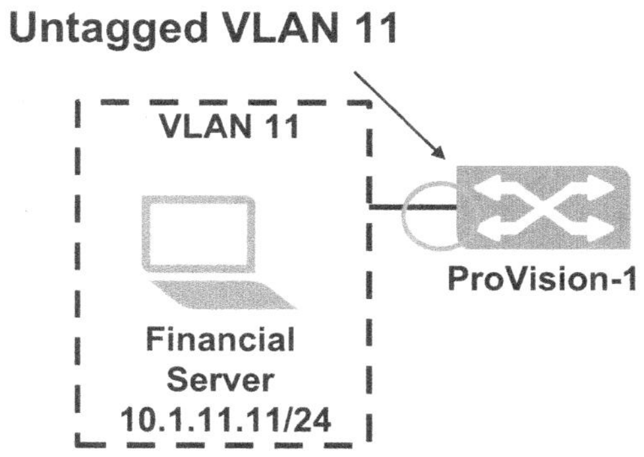


Figure 5-2: Assign an endpoint to a VLAN

On Comware switches, you make the port that connects to the endpoint an access port in the desired VLAN, shown in Figure 5-3. Again, all untagged traffic is assigned to the access port VLAN; tagged traffic is not supported.

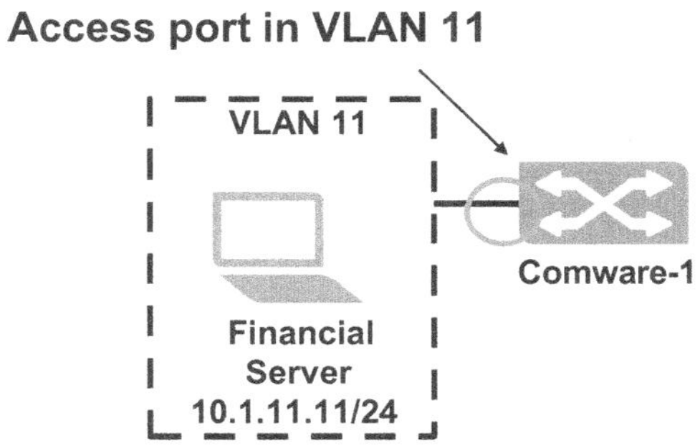


Figure 5-3: Assigning an endpoint to a VLAN (Comware)

## Extending the VLAN across multiple switches

You might want to place endpoints that connect to different switches in the same VLAN. For example, in Figure 5-4, you can place the server, which is connected to the ProVision-1 switch, and the client, which is connected to the ProVision-2 switch, in VLAN 11 so that they can communicate with each other at Layer 2.

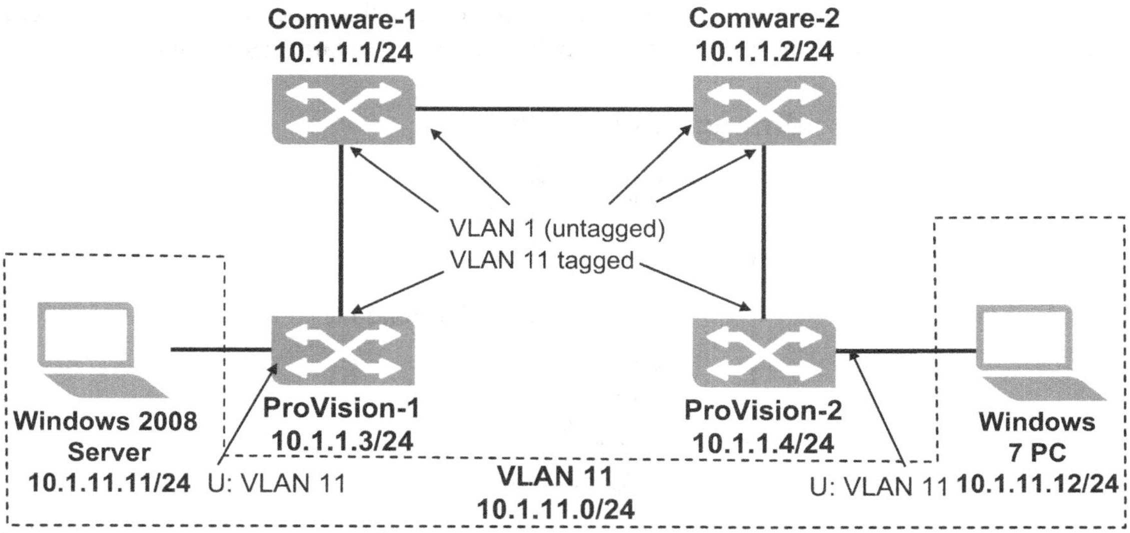


Figure 5-4: Extending the VLAN across multiple switches

You already learned how to assign the server's port to VLAN 11 on ProVision-1 and the client's port to VLAN 11 on ProVision-2. But, in order for the server and client to reach each other, you also need to extend the VLAN across the switch topology. The switches do not necessarily need IP addresses in this VLAN, but they need to *support* the VLAN so that the client and server can communicate.

Examine the switch-to-switch links in this topology. They already support VLAN 1, which the switches still use for their management IP addresses. Now they also need to support VLAN 11. In fact, switch-to-switch links typically need to support multiple VLANs.

How can you set up the links to support multiple VLANs? Do you need to add a link for each new VLAN? The cost of that solution would be prohibitively high.

Fortunately, you can set up a single switch-to-switch link to carry multiple VLANs. You simply need to configure the switch ports to distinguish traffic in one VLAN from another.

The most common way is through the IEEE 802.1Q standard for VLAN tagging. This standard defines a 4-byte field that can be inserted into an Ethernet frame (see Figure 5-5 and Figure 5-6). This field allows each Ethernet frame to be identified as part of a particular VLAN. An Ethernet frame can carry a payload of 1500 bytes. An untagged Ethernet frame has a maximum size of 1518 bytes while a tagged frame has a maximum size of 1522 bytes.

Because the VLAN ID component in the tag is 12 bits, valid VLAN IDs range from 1 to 4094. (VLAN 0 and VLAN 4095 are reserved and not permitted as valid VLAN IDs.)

**Note**

You can enable a switch to carry jumbo frames over this size. Comware switches support jumbo frames by default on interfaces with Gbps speeds or higher. On ProVision switches, jumbo frames are not supported by default, but you can enable them on VLANs. Check your switch documentation for the specific command.

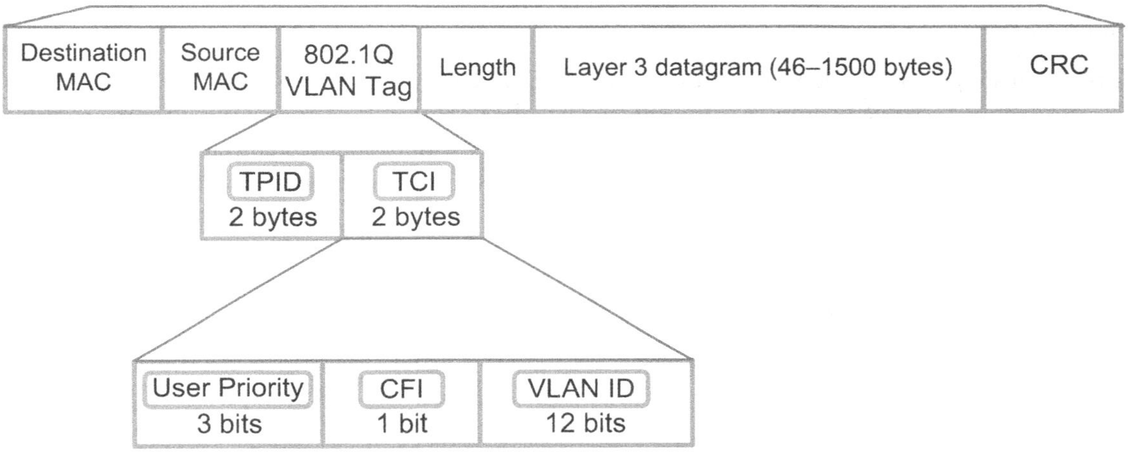


Figure 5-5: 802.1Q field (VLAN tag)

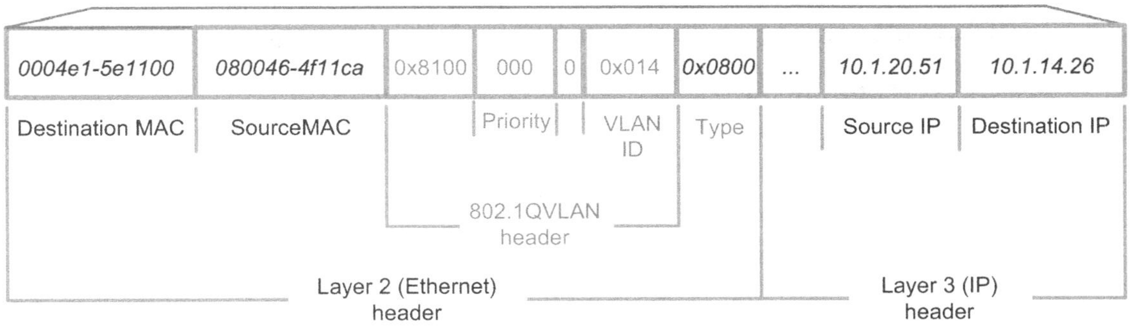


Figure 5-6: Example Ethernet header

The 802.1Q field is often called a VLAN tag, and when the 802.1Q field is inserted into an Ethernet frame, the frame is "tagged." In other words, it has VLAN ID information in it. If an Ethernet frame does not include this field, the frame is "untagged"; it does not give any information about the VLAN ID. The switch must classify the traffic on its own—for example, using a port-based assignment, as you learned earlier.

Setting up a switch port to support VLAN tagging is simple, but you take a different approach on Comware and ProVision switches.

## HP ProVision: Supporting multiple VLANs

On a ProVision port, you can simply add tagged VLAN assignments to any port. The port is then allowed to send and receive traffic that has a tag with that specific VLAN ID (remember that tag is another term for the 802.1Q field). The port drops tagged traffic with a VLAN ID that is not tagged on the port.

A port can support multiple tagged VLANs, as shown in Figure 5-7. It can also continue to support a *single untagged* VLAN assignment. You can remove the untagged VLAN assignment if the port has at least one tagged assignment. In that case, the port supports only tagged traffic and drops untagged traffic. If you try to remove the untagged assignment without giving the port a tagged assignment first, you receive an error. A port must be a member of at least one VLAN.

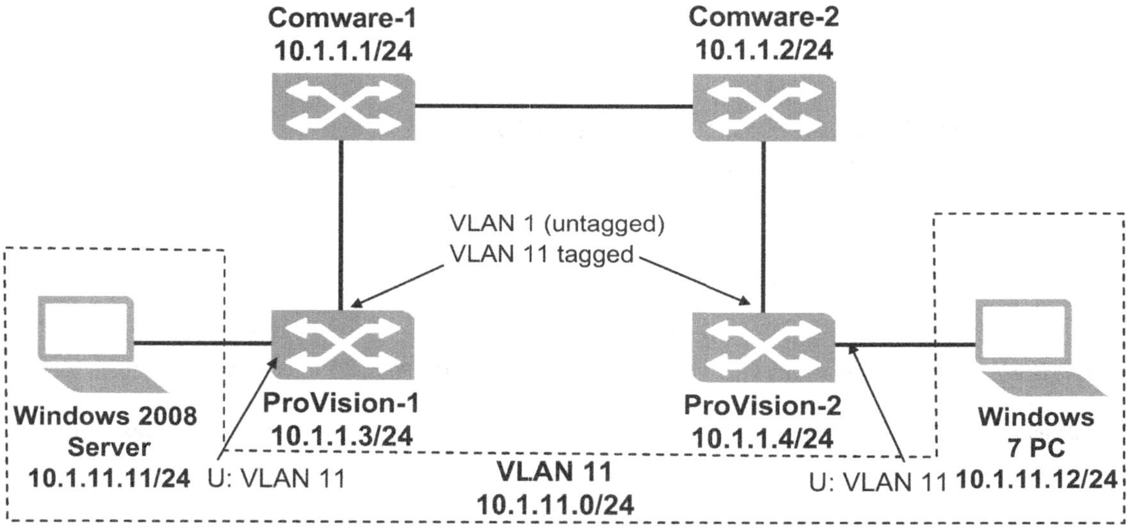


Figure 5-7 ProVision tagged VLAN assignments

HP ExpertOne If you want to view a video about VLANs, access the HP Aurasma application on your smartphone and use the application to focus on Figure 5-7. This action will launch the video.

## HP Comware: Supporting multiple VLANs

A Comware access port, which you looked at previously, does not accept tagged traffic. You must configure a port that you want to support multiple VLANs as a trunk port. The trunk port has two properties, which are highlighted in Figure 5-8:

- Port VLAN Identifier (PVID)—The PVID is the VLAN ID for untagged traffic. In other words, if the switch port sends traffic in this VLAN, it sends the traffic untagged. If the port receives untagged traffic, it assigns it to this VLAN.
- Permit—If a VLAN is permitted on a trunk port, the switch accepts and transmits traffic in that VLAN. If the PVID is permitted, untagged traffic is permitted. Otherwise, untagged traffic is dropped. If other VLANs are permitted, the traffic associated with those VLANs is tagged. The port drops traffic that is tagged for a VLAN that is not permitted on the port.

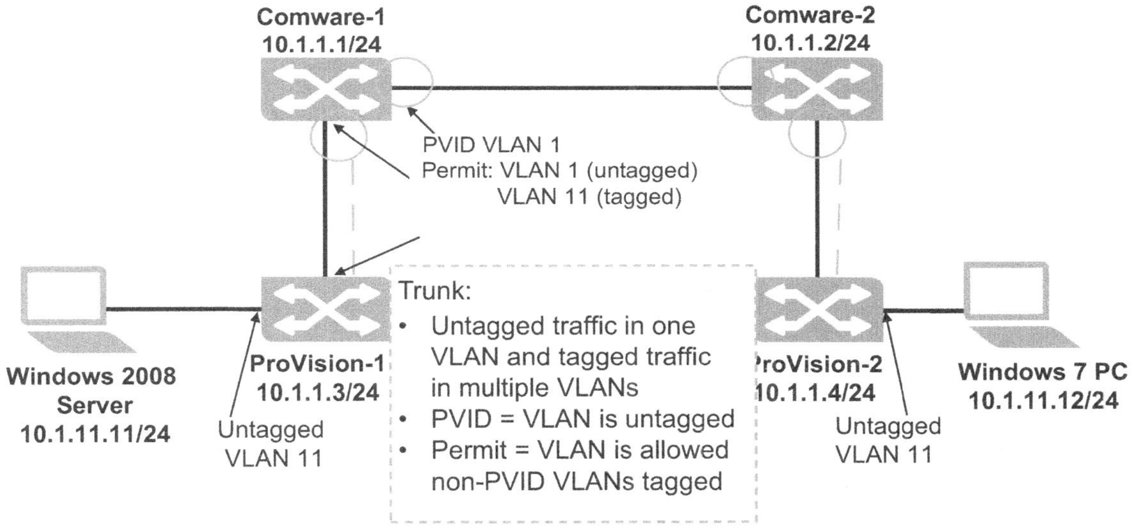


Figure 5-8: Comware trunk port settings

## Why use VLANs?

Take a moment to consider two questions:

- What are benefits of placing endpoints in a new broadcast domain/subnet (in other words, in a VLAN)?
- What are benefits of using VLANs to create broadcast domains/subnets instead of the traditional method? The traditional method required a router to separate devices in different subnets, as shown in Figure 5-9.

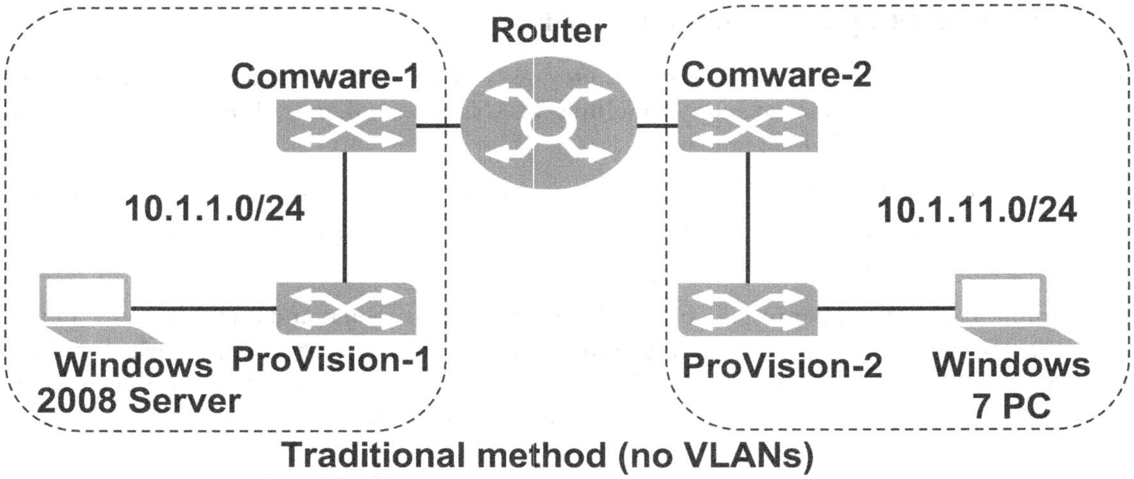


Figure 5-9: Traditional method for dividing subnets

If you want to view a video about the way VLANs are implemented on HP switches, access the HP Aurasma application on your smartphone and use the application to focus on Figure 5-9. This action will launch the video.

## Answers

Creating a separate broadcast domain, or VLAN, for your endpoints provides a number of benefits:

- You can improve security. Endpoints are now in a separate broadcast domain from the network infrastructure devices. To reach the devices, they must communicate at Layer 3 through a Layer 3 switch or router. At that point, the traffic can be better controlled.

  You might have suggested that placing servers and users in different VLANs will improve security even more. This can be true. You will look at this type of solution later in this module.

- Broadcast traffic is confined within the VLAN. Isolating the broadcasts can help to reduce congestion and CPU usage on individual devices. In this way, you can use bandwidth more efficiently and improve network performance. VLANs can also help to confine issues (such as excessive broadcasts) to a smaller number of devices.

Using VLANs rather than routers to create broadcast domains provides additional benefits. You have better flexibility in defining the broadcast domains. For example, in this scenario, you want to place endpoints on ProVision-1 and ProVision-2 in the same VLAN, even though they are in different locations. VLANs let you do so. You'll see how in more detail as the module progresses.

You could also create more than one broadcast domain on a switch. For example, in a later scenario, you will place servers and clients in different VLANs. You might have a switch that connects to both types of devices, and you can use VLANs to isolate them. Thus network construction and maintenance is much easier and more flexible.

## Example configuration: Configure VLANs on HP switches

This section provides step-by-step instructions for configuring VLANs on HP ProVision and Comware switches. Specifically, it explains how to create VLANs, implement tagged and untagged ports on ProVision switches, and configure trunk ports on Comware switches.

To illustrate how VLANs can be extended over a network, this section uses an example configuration, based on the network topology shown in Figure 5-10.

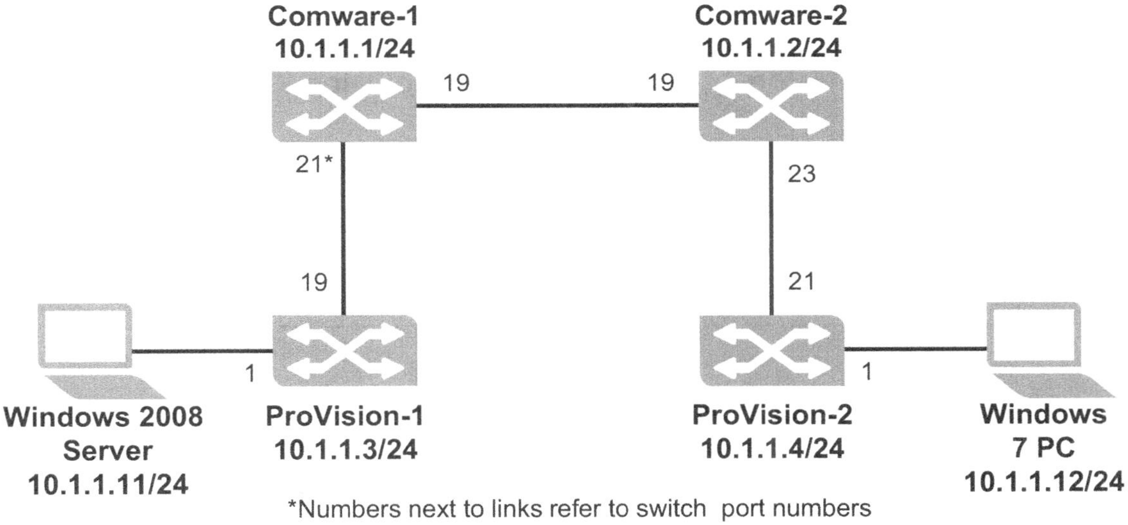


Figure 5-10: Beginning network topology for the example configuration

The four switches shown in Figure 5-10, the server, and the client are configured with the VLAN 1 IP addresses listed in Table 5-1.

Table 5-1: VLAN IP addressing scheme

| Switch | VLAN 1 | VLAN 11 |
|---|---|---|
| Comware-1 | 10.1.1.1 | |
| Comware-2 | 10.1.1.2 | |
| ProVision-1 | 10.1.1.3 | |
| ProVision-2 | 10.1.1.4 | |
| Windows 2008 | | 10.1.11.11 |
| Windows 7 | | 10.1.11.12 |

In this example configuration, a new VLAN—VLAN 11—is added to the network. The server and client in the network topology are placed in this VLAN and configured with the IP addresses shown in Table 5-1. Figure 5-11 shows network settings that will be configured for the new VLAN.

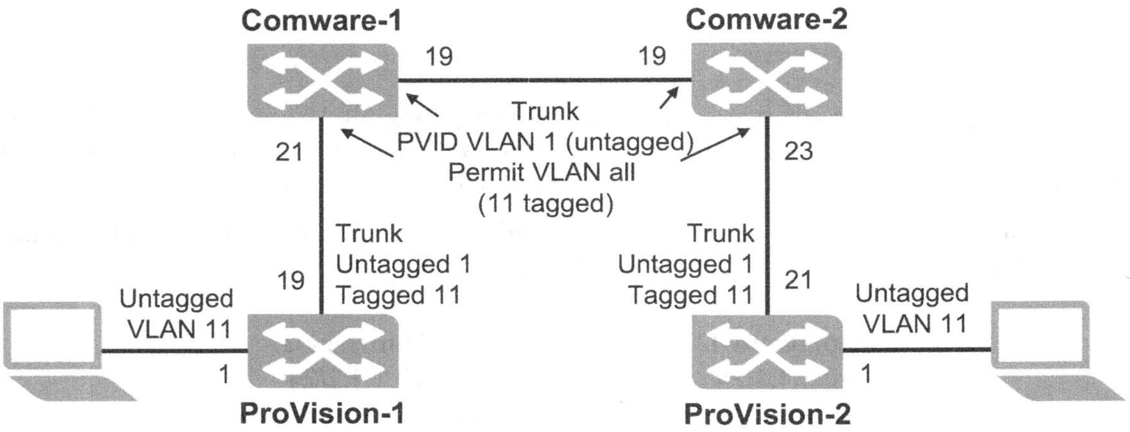


Figure 5-11: Final network topology for the example configuration

> **Important**
> These instructions do not cover the steps for setting up access to the switch and accessing it. It is assumed that you have access to the CLI on all switches.

## Configure VLAN 11 on HP ProVision-1

1. From the global configuration on the ProVision-1 switch, create VLAN 11 and then make port 1 (the Windows 2008 server connection) an untagged member of VLAN 11.

```
ProVision-1(config)# vlan 11

ProVision-1(vlan-11)# untagged 1

ProVision-1(vlan-11)# exit
```

2. Verify that the interface is an untagged member of VLAN 11.

```
ProVision-1(config)# show vlan 11

 Status and Counters - VLAN Information - VLAN 11

  VLAN ID : 11

  Name : VLAN11
```

```
Status : Port-based

Voice : No

Jumbo : No

Port Information Mode     Unknown VLAN Status

---------------- -------- ------------ ----------

1                Untagged Learn        Up
```

## Windows server

3. Access the Windows 2008 server desktop and configure its IP address. (If you are familiar with this process, you can skip to the "Configure VLAN 11 on HP ProVision-2" section on page 128.)
4. Access the **Properties** window, select **Internet Protocol version 4 (TCP/IPv4)**, and click **Properties** (see Figure 5-12).

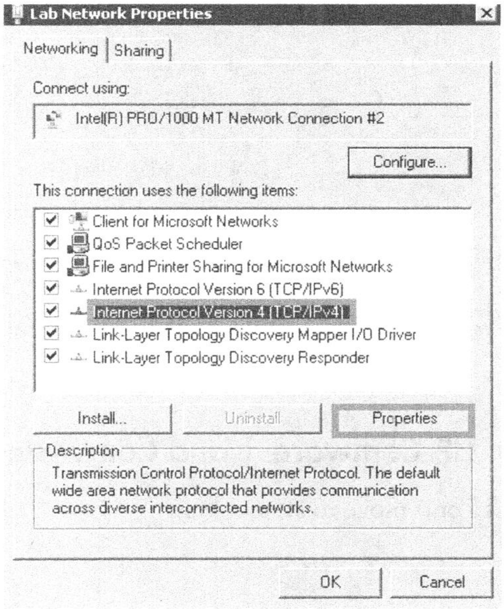


Figure 5-12: Microsoft Windows Network Properties

5. Enter these settings:
   - IP address = 10.1.11.11
   - Mask = 255.255.255.0
   - Gateway = 10.1.11.1

   For a test network such as this, you would probably set the DNS server address to 127.0.0.1.

6. Click **OK** and then click **Close**.
7. Open a command prompt on the server. Validate that you have set the IP address correctly.

```
ipconfig
```

## Configure VLAN 11 on HP ProVision-2

8. Access the ProVision-2 CLI and configure port 1 (the Windows 7 PC connection) as an untagged member of VLAN 11.

```
ProVision-2(config)# vlan 11
ProVision-2(vlan-11)# untagged 1
ProVision-2(vlan-11)# exit
```

9. View the VLANs that are configured on interface 1 and verify that the untagged VLAN is 11.

```
ProVision-2(config)# show vlan port 1 detail
 Status and Counters - VLAN Information - for ports 1
  VLAN ID Name                 | Status     Voice Jumbo Mode
  ------- -------------------- + ---------- ----- ----- --------
  11      VLAN11               | Port-based No    No    Untagged
```

## Windows client

10. Access the Windows 7 PC desktop and configure its IP address, 10.1.11.12/24. Configure the gateway as 10.1.11.1.

## Configure VLAN 11 on HP Comware-1 and Comware-2

11. Access the Comware-1 CLI and move to system view.

```
<Comware-1> system
```

12. Add VLAN 11.

```
[Comware-1] vlan 11
```

13. Access the Comware-2 CLI and move to system view.
14. Add VLAN 11.

```
[Comware-2] vlan 11
```

## Test connectivity

15. Access the server desktop and open a command prompt. Attempt to ping the client.

```
ping 10.1.11.12
```

16. The ping fails. Adding a VLAN to the switches is not enough to establish connectivity. You need to extend the VLAN on the switch-to-switch links.

## Configure switch-to-switch links on HP ProVision-1

17. Access the ProVision-1 CLI and configure the interface that connects to Comware-1 as a tagged member of VLAN 11. (You should already be in the VLAN 11 context.)

```
ProVision-1(config)# vlan 11 tagged 19
```

18. Verify the port membership for VLAN 11. Notice that port statuses are UP since the interfaces are currently up.

```
ProVision-1(config)# show vlan 11

 Status and Counters - VLAN Information - VLAN 11

  VLAN ID : 11

  Name : VLAN11

  Status : Port-based

  Voice : No

  Jumbo : No

  Port Information Mode     Unknown VLAN Status

  ---------------- -------- ------------ ----------

  1                Untagged Learn        Up

  19               Tagged   Learn        Up
```

## Configure switch-to-switch links on Comware-1

19. Access the Comware-1 CLI.

20. Move to the Comware-1 port that connects to ProVision-1. Define the port as a trunk and permit the trunk to carry all VLANs configured on the switch.

```
[Comware-1-vlan11] interface g1/0/21
```

```
[Comware-1-GigabitEthernet1/0/21] port link-type trunk
[Comware-1-GigabitEthernet1/0/21] port trunk permit vlan all
```

21. Move to the Comware-1 port that connects to Comware-2 and complete the same steps.

```
[Comware-1-GigabitEthernet1/0/21] interface g1/0/19
[Comware-1-GigabitEthernet1/0/19] port link-type trunk
[Comware-1-GigabitEthernet1/0/19] port trunk permit vlan all
```

22. View VLAN 11 and verify that both interfaces carry this VLAN.

```
[Comware-1] display vlan 11
 VLAN ID: 11
 VLAN Type: static
 Route Interface: Not configured
 Description: VLAN 0011
 Name: VLAN 0011
 Tagged   Ports:
    GigabitEthernet1/0/19    GigabitEthernet1/0/21
 Untagged Ports: none
```

## Configure switch-to-switch links on Comware-2

23. Access the terminal session for the Comware-2 switch.
24. Move to the Comware-2 port that connects to Comware-1. Define the port as a trunk and permit the trunk to carry all VLANs configured on the switch.

```
[Comware-2-vlan11] interface g1/0/19
[Comware-2-GigabitEthernet1/0/19] port link-type trunk
[Comware-2-GigabitEthernet1/0/19] port trunk permit vlan all
```

25. Move to the Comware-2 port that connects to ProVision-2 and complete the same steps.

```
[Comware-2-GigabitEthernet1/0/19] interface g1/0/23
[Comware-2-GigabitEthernet1/0/23] port link-type trunk
[Comware-2-GigabitEthernet1/0/23] port trunk permit vlan all
```

26. View trunk ports and verify that the two interfaces are trunks that permit VLAN 11. Note the PVID.

```
[Comware-2] display port trunk

Interface                PVID  VLAN passing

GE1/0/19                    1     1, 11

GE1/0/23                    1     1, 11
```

## Configure switch-to-switch links on ProVision-2

27. Access the ProVision-2 CLI.

28. Define the ProVision-2 port that connects to Comware-2 as a tagged member of VLAN 11. Remember to specify VLAN 11 if you are not already in the VLAN 11 context.

```
ProVision-2((config)# vlan 11 tagged 21
```

Verify the port membership for VLAN 11. Notice that port statuses are UP, since the interfaces are currently up.

```
ProVision-2(config)# show vlan 11

 Status and Counters - VLAN Information - VLAN 11

  VLAN ID : 11

  Name : VLAN11

  Status : Port-based

  Voice : No

  Jumbo : No

  Port Information Mode     Unknown VLAN Status

  ---------------- -------- ------------ ----------

  1                Untagged Learn        Up

  21               Tagged   Learn        Up
```

## Windows server

You have now extended VLAN 11 across your topology. The server and client should be able to reach each other. Validate now.

29. Access the server desktop and open a command prompt.

30. Ping the client and verify success.

```
ping 10.1.11.12
```

## Troubleshooting tip

If your ping succeeds, move directly to "Save the configuration" on page 133.

If the ping does not succeed, complete the following steps. You have probably missed adding VLAN 11 as a tagged VLAN on one of the interfaces. You can use this tip to troubleshoot:

31. Assign each switch an IP address in VLAN 11.

```
[Comware-1] interface vlan 11

[Comware-1-Vlan-interface11] ip address 10.1.11.1 24

[Comware-2] interface vlan 11

[Comware-2-Vlan-interface11] ip address 10.1.11.2 24

ProVision-1(config)# vlan 11 ip address 10.1.11.3/24

ProVision-2(config)# vlan 11 ip address 10.1.11.4/24
```

32. Access the server desktop and open a command prompt. Ping the IP addresses is the order shown below.

```
ping 10.1.11.3

ping 10.1.11.1

ping 10.1.11.2

ping 10.1.11.4
```

33. The first ping that fails indicates where the tagging is missing. For example, if the ping to ProVision-1 succeeds, but the ping to Comware-1 fails, the tagging is missing on the link between these switches. See Table 5-2 for help in pinpointing the problem.

34. Check the tagging on both sides of the link.

    - To check the tagging on a Comware switch, use the `display port trunk` command.
    - To check the tagging on a ProVision switch, use the `show vlan port <ID>` command.

35. Fix the problem.

36. Remove the IP addresses for now.

```
[Comware-1] undo interface vlan 11

[Comware-2] undo interface vlan 11

ProVision-1(config)# no vlan 11 ip address

ProVision-2(config)# no vlan 11 ip address
```

Table 5-2: Troubleshooting table

| Failed ping | Link with the issue |
|---|---|
| 10.1.11.3 | Link between the server and ProVision-1 |
| 10.1.11.1 | Link between the ProVision-1 and Comware-1 |
| 10.1.11.2 | Link between Comware-1 and Comware-2 |
| 10.1.11.4 | Link between Comware-2 and ProVision-2 |
| 10.1.11.12 | Link between ProVision-2 and the PC |

## Save the configuration

37. Save the configuration on each switch. For ProVision, use the `write memory` command. For Comware, use the `save [force]` command.

# Learning check 1

Answer the following questions to test your understanding of VLANs.

1. Which command can you use to see all interfaces that are part of a VLAN?
   a. On Comware switches?
   b. On ProVision switches?

2. Which command or commands can you use to see all the VLANs that are carried on an interface?
   a. On Comware switches?
   b. On ProVision switches? Which option do you need to use to see whether the VLANs are tagged or untagged?

## Answers to learning check 1

1. Which command can you use to see all interfaces that are part of a VLAN?

   a. On Comware switches?

      `display vlan <vid>`

   b. On ProVision switches?

      `show vlan <vid>`

2. Which command or commands can you use to see all the VLANs that are carried on an interface?

   a. On Comware switches?

      `display port trunks`

      `display interface <int-id>`

   b. On ProVision switches? Which option do you need to use to see whether the VLANs are tagged or untagged?

      `show vlan port <int-id> [detail]`

      Use the `detail` option to see whether the VLANs are tagged or untagged.

## Review: Trace tagging across the topology

Figure 5-13 shows a frame that is sent from the Windows 7 PC to the server in the example. The path is divided into five segments.

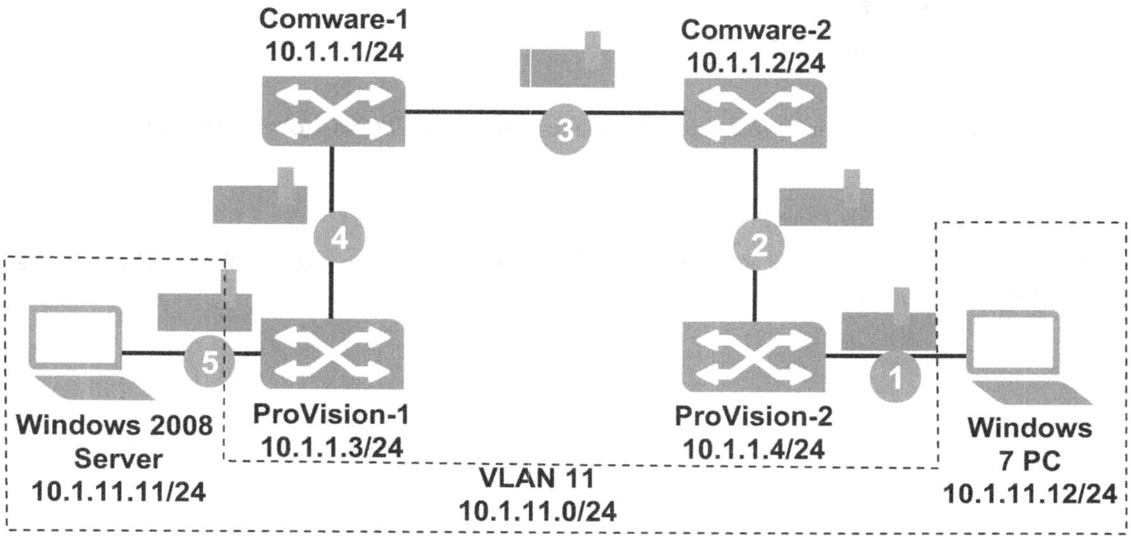


Figure 5-13: Review: Trace tagging across the topology

Table 5-3 includes a row for each segment. For each segment, decide the VLAN to which the frame belongs and whether the frame is tagged or untagged. Then fill in the second column for the row that corresponds to that segment. For example, for the first row, the PC transmits the frame to ProVision-2:

- To which VLAN does ProVision-2 assign the frame?
- Is the frame untagged and tagged?

Table 5-3: Tracing a frame across the network topology

| Segment | VLAN ID Tagged or untagged | Source MAC Address | Destination MAC Address |
|---|---|---|---|
| 1—PC to ProVision-2 | | Client | Server |
| 2—ProVision-2 to Comware-2 | | Client | Server |
| 3—Comware-2 to Comware-1 | | Client | Server |
| 4—Comware-1 to ProVision-1 | | Client | Server |
| 5—ProVision-1 to Server | | Client | Server |

Table 5-4 shows the VLAN tagging for the frame as it is sent from the server to the client.

Table 5-4: Answers for Tracing a frame across the topology

| Segment | VLAN ID tagged or untagged | Source MAC address | Destination MAC address |
|---|---|---|---|
| 1—PC to ProVision-2 | 11 untagged | Client | Server |
| 2—ProVision-2 to Comware-2 | 11 untagged | Client | Server |
| 3—Comware-2 to Comware-1 | 11 untagged | Client | Server |
| 4—Comware-1 to ProVision-1 | 11 untagged | Client | Server |
| 5—ProVision-1 to Server | 11 untagged | Client | Server |

## Add another VLAN

Sometimes you do not want all of the endpoints in the same VLAN. In the example configuration, for example, you may want to place the servers and the user devices in different VLANs:

- Servers in VLAN 11
- User devices in VLAN 12

Both ProVision-1 and ProVision-2 will support devices in both VLANs, and the VLANs will be extended across the switch topology, as illustrated in Figure 5-14.

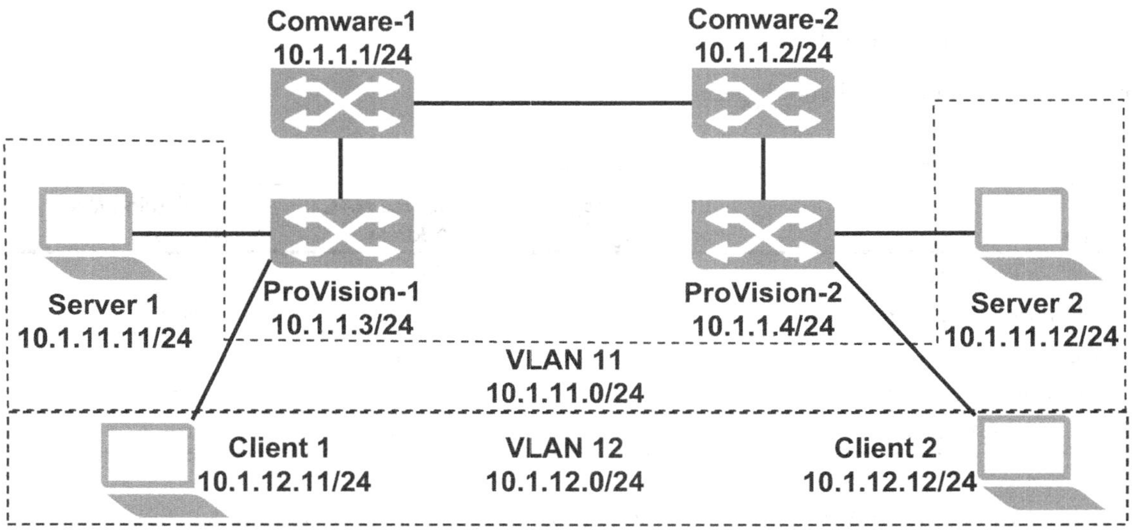


Figure 5-14: Add another VLAN

As Figure 5-15 illustrates, the VLAN 11 and VLAN 12 devices are logically separated as if by different routers. Physically, though, the devices are located wherever they need to be, as illustrated in Figure 5-14. The ability to create a solution like this is a benefit of using VLANs.

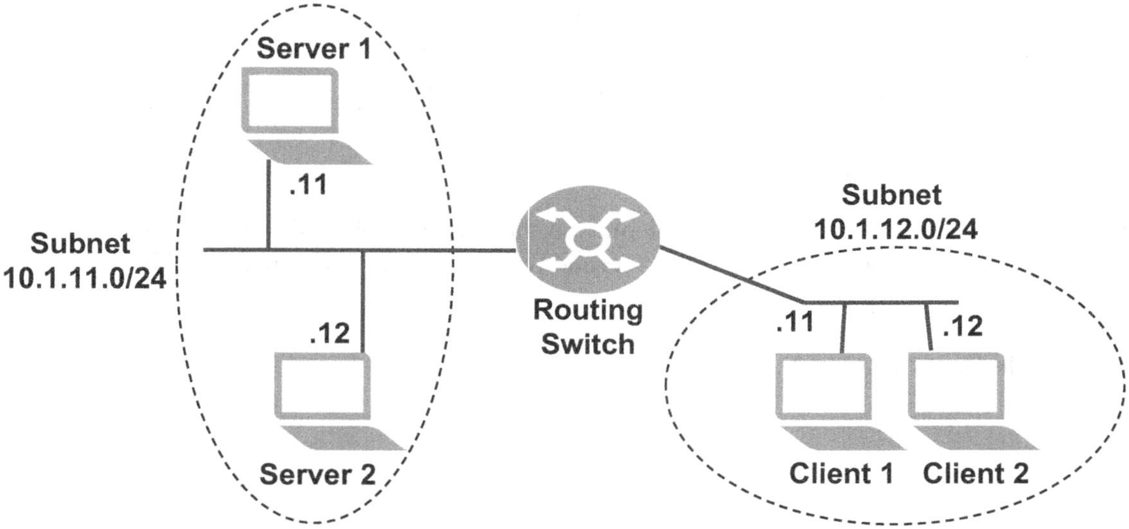


Figure 5-15: Logical topology

## Review: Adding another VLAN

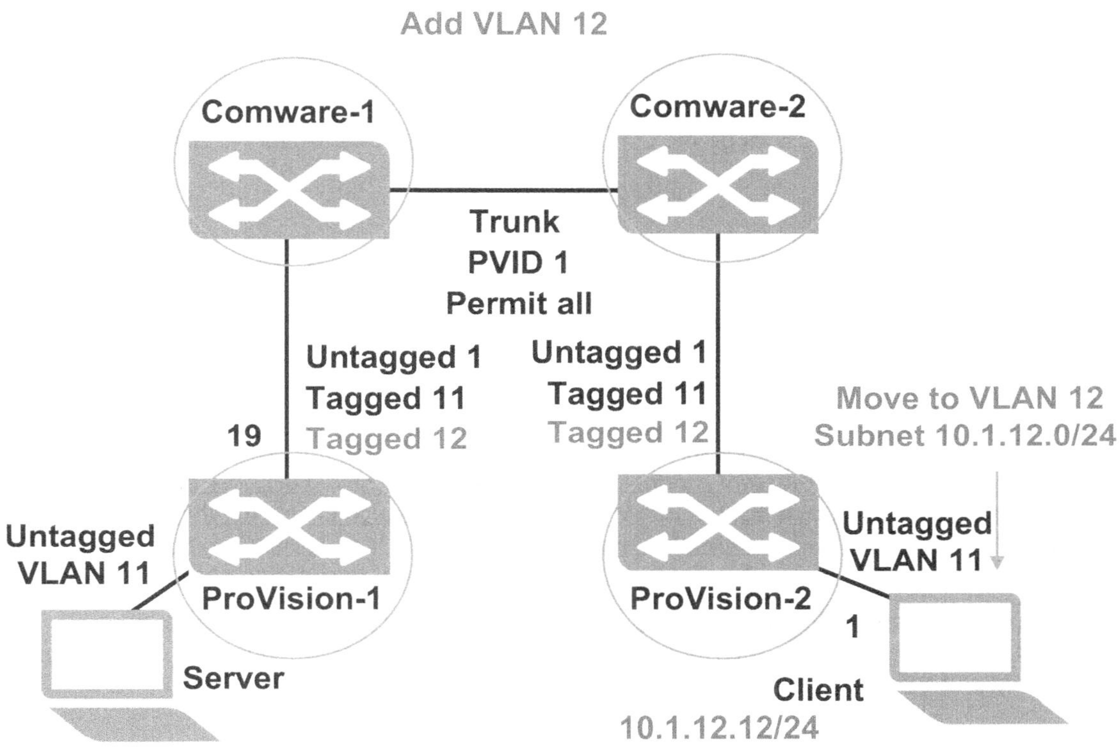


Figure 5-16: Adding another VLAN to the network topology

Take a few minutes to consider the following questions. Refer to Figure 5-16 as you answer the questions.

1. Why would you want to put different endpoints in different VLANs?

2. In the next section, you will return to the example configuration and review the steps to move the client to VLAN 12 and extend VLAN 12 across the entire network topology, as Figure 5-16. For this example, you want to set up a solution in which both ProVision-1 and ProVision-2 can connect to clients that should be in VLAN 12.

   How do you change the Windows PC's IP address?

3. On which switches, do you need to add VLAN 12?

4. How do you need to change the configuration for the ProVision-2 interface that connects to the Windows PC?

5. How do you need to change the configuration for the ProVision switch-to-switch interfaces?

6. Why is it unnecessary to change the configuration on the Comware interfaces?

## Answers

1. Why would you want to put different endpoints in different VLANs?

   Ideas include:

   You want to treat the endpoints in a different way and set up controls between their communications. By placing the server and the client in different VLANs, you ensure that they have IP addresses with different ranges and that the clients' traffic must pass through a router to reach the server. Access control lists (ACLs) can control which clients in which VLANs can reach which servers in which VLANs. To help you to impose these controls, you might assign different users to different VLANs based on identity or identity and endpoint type.

   Assigning different endpoints to different VLANs also helps to create reasonably sized broadcast domains, which in turn improves performance. Therefore, you might also assign different endpoints to different VLANs based on their location.

2. How do you change the Windows PC's IP address?

   You need to change the IP properties for the NIC. The new address is in the VLAN 12 subnet, 10.1.12.0/24.

   (You might also suggest using DHCP to assign addresses automatically. DHCP is a great way to assign addresses to clients more easily, and you will cover DHCP in the next module.)

3. On which switches, do you need to add VLAN 12?

   All of the switches require VLAN 12.

   The scenario stated that both ProVision-1 and ProVision-2 need to support endpoints in VLAN 12. Therefore, both ProVision switches need to support VLAN 12, as do both of the Comware switches, which connect the ProVision switches to each other.

   You might also know that a subnet needs a default router and that the VLAN must extend to that router.

4. How do you need to change the configuration for the ProVision-2 interface that connects to the Windows PC?

   You must make this interface an untagged member of VLAN 12.

5. How do you need to change the configuration for the ProVision switch-to-switch interfaces?

   You must add VLAN 12 as a tagged membership. The interfaces continue to have the same untagged membership and to keep the tagged membership in VLAN 11.

6. Why is it unnecessary to change the configuration on the Comware interfaces?

   The Comware switch-to-switch interfaces also must support VLAN 12 as a tagged VLAN. However, you do not have to change the configuration on these interfaces because they permit all VLANs on the switch. As soon as you add VLAN 12 to the switch, the interface will begin to accept and transmit tagged VLAN 12 traffic.

## Example configuration: Add a VLAN 12

This section builds on the example configuration, adding VLAN 12 as shown in Figure 5-17. The network topology includes VLANs 1 and 11.

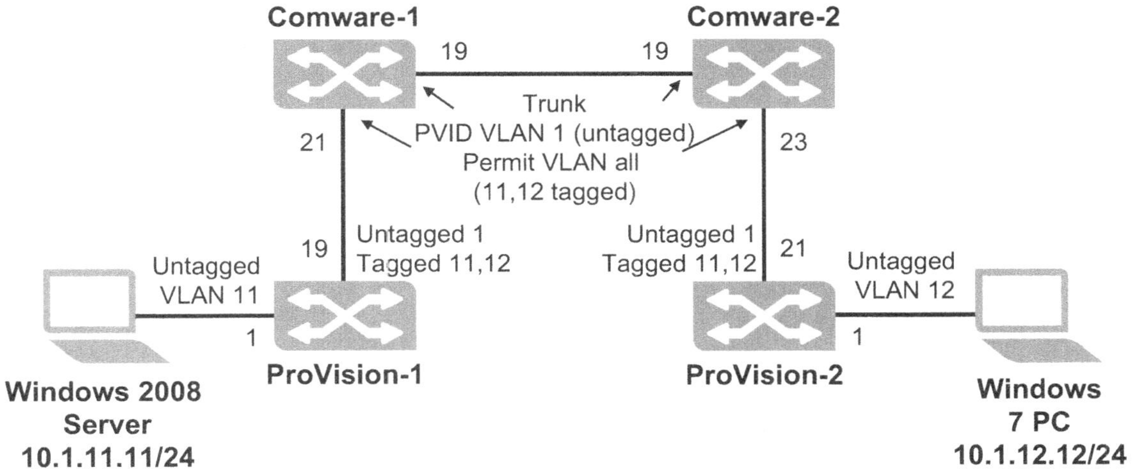


Figure 5-17: Final network topology

Table 5-5 shows the VLANs and IP addresses that will be established in the network topology. (Changes are indicated in boldface type.)

Table 5-5: VLAN IP addressing scheme

| Switch | VLAN 1 | VLAN 11 | VLAN 12 |
|---|---|---|---|
| Comware-1 | 10.1.1.1 | | |
| Comware-2 | 10.1.1.2 | | |
| ProVision-1 | 10.1.1.3 | | |
| ProVision-2 | 10.1.1.4 | | |
| Windows 2008 | | 10.1.11.11 | |
| Windows 7 | | | **10.1.12.12** |

1. Access the Windows 7 PC desktop. Change its IP address to 10.1.12.12/24 and set the default gateway address to 10.1.12.1.
2. After you have set the address, remember to click **OK** in the Properties window and then **Close**.
3. Enter the following commands to add VLAN 12 to each switch.

```
[Comware-1] vlan 12
[Comware-2] vlan 12
ProVision-1(config)# vlan 12
ProVision-2(config)# vlan 12
```

4. Make sure that you are at the ProVision-2 CLI and then make sure that you are in the VLAN 12 context.
5. Enter this command:

```
ProVision-2(vlan-12)# untagged 1
```

6. Make the ProVision-2 port that connects to Comware-2 a tagged member of VLAN 21.

```
ProVision-2(vlan-12)# tagged 21
```

7. Access the ProVision-1 CLI.
8. Make sure that you are in the VLAN 12 context.
9. Add VLAN 12 as a tagged VLAN on the ProVision-1 port that connects to Comware-2.

```
ProVision-1(vlan-12)# tagged 19
```

## Verify the VLAN topology

The Comware switch-to-switch interfaces also must carry VLAN 12 as a tagged VLAN. However, you did not have to change the configuration on these interfaces because they are permitted to carry all VLANs on the switch. As soon as you added VLAN 12 to the switch, the switch-to-switch interfaces began to accept and transmit tagged VLAN 12 traffic. This is because the trunk ports are configured to permit all VLANs. If the trunk ports only permitted specific VLANs, you would need to add VLAN 12 to the list.

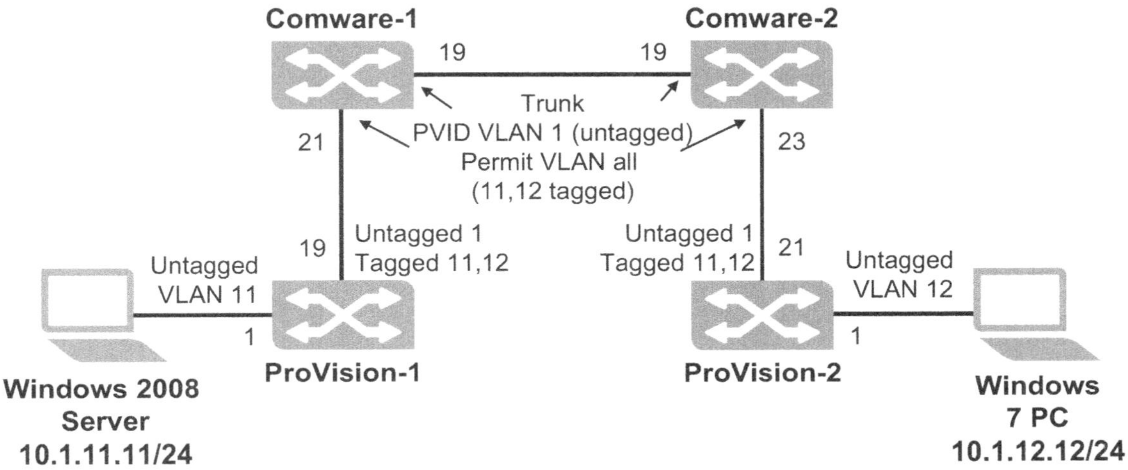


Figure 5-18: Final network topology

You can now use the following `show` and `display` commands to verify that the VLAN topology matches that shown Figure 5-18.

```
ProVision-1# show vlan 12

 Status and Counters - VLAN Information - VLAN 12

<-output omitted->

  Port Information Mode     Unknown VLAN Status

  ---------------- -------- ------------ ----------

  19               Tagged   Learn        Up

<Comware-1> display port trunk

Interface             PVID    VLAN passing

GE1/0/19                 1      1, 11-12,

GE1/0/21                 1      1, 11-12,

<Comware-2> display port trunk

Interface             PVID    VLAN passing

GE1/0/19                 1      1, 11-12,
```

```
GE1/0/23                1      1, 11-12,

ProVision-2# show vlan 12

 Status and Counters - VLAN Information - VLAN 12

<-output omitted->

  Port Information Mode     Unknown VLAN Status
  ---------------- -------- ------------ ----------
  1                Untagged Learn        Up
  21               Tagged   Learn        Up
```

## Save

Save the current configuration on each of the four switches. For ProVision, use the `write memory` command. For Comware, use the `save` command.

## Learning check 2

You will answer some questions about the example configuration.

1. Can the Windows 7 PC reach the server?

2. When you attempt to ping the Windows 7 PC from the server, for which IP address does the server send an ARP request?

## Answers to learning check 2

1. Can the Windows 7 PC reach the server?

   No. The PC and the server cannot communicate because they are in different subnets, and they need a router to route their traffic.

2. When you attempt to ping the Windows 7 PC from the server, for which IP address does the server send an ARP request?

   The server sent an ARP message, not to the client, but to its default gateway address. It did this because it recognized that the client's IP address was not part of its own subnet; therefore, it attempted to find its default gateway, which is within its own subnet. (You set the default gateway address as 10.1.11.1 when you set the server's IP address.)

Currently, the server is not receiving a response because you have not set up the default gateway. You will complete that task in a minute.

## Routing between VLAN 11 and 12

In the example configuration, endpoints in different VLANs can no longer reach each other at Layer 2. Nor can they exchange broadcasts such as ARP messages. When a device needs to reach a device in another subnet, the traffic needs to be routed. Endpoints that cannot route their own traffic send the traffic to a default gateway for routing.

If you attempted to ping the PC from the server in the example network, the server would send an ARP message, not to the client, but to its default gateway address. The server would do this because it recognized that the client IP address was not part of its own subnet; therefore, it would attempt to find its default gateway, which is within its own subnet.

To allow an endpoint in VLAN 11 and VLAN 12 to communicate, you need to set up the default gateway in each VLAN (see Figure 5-19).

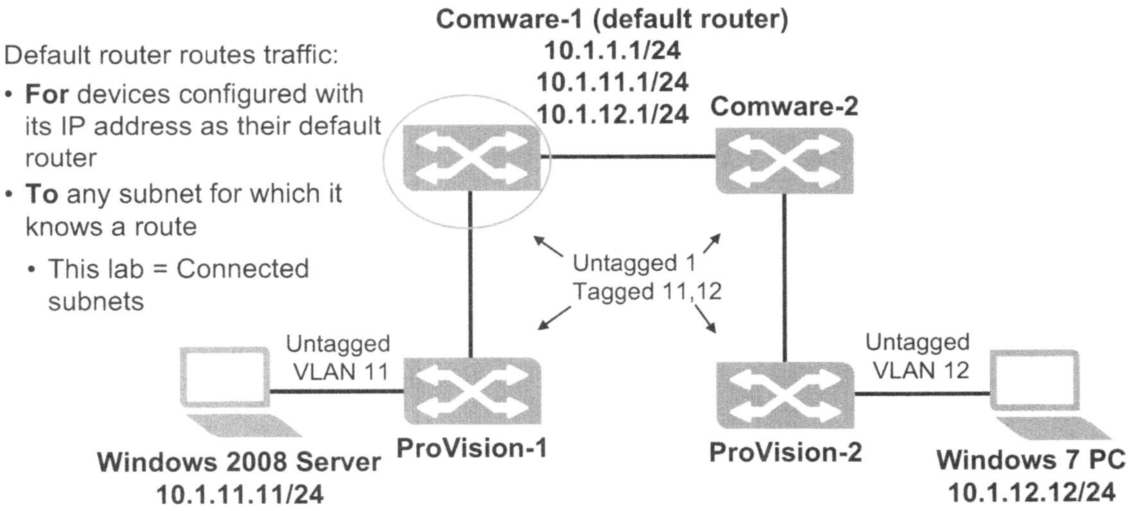


Figure 5-19: Routing between VLAN 11 and 12

In this example network, Comware-1 acts as the default gateway. You simply need to assign Comware-1 the default gateway address for each subnet on the associated VLAN. Table 5-6 lists these IP addresses.

Table 5-6: VLAN IP addresses

| VLAN paragraph style) | IP subnet | IP default gateway address |
|---|---|---|
| 11 | 10.1.11.0/24 | 10.1.11.1/24 |
| 12 | 10.1.12.0/24 | 10.1.12.1/24 |

After you assign the IP addresses to Comware-1, the Comware switch will automatically route traffic for other devices, such as endpoints configured to use its IP address as their gateway. Remember the server and PC were configured manually with these default gateway addresses.

**Important**

To achieve the same effect on a ProVision switch, you must also enable IP routing, which is disabled by default. The command is: `ip routing`.

The switch will route traffic to any network for which it knows a route. It automatically knows a route for any subnet on which it has an IP address. This type of subnet is called a direct or connected route. For this activity, you simply want to route between VLAN 11, 10.1.11.0/24 and VLAN 12, 10.1.12.0/24. Comware-1 will have an IP address on both of those VLANs, so you do not need to set up any additional routing.

**Note**

You will learn about setting up IP routing for more complicated topologies in "Chapter 9: IP Routing."

## Configuring a default route for switches

In this section, you will learn how to set up a default router for your switches. Comware-2, ProVision-1, and ProVision-2 each have only one IP address on one VLAN. They are acting essentially as Layer 2 devices. Just like the client and the server, the switches need a default router if they need to communicate with devices outside of their VLAN. Comware-1 will be their default router on VLAN 1 using its 10.1.1.1 IP address.

To complete the configuration on the HP switches, you create a default route with the default router IP address (10.1.1.1, in this scenario) as the next hop, as illustrated in Figure 5-20.

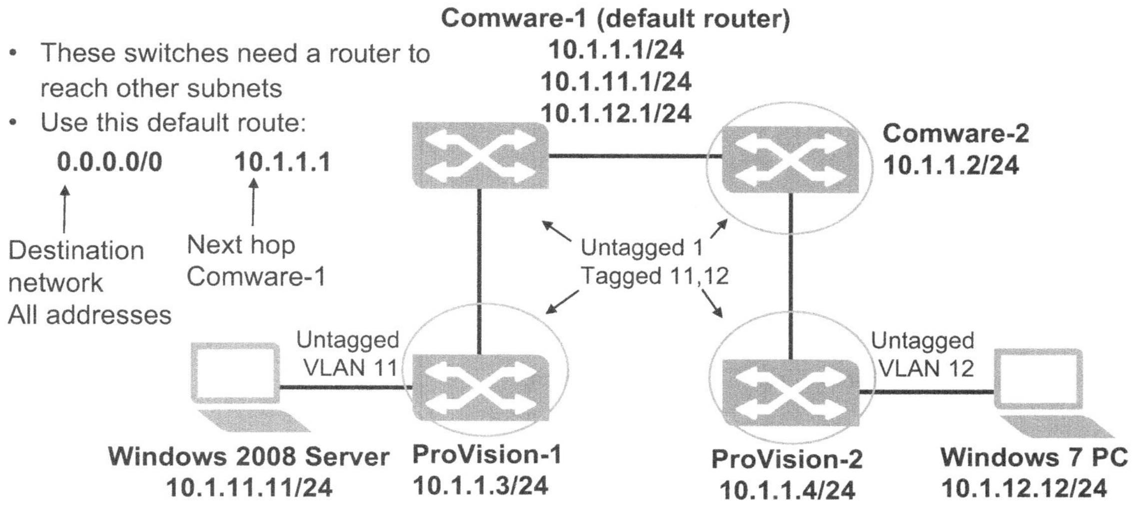


Figure 5-20: Configuring a default route for switches

The default route is a route to network 0.0.0.0/0. The 0 prefix length means that this network address matches *all* IP addresses. In other words, whenever the switch needs to reach any IP address that is not in its same subnet, the switch sends the traffic to the default router.

---

**Note**

When a ProVision switch has IP routing disabled (the default), you can use an alternative command to set the default router: `ip default-gateway <default router IP address>`. This command only works when IP routing is disabled. The `ip route` command for creating the default route, on the other hand, works whether or not routing is enabled, so you will use it in Activity 5.3.

---

## Example configuration: Set up basic IP routing

In this section, the example configuration will be modified to:

- Enable tracerouts on a Comware switch
- Set up a Comware switch as the default router
- Configure default routes on the other switches

### Prepare for traceroutes

The `traceroute` command can be used to trace the route between endpoints. When a device traces a route, it sends an Internet Control Message Protocol (ICMP) message (a ping) with a time to live (TTL) of 1. When the next hop in the IP route receives that message, it detects that the TTL has expired and returns an ICMP unreachable response with its IP address. Then the device tracing the route sends an ICMP message with a TTL of 2, which receives a response from the next hop, and so on, until the ICMP message reaches the final destination. In this way, the device can detect every hop in the route.

By default, Comware devices ignore ICMP messages with TTLs and do not respond with ICMP unreachable messages. Therefore, a traceroute fails when it reaches a Comware device. To allow your traceroutes to succeed, you must enable these functions on the Comware switches.

1. Access the Comware-1 CLI and move to system view.
2. Enable these functions on the Comware-1 switch.

```
[Comware-1] ip ttl-expires enable
[Comware-1] ip unreachables enable
```

## Set up Comware-1 as the default router for VLAN 11 and 12

To set up Comware-1 as the default router for VLAN 11 and VLAN 12, you simply need to assign the switch the default router IP address. You set this address manually on the server and Windows PC when you configured their IP settings.

3. On the Comware-1 switch, create a Layer 3 interface associated with VLAN 11.

```
[Comware-1] interface vlan 11
```

4. Assign Comware-1 the default router IP address on this interface.

```
[Comware-1-VlanInterface11] ip address 10.1.11.1 24
```

5. Access the Windows 2008 server desktop and open a command prompt.
6. Perform a traceroute to 10.1.1.1. (The `-d` option disables reverse DNS lookups, which speeds up the process).

```
tracert -d 10.1.1.1

Tracing route to 10.1.1.1 over a maximum of 30 hops

  1     1 ms     2 ms     1 ms  10.1.1.1
```

Instead of using the `tracert` command, you can use the `pathping` command (`pathping 10.1.1.1`), which also shows routing hops but also includes the hostname of the client. If you prefer the `pathping` command, you can use it in the following steps.

7. Perform a traceroute to the Windows 7 PC.

```
tracert -d 10.1.12.12

Tracing route to 10.1.12.12 over a maximum of 30 hops

  1     10.1.11.1 reports: Destination net unreachable.
```

The trace failed for the IP address in VLAN 12, but it was successful for the IP address in VLAN 1. Comware-1 has an IP address in the 10.1.1.0/24 subnet and knows a route to this subnet. However, it does not know a route to the VLAN 12 subnet. You will add a connected route now by assigning Comware-1 an IP address on VLAN 12.

8. On the Comware-1 switch, create a Layer 3 interface for VLAN 12.

```
[Comware-1] interface vlan 12
```

9. You will first conduct an experiment by assigning Comware-1 a different IP address from the VLAN 12 default router address. Assign Comware-1 IP address 10.1.12.2/24 on this interface.

```
[Comware-1-VlanInterface12] ip address 10.1.12.2 24
```

10. Access a command prompt on the Windows server. Trace a route to Comware-1 at 10.1.12.2.

```
tracert -d 10.1.12.2

Tracing route to 10.1.12.2 over a maximum of 30 hops

  1     1 ms     2 ms     1 ms  10.1.12.2
```

11. Trace a route to the Windows 7 PC.

```
tracert -d 10.1.12.12

Tracing route to 10.1.12.12 over a maximum of 30 hops

  1     1 ms     2 ms     1 ms  10.1.11.1

  2     *        *        *     Request timed out.

  3     *        *        *     Request timed out.
```

The first traceroute succeeded, but the second one timed out. Why?

When you assigned Comware-1 an IP address in the 10.1.12.0/24 subnet, it can route traffic to this subnet. The traceroute to its own IP address succeeds. The traceroute to the Windows 7 PC reaches the PC. However, the PC cannot reach its default gateway, so it cannot send the ICMP message back to the server. Therefore, the request times out.

You will fix the problem now.

12. Set the Comware-1 VLAN 12 IP address to the correct IP address for the default router.

```
[Comware-1-VlanInterface12] ip address 10.1.12.1 24
```

13. Access a command prompt on the Windows server. Trace a route to the Windows 7 PC and verify success.

```
tracert -d 10.1.12.12

Tracing route to 10.1.12.12 over a maximum of 30 hops

  1     3 ms     2 ms     1 ms  10.1.11.1

  2     5 ms    <1 ms    <1 ms  10.1.12.12
```

## Set up default routes on the Layer 2 switches

You want the server to be able to reach all of the switches. Test to determine which switches the server can reach.

14. Access a command prompt on the server. Ping these four IP addresses and observe the results. (Some pings will be unsuccessful.)

```
ping 10.1.1.1

ping 10.1.1.2

ping 10.1.1.3

ping 10.1.1.4
```

Why does the ping fail for Comware-2 and the ProVision switches? The switches are in VLAN 1, subnet 10.1.1.0/24, and have no route for reaching the server in 10.1.11.0/24. The switches need a default route to reach devices outside of their subnet. You will now configure the switches to use Comware-1 as the next hop in this route; in other words, Comware-1 is their default router.

15. Access the Comware-2 CLI and move to the system view.
16. Configure the default route.

```
[Comware-2] ip route 0.0.0.0 0 10.1.1.1
```

17. Access the ProVision-1 CLI and move to the global configuration context.
18. Configure the default route.

```
ProVision-1(config)# ip route 0.0.0.0/0 10.1.1.1
```

19. Complete the same steps on ProVision-2.

```
ProVision-2(config)# ip route 0.0.0.0/0 10.1.1.1
```

20. Access a command prompt on the server. Ping these four IP addresses and verify success.

```
ping 10.1.1.1

ping 10.1.1.2
```

```
ping 10.1.1.3
ping 10.1.1.4
```

## Save the configuration

Save the current configuration on each of the four switches. For ProVision, use the `write memory` command. For Comware, use the `save [force]` command.

## Learning check 3

You will now answer questions about basic routing concepts.

1. An admin is managing VLANS in the network. VLAN 1 is, by default, the default VLAN and the primary VLAN. On ProVision-2, the admin adds VLAN 4000.

   ```
   ProVision-2(config)# vlan 4000
   ProVision-2(vlan-4000)# exit
   ```

   The admin then changes the primary VLAN to 4000. VLAN 1 remains the default VLAN.

   ```
   ProVision-2(config)# primary-vlan 4000
   ```

   The admin defines VLAN 13 as the untagged VLAN for an unused interface, such as interface 4.

   ```
   ProVision-2(config)# vlan 13 untagged 4
   ```

   And the admin deletes VLAN 13.

   ```
   ProVision-2(config)# no vlan 13
   ```

   The admin examines the VLAN assigned to the interface and sees:

   ```
   ProVision-2(config)# show vlan port 4
    Status and Counters - VLAN Information - for ports 4
     VLAN ID Name                             | Status     Voice Jumbo
     ------- -------------------------------- + ---------- ----- -----
     1       DEFAULT_VLAN                     | Port-based No    No
   ```

   What has the admin discovered about the default VLAN? What is the role of the primary VLAN?

2. Assume that the admin wants to change the port 3 VLAN assignment to VLAN 13. What is the correct procedure? Should the admin remove the current untagged assignment, or simply assign port 3 to VLAN 13?

3. The admin tries removing the current untagged VLAN assignment.

```
ProVision-1(config)# no vlan 1 untagged 3
```

The admin receives an error. To change the untagged VLAN assignment, the admin enters the `vlan <ID> untagged <port ID>` command. The admin then defines a range of VLANs on Comware-1.

```
<Comware-1> vlan 13 to 20
```

The admin views the trunk ports and sees that all of the new VLANs are carried on the ports.

```
<Comware-1> display port trunk
```

What are reasons that the admin might want to permit all VLANs on trunk ports? What are reasons that the admin might want to permit only specific VLANs?

## Answers

1. What have you discovered about the default VLAN? What is the role of the primary VLAN?

   The default VLAN is always VLAN 1, and the ID cannot be changed. This is the VLAN to which all interfaces are assigned by default. When you deleted VLAN 13, interface 4 did not have a VLAN assigned to it anymore, so it was moved to the default VLAN, VLAN 1.

   The primary VLAN is the VLAN in which protocol data units and packets are processed. It is VLAN 1 by default, but you can change the primary VLAN ID.

2. Assume that you want to change the port 3 VLAN assignment to VLAN 13. What is the correct procedure? Should you remove the current untagged assignment or simply assign port 3 to VLAN 13?

   You should assign the port to VLAN 13. If you try to remove the current untagged assignment, you receive an error message that the port would be orphaned.

3. What are reasons that you might want to permit all VLANs on trunk ports? What are reasons that you might want to permit only specific VLANs?

   If you permit all VLANs on trunk ports, you are sure that the ports will always support a VLAN as soon as you create it on the switch. This approach can help to eliminate misconfigurations and disconnected VLANs. In some topologies, though, you do not need or want to support all VLANs on all trunk ports. In that case, you might want more control over which trunk ports support which VLANs.

## Review: Tracing a frame across the routed topology

The example network has been modified so that traffic can be routed. You will trace how a frame is forwarded across this routed topology. Figure 5-21 divides the path from the client to the server into several segments, and Table 5-7 includes a row for each segment.

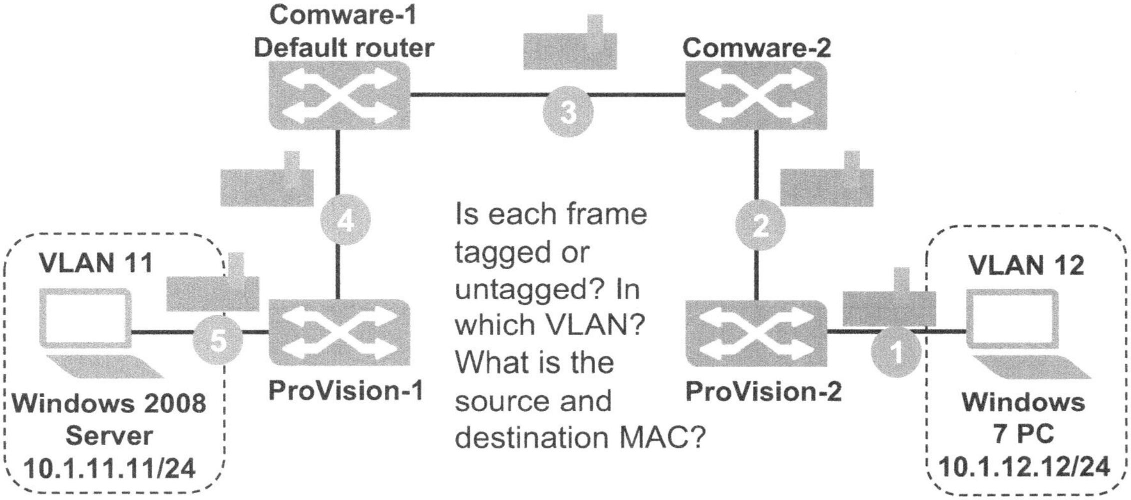


Figure 5-21: Review: Tracing a frame across the routed topology

In each row of Table 5-7, define:

- The frame's VLAN and whether the frame will be tagged using 802.1Q
- The frame's source MAC address
- The frame's destination MAC address

You do not need to write the specific MAC address. Just write server, PC, VLAN 11 default gateway, or VLAN 12 default gateway.

Table 5-7: Tracing a frame across the routed topology

| Segment | VLAN ID tagged or untagged | Source MAC address | Destination MAC address |
|---|---|---|---|
| 1—PC to ProVision-2 | | | |
| 2—ProVision-2 to Comware-2 | | | |
| 3—Comware-2 to Comware-1 | | | |
| 4—Comware-1 to ProVision-1 | | | |
| 5—ProVision-1 to Server | | | |

Table 5-8 shows the information about the routed topology.

Table 5-8: Tracing a frame across the routed topology

| Segment | VLAN ID tagged or untagged | Source MAC address | Destination MAC address |
|---|---|---|---|
| 1—PC to ProVision-2 | VLAN 12 untagged | PC | VLAN 12 default gateway |
| 2—ProVision-2 to Comware-2 | VLAN 12 tagged | PC | VLAN 12 default gateway |
| 3—Comware-2 to Comware-1 | VLAN 12 tagged | PC | VLAN 12 default gateway |
| 4—Comware-1 to ProVision-1 | VLAN 11 tagged | VLAN 11 default gateway | server |
| 5—ProVision-1 to Server | VLAN 11 untagged | VLAN 11 default gateway | server |

## HP Comware: Hybrid ports

You have learned about two types of ports on Comware switches:

- Access ports
- Trunk ports

These switches support a third type of port: hybrid ports. A hybrid port is like a trunk port in that it can support multiple VLANs. But the hybrid port can support multiple tagged VLANs *and multiple untagged VLANs.*

How can a hybrid port support multiple untagged VLANs? Without a tag, how can the switch determine to which VLAN a frame belongs? It would not be able to if it were only using port-based VLANs—the type of VLANs on which you have focused in this chapter. Therefore, a hybrid port is typically configured with another type of VLAN. These special VLAN types and their use cases are briefly described below.

If you want to learn more about these use cases and how to set up hybrid ports to support them, you should attend other HP training.

### MAC-based VLAN (RADIUS-assigned VLANs for multiple endpoints)

When a hybrid port is configured to support MAC-based VLANs, it examines the inbound MAC address of incoming traffic to determine the correct VLAN for that traffic. You can set up static MAC-to-VLAN mappings, but typically the switch learns the mappings dynamically from a RADIUS server. Any untagged traffic that is not associated with a MAC-based VLAN is assigned to the hybrid port's PVID. Outgoing traffic to these MAC addresses is sent untagged. Broadcasts for all VLANs untagged on the port are also sent untagged.

In the typical use case for MAC-based VLANs, a port connects to multiple downstream endpoints—perhaps through a less intelligent switch—and enforces authentication. The endpoints authenticate to a RADIUS server, which assigns them to VLANs based on the user identity, device type, or other criteria. The RADIUS server informs the switch of the assigned VLAN for each endpoint, and the switch port keeps track of all the source-MAC-address-to-VLAN mappings. In this way, the switch and centralized authentication server can work together to provide flexible, secure access for a variety of devices, including PCs, laptops, phones, and so on.

## Voice VLAN (VoIP phones)

Voice VLANs are used with voice over IP (VoIP) phones. A hybrid port is required when both a phone and a computer connect to the same port, and the phone sends untagged traffic. The voice VLAN setting helps the port to recognize the untagged VoIP traffic and assign it to a different VLAN from the computer traffic. The switch has a vendor OID list so that it can recognize the source MAC address of VoIP traffic.

Often VoIP phones tag their traffic. Although you can use a hybrid port in this case, using a trunk port often works just as well and is simpler. The trunk port also supports voice VLAN settings, which might be required for assigning the correct priority to the VoIP traffic or for informing the phone of the correct VLAN ID.

## IP-based VLAN

When configured for IP-subnet-based VLANs, the hybrid switch port examines incoming untagged traffic and assigns it to a VLAN based on the source IP address. Templates that map subnets to VLANs are created statically on the switch and activated on the hybrid port. The port is also configured to support the proper VLANs as untagged. Any untagged frame that does not match a template is assigned to the PVID.

This feature is not as often required in modern networks. A use case is a provider switch that connects to multiple customer switches that do not support tagging. Traffic from the customer switches arrives on the provider switch untagged, and the provider switch uses the subnets to isolate each customer switch in its own VLAN.

## Protocol-based VLAN

The setup for a protocol-based VLAN is similar to that for an IP-based VLAN; however, templates associate *protocols* with VLAN IDs. This type of VLAN is used to isolate non-IP protocols, such as

IPX, in their own VLAN when devices connected to a port send traffic that use multiple protocols. Comware switches can use a variety of network protocols carried by Ethernet to define protocol-based VLANs. Refer to your switch documentation for a complete list.

## Special VLAN types on ProVision switches

ProVision switches also support special VLAN types. However, you do not need to use a special port type to use the VLAN. The focus is on the use case; detailed configuration is beyond the scope of this study guide.

### MAC-based VLAN (RADIUS-assigned VLANs for multiple endpoints)

Many ProVision switches support MAC-based VLANs for the same use case discussed for Comware switches. On these switches, you do not need to configure the feature. The switch automatically applies it when multiple devices authenticate to the same port and receive different VLAN assignments from a RADIUS server. In order to apply the feature, though, the switch hardware must support it. For example, HP 3800 Series switches and version 2 modules for ProVision 5400, 5400R, and 8200 Series switches support MAC-based VLANs, but version 1 modules do not. You can refer to the switch documentation to determine if its hardware supports the feature.

### Voice VLANs (VoIP phones)

ProVision switches also support voice VLANs. On a ProVision switch, you simply define the VLAN as a voice VLAN. You assign that VLAN to ports as normal, using a tagged or an untagged assignment based on the type of traffic sent by the VoIP phone. The voice VLAN definition does *not* play a role in identifying traffic for the VLAN. The port assignment does that. Instead the voice VLAN helps the switch to apply the correct priority to traffic. It can also help the switch to communicate the correct VLAN ID to a VoIP phone using LLDP-MED, a protocol for communicating settings between switches and phones.

### Protocol-based VLANs

ProVision switches can also classify traffic into a VLAN based on the traffic's protocol. You simply create a VLAN and then define the protocol for that VLAN. For example, you can define the protocol as IPv6 or IPX. Refer to your switch documentation for a list of supported protocols.

Defining the protocol transforms the VLAN into a protocol-based VLAN. (VLANs are port-based VLANs by default.) You can then assign the protocol-based VLAN to switch ports using the untagged and tagged commands. Unlike port-based VLANs, you can assign multiple untagged protocol-based VLANs to a port. The switch examines the network protocol (Ethernet frame type) for incoming frames and uses that to assign them to the correct VLAN. Untagged frames that do not match an untagged protocol-based VLAN on the port are assigned to the untagged port-based VLAN.

## Summary

In this chapter, you learned why you would add VLANs to a topology and how to add those VLANs. You now understand how to assign edge ports to the correct VLAN, as well as how to extend a VLAN across switch-to-switch ports. You can use the appropriate ProVision tagged or untagged memberships and Comware port types to complete these configurations.

Finally, you learned how to set up a switch as a default router in multiple VLANs so that it can route traffic between those VLANs.

## Learning check 4

The following questions will help you measure your understanding of the concepts presented in this chapter. Read all the choices carefully, as there may be more than one correct answer. Choose all correct answers for each question.

1. What is the maximum number of usable VLANs that 802.1Q supports?

   a. 512
   b. 1023
   c. 1024
   d. 2047
   e. 2048
   f. 4094
   g. 4096

2. What ProVision or Comware command creates VLAN 10?

3. What ProVision command(s) assign port 11 as an untagged member of VLAN 10?

4. What Comware command(s) configure interface g1/0/1 to support VLAN 1 (untagged) and VLANs 10 and 11 (tagged)?

## Answers to learning check 4

1. What is the maximum number of usable VLANs that 802.1Q supports?

   ☑ **F**. 4,094 is correct. The VLAN ID can have up to 4096 values, two of which (0 and 4095) are reserved.

   ☒ **A**, **B**, **C**, **D**, **E**, and **G** are incorrect

2. What ProVision or Comware command creates VLAN 10?

   VLAN 10

3. What ProVision command(s) assign port 11 as an untagged member of VLAN 10?

   VLAN 10 untagged 11

4. What Comware command(s) configure interface g1/0/1 to support VLAN 1 (untagged) and VLANs 10 and 11 (tagged)?

   ```
   interface g1/0/1
   port link-type trunk
   port trunk permit vlan 10 11
   ```

   These two commands would be required if the interface is not operating at default settings:

   ```
   port trunk pvid vlan 1
   port trunk permit vlan 1
   ```

# 6 Dynamic Host Configuration Protocol (DHCP) Services

## EXAM OBJECTIVES

✓ Configure HP switches to act as a Dynamic Host Configuration Protocol (DHCP) server

✓ Configure HP switches to relay client requests to a network DHCP server

## ASSUMED KNOWLEDGE

- IP addressing
- VLANs

## INTRODUCTION

DHCP, an important network service, helps to automate the configuration of client IP settings. This chapter teaches you how to configure HP switches to support your network's DHCP needs.

## DHCP use case

Up to this point, you have configured your client with a static IP address. You can imagine how time-consuming this process would be in a network with thousands of clients. You would find it difficult to manage changes in the event of mergers and new acquisitions. Duplicate IP addresses could easily occur.

In most networks, clients use DHCP to acquire their IP addresses and other IP settings dynamically. Originally defined in RFC 2131 and updated in 2939, DHCP was based on the Bootstrap Protocol (BOOTP).

DHCP provides the following advantages:

- It reduces the amount of manual configuration on devices.
- It reduces likelihood of configuration errors in devices' IP settings.
- It gives you more administrative control by centralizing IP addressing information and management.

With DHCP, when you need to make a change to your company's IP addressing, you do not need to make the change on every client. Instead, you only change the configuration on DHCP servers. When the clients either reboot or renew their addressing information, they will acquire the addressing information from the new addressing scheme. Of course, you would also have to update network infrastructure addressing on switches and routers; you might need to adjust access control lists (ACLs), DNS services, routing configurations. DHCP simply makes the client reconfiguration simpler.

HP switches can also act as DHCP clients on a VLAN interface and obtain their IP addresses automatically from a DHCP server. You might find this a simple way to set up IP addressing on many network infrastructure devices. However, often you would want to keep closer control over network infrastructure device IP addresses. In that case, you should generally assign the addresses statically.

## DHCP process

DHCP is built on a client–server model. The server stores IP settings and assigns them to clients. The client requests and accepts the settings.

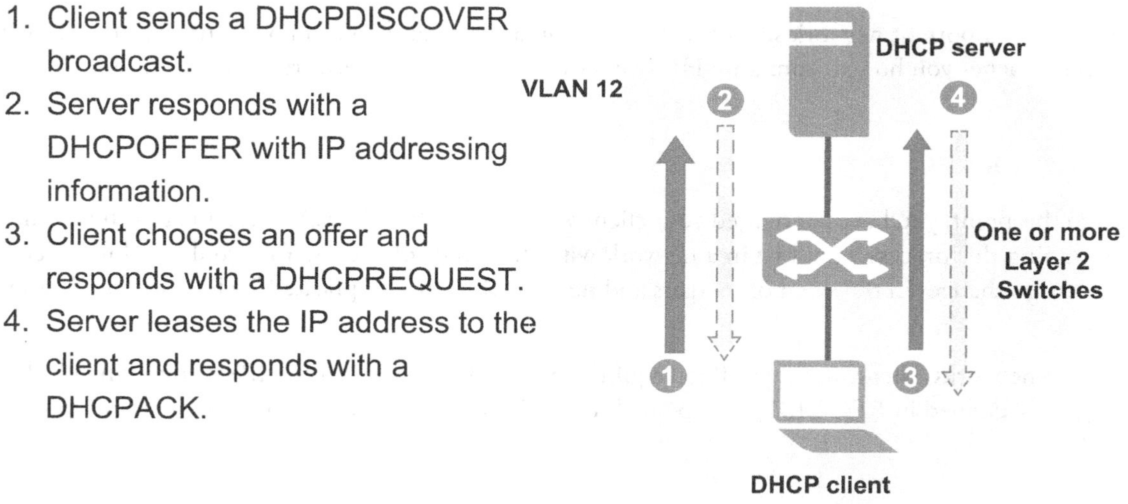


Figure 6-1: DHCP process

A client obtains its settings through this process, which is illustrated in Figure 6-1.

The client generates a DHCPDISCOVER local broadcast to UDP port 67. The broadcast will reach any DHCP server in its broadcast domain or VLAN.

The DHCP server responds with a DHCPOFFER message to UDP port 68. Generally, the server unicasts the offer to the client's MAC address (and the offered IP address). Most clients will process this message because it is to their MAC address. But some clients will not process unicasts until they have an IP address; in that case, they set a broadcast flag in the discover message, and the server broadcasts the DHCPOFFER.

In any case, the DHCPOFFER message offers IP addressing information to the client. The server can take several approaches to selecting an IP address for the client:

- Dynamic allocation—The server chooses any available IP address from a pool of addresses. It assigns this address to a client for a limited period of time, which is called a lease. Most DHCP clients obtain their addresses in this way.
- Automatic allocation—The server chooses an available IP address from its pool, but it assigns the address to the client on a more permanent basis (but the server can still recycle IP addresses in some cases).
- Static allocation—The network administrator configures the DHCP server to associate a specific IP address to the client, perhaps based on the client's MAC address. The DHCP server always assigns this address to the client.

In addition to the IP address, the DHCPOFFER also includes other settings.

**Note**

If the broadcast domain includes multiple servers, all servers can respond. If a client receives messages from multiple servers, it chooses one (typically the first one).

The client chooses an offer. It responds to the server with a DHCPREQUEST message, indicating that it wants to use the settings that the server offered.

**Note**

Sometimes a client receives a DHCPOFFER that conflicts in some way with its configuration and that the client cannot accept. If the client receives only this offer, it sends a DHCPDECLINE message to the server. The server might be able to send a different offer; otherwise, the client's DHCP process will fail.

The DHCP server responds with a DHCPACK, which acknowledges to the client that it received the DHCPREQUEST message and has leased that IP address to the client.

**Note**

If a client is late in responding with its DHCPREQUEST, the IP address might no longer be available. In that case, the server sends a DHCPNACK, indicating that the offer is no longer valid and the client should request new information.

The client now has its IP address and other settings. It retains these settings for the lease time. Before reaching the end of this time limit, the client must renew its lease with the current server or get new IP addressing information.

When a client shuts down gracefully, it can generate a DHCPRELEASE message, telling the server it no longer needs its assigned IP address.

**Note**

If a server does not respond to the client, a Microsoft client's TCP/IP protocol stack will automatically pick an IP address from the range of this Class B network: 169.254.0.1–169.254.255.254, based on RFC 3927. This process is referred to as Automatic Private IP Addressing (APIP). Other operating systems, such as HP ProVision, HP Comware, or Linux do not enable the interface if it cannot obtain IP addressing.

In Figure 6-2, you can see the general DHCP process:

1. The client sends a DHCP Discover message.
2. The server sends a DHCP Offer.
3. The client sends a DHCP Request.
4. The server sends a DHCP ACK.

You can also see some of the accompanying processes. For example, the two ARP messages in between the client's two DHCP Discover messages indicate that the DHCP server is trying to send a ping to 10.1.12.21. Remember that you configured Comware-2 to send this ping to ensure that the IP address that it wants to assign to the client is not in use. Only then does the server send the DHCP Offer message with IP address 10.1.12.21.

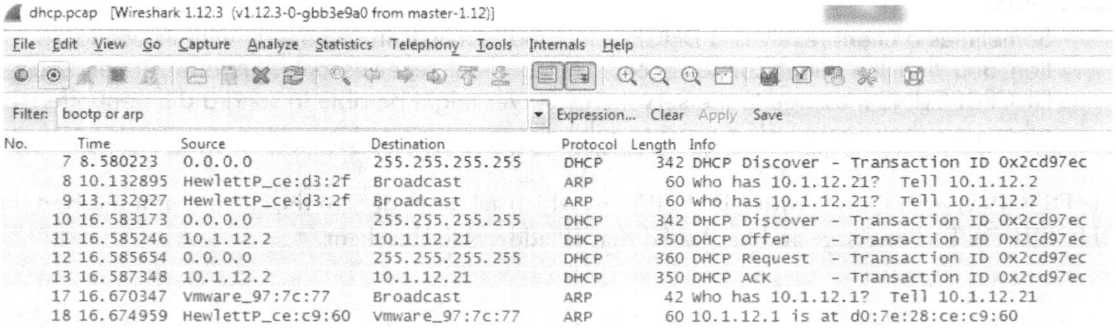


Figure 6-2: DHCP process shown in packet capture

Figure 6-3 shows the packet information in more detail. You can see that, for example, a DHCP Discover message includes various information about the client, such as its MAC address, its client identifier, its host name, and so on. The server needs the MAC address so that it can target the DHCP Offer to the correct destination. Depending on the server and its configuration, the server might use other information to decide the correct settings for the client. The client can also request specific settings in a parameter request list.

```
  Hops: 0
  Transaction ID: 0x02cd97ec
+ Seconds elapsed: 11
+ Bootp flags: 0x0000 (Unicast)
  Client IP address: 0.0.0.0 (0.0.0.0)
  Your (client) IP address: 0.0.0.0 (0.0.0.0)
  Next server IP address: 0.0.0.0 (0.0.0.0)
  Relay agent IP address: 0.0.0.0 (0.0.0.0)
  Client MAC address: Vmware_97:7c:77 (00:50:56:97:7c:77)
  Client hardware address padding: 00000000000000000000
  Server host name not given
  Boot file name not given
  Magic cookie: DHCP
+ Option: (53) DHCP Message Type (Discover)
+ Option: (61) Client identifier
+ Option: (50) Requested IP Address
+ Option: (12) Host Name
+ Option: (60) Vendor class identifier
+ Option: (55) Parameter Request List
+ Option: (255) End
  Padding
```

Figure 6-3: DHCP Discover details

The server does not necessarily have to send parameters for all of these settings. In Figure 6-4, you see the settings that the server does send in the DHCP Offer. As you remember, you configured these in the DHCP IP pool on Comware-2.

```
  Relay agent IP address: 0.0.0.0 (0.0.0.0)
  Client MAC address: Vmware_97:7c:77 (00:50:56:97:7c:77)
  Client hardware address padding: 00000000000000000000
  Server host name not given
  Boot file name not given
  Magic cookie: DHCP
+ Option: (53) DHCP Message Type (Offer)
+ Option: (54) DHCP Server Identifier
- Option: (51) IP Address Lease Time
    Length: 4
    IP Address Lease Time: (86400s) 1 day
+ Option: (58) Renewal Time Value
+ Option: (59) Rebinding Time Value
- Option: (1) Subnet Mask
    Length: 4
    Subnet Mask: 255.255.255.0 (255.255.255.0)
- Option: (3) Router
    Length: 4
    Router: 10.1.12.1 (10.1.12.1)
+ Option: (255) End
  Padding
```

Figure 6-4: DHCP Offer details

## DHCP server settings on HP switches

DHCP servers group settings for clients into scopes. Typical settings are listed in Figure 6-5.

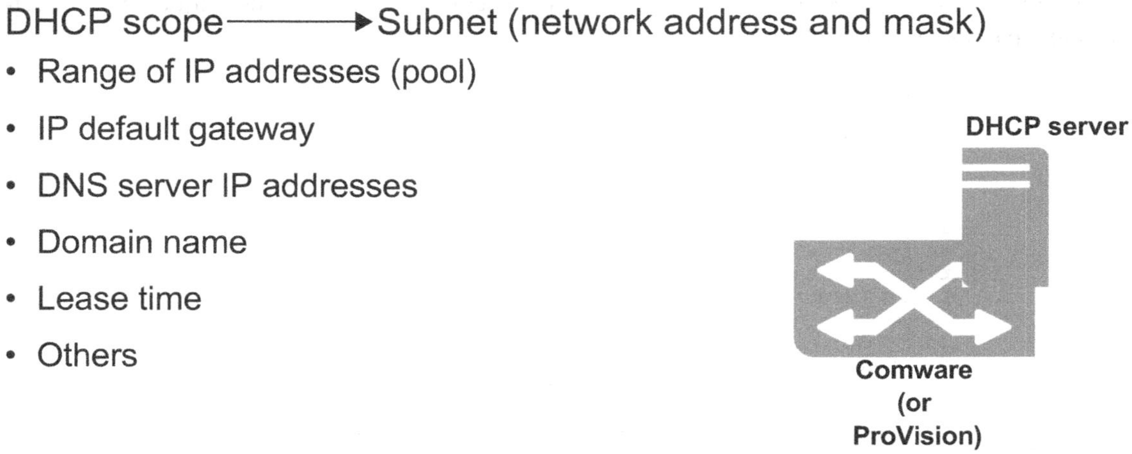


Figure 6-5: DHCP server settings on HP switches

Each scope is associated with an IP subnet: a network address and a subnet mask. The scope includes a range of IP addresses available for assignment to clients, called a pool. You can also exclude addresses in the subnet from the range.

The scope can also have other settings, which the server sends in the DHCPOFFER. These include:

- **An IP default gateway address**—As you learned in the previous chapter, clients need this address so that they can reach devices in other subnets.
- **DNS server IP addresses**—In the real-world, users access resources through DNS names. They need the IP addresses of DNS servers so that those servers can resolve names to IP addresses.
- **Domain name**—The DHCP server can give clients the corporate domain name. The clients automatically use that domain name as the suffix when they need to resolve a hostname without a domain name. For example, the domain name is "hp.com," and a client attempts to reach "server." The DNS request is for "server.hp.com."
- **Lease time**—The length of time for which the server leases the IP address to the client.

On HP Comware switches, the DHCP scope is called a pool. Comware switches take the range of IP addresses from the IP subnet. If you do not want to include any addresses within the network in the pool, you must exclude them globally. Comware switches support all of the options listed above, as well as other options.

HP ProVision switches also support DHCP services in software K/KA/KB15.16 and later. ProVision switches support dynamic DHCP pools with settings such as the ones listed above. ProVision switches also support static pools, which bind a specific MAC address to a specific IP address.

## Configure DHCP server settings on HP ProVision switches

ProVision switches that run software K/KA/KB15.16 and later support DHCP server functions, so you can configure a ProVision switch as a server.

You first create a DHCP pool.

```
ProVision(config)# dhcp-server pool mypool
```

You assign the pool a network.

```
ProVision(mypool)# network 10.1.12.0 255.255.255.0
```

You then set the IP range. In this example, the range is set from 10.1.12.21 to 10.1.12.254.

```
ProVision(mypool)# range 10.1.12.21 10.1.12.254
```

You set the default router IP address

```
ProVision(mypool)# default-router 10.1.12.1
```

Set the lease in DD:HH:MM.

```
ProVision(mypool)# lease 01:00:00
```

If the switch is the only DHCP server for the subnet, you set the pool as authoritative. This option allows the server to respond to DHCP Inform messages. DHCP clients send these messages to inform the DHCP server of the IP address that it obtained elsewhere (often statically) and request other dynamic options.

```
ProVision(mypool)# authoritative
```

You set the number of ping packets that the server sends to test an IP address.

```
ProVision(mypool)# exit
```

```
ProVision(config)# dhcp-server ping packets 2
```

You can also configure the timeout for how long the server waits for a response to a ping in seconds.

```
ProVision(config)# dhcp-server ping timeout 1
```

You must also ensure that the switch has an IP address on the VLAN and enable the DHCP server on it. In this example, the switch receives the requests on VLAN 12.

```
ProVision(config)# vlan 12
```

```
ProVision(vlan12)# ip address 10.1.12.3/24
ProVision(vlan12)# dhcp-server
```

Finally, you enable the DHCP server.

```
ProVision(vlan12)# exit
ProVision(config)# dhcp-server enable
```

## Configure DHCP server settings on HP Comware switches

You will now examine an example scenario based on the network topology shown in Figure 6-6. In this example, the Comware-2 switch will be configured to act as a DHCP server on VLAN 12, subnet 10.1.12.0/24.

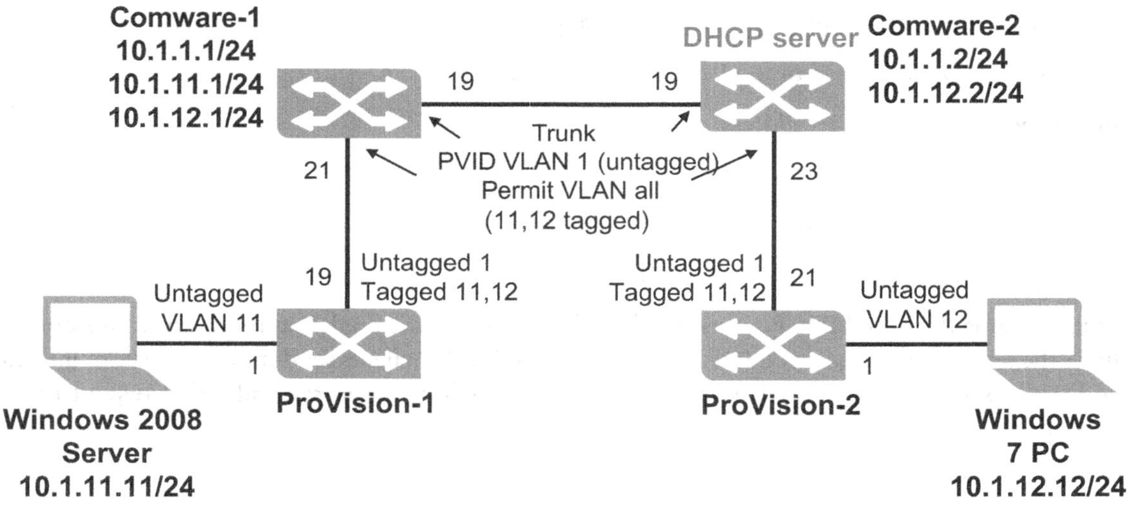


Figure 6-6: Example network topology

As Figure 6-7 shows, Comware-2 has an IP address in VLAN 12 so that it can respond to broadcasts from DHCP clients. Figure 6-7 also shows other settings that are configured on the switch. For example, the port on Comware-2 that connects to ProVision-2 is a trunkport, which supports VLANs 1, 11, and 12. VLAN 1 is the PVID and is, therefore, untagged. VLANs 11 and 12 are tagged.

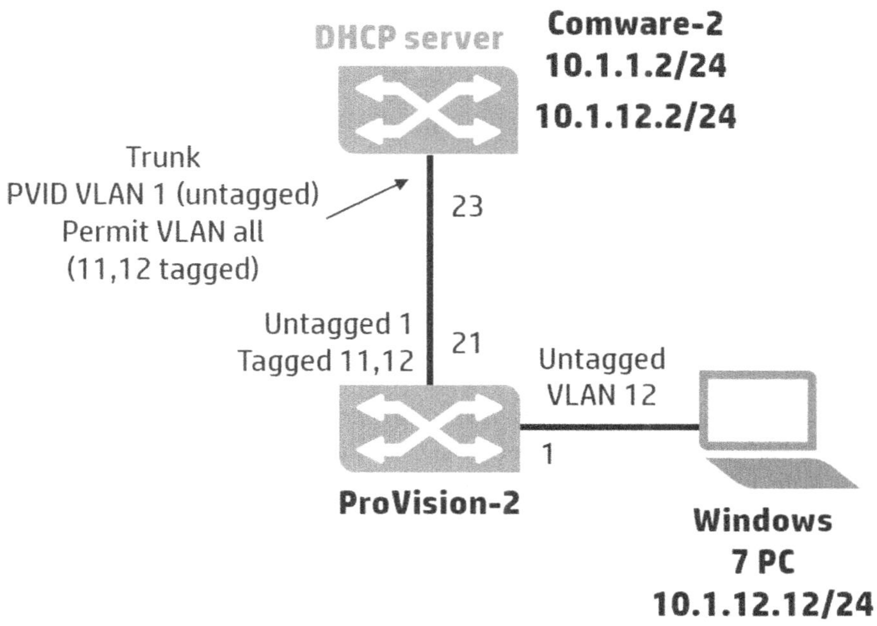


Figure 6-7: Settings required on the Comware-2 switch

Table 6-1 shows the settings for the DHCP pool. This pool contains IP addresses and other IP settings that the DHCP server on the switch will assign to clients.

Table 6-1: DHCP pool

| Parameter | Setting |
|---|---|
| ID | 1 |
| Subnet | 10.1.12.0/24 |
| Default gateway | 10.1.12.1 |
| Lease time | 1 day |

To create the DHCP server pool with an ID of 1, enter:

```
[Comware-2] dhcp server ip-pool 1
[Comware-2-dhcp-pool-1]
```

The switch prompt shows you move to the DHCP pool context. You configure the other settings listed in Table 6-2. You first configure the subnet, which is 10.1.12.0/24 in this example:

```
[Comware-2-dhcp-pool-1] network 10.1.12.0 24
```

You then configure the default gateway, which is 10.1.12.1 in this example.

```
[Comware-2-dhcp-pool-1] gateway-list 10.1.12.1
```

Finally, you configure the lease time. In this example, the lease time is 1 day.

```
[Comware-2-dhcp-pool-1] expired day 1
```

Then you exit the DHCP pool context.

```
[Comware-2-dhcp-pool-1] quit
```

Next, you configure several settings that will help the DHCP server function better, and you enable the server.

Often subnets include a few devices with fixed IP addresses assigned to them statically. In this example network, 10.1.12.0/24 includes a default router and DHCP server. It is important to exclude these addresses from DHCP. To exclude theses settings, you must be at the *system view*. For this example, you exclude the first 20 addresses in 10.1.12.0/24 from the DHCP server pool.

```
[Comware-2] dhcp server forbidden-ip 10.1.12.1 10.1.12.20
```

For the example configuration, you also configure the DHCP server to ping an IP address to ensure that it is not in use before it assigns that address to a client. You configure the server to perform two pings with a timeout of 750 milliseconds.

```
[Comware-2] dhcp server ping packets 2

[Comware-2] dhcp server ping timeout 750
```

You then enable the DHCP server.

```
[Comware-2] dhcp enable

 DHCP is enabled successfully!
```

You can view statistics about the DHCP server and verify that a pool is configured by entering:

```
[Comware-2] display dhcp server statistics

    Global Pool:

      Pool Number:                          1

<-output omitted->
```

You can view the IP addresses that the server currently has free to assign to clients.

```
[Comware-2] display dhcp server free-ip
```

```
IP Range from 10.1.12.21          to  10.1.12.254
```

You can also view the IP addresses the server has assigned clients.

```
[Comware-2] display dhcp server ip-in-use all

Pool utilization: 0.42%

 IP address   Client-identifier/   Lease expiration      Type
              Hardware address

 10.1.12.21       0050-5697-35ee  Apr 28 2000 13:11:50  Auto:COMMITTED

 --- total 1 entry ---
```

The output from this command also lists the MAC address of the client and the lease expiration.

When you configure DHCP server settings on a switch, remember to save the configuration by entering `save force`.

## Learning check 1

You will now answer several questions about DHCP services on HP switches.

1. Why is it important to configure excluded IP addresses on the DHCP server?

2. In the example configuration, why does Comware-2 need an IP address in VLAN 12?

## Answers to learning check 1

1. Why is it important to configure excluded IP addresses on the DHCP server?

   A Comware switch uses all the IP addresses in the pool's subnet as available addresses in the pool. Often subnets have a few devices that are assigned static IP addresses. For example, the default router should have a static address. You must exclude these IP addresses to prevent the server from assigning them to clients and causing IP duplications. A client using the default router IP address would cause the subnet to become unusable.

2. In the example configuration, why does Comware-2 need an IP address in VLAN 12?

   DHCP clients use broadcasts to discover the server. The broadcasts do not cross VLAN boundaries. Comware-2, acting as a DHCP server, can only listen for discovery messages on its own IP interfaces. Therefore, Comware-2 needs an IP address on the client's VLAN.

## DHCP relay use case

Often networks include many VLANs and subnets. Deploying a DHCP server on each of those subnets would be expensive and time-consuming. Companies need a centralized DHCP server in a server subnet, which can provide IP addresses to clients in multiple different subnets. However, clients' DHCPDISCOVER broadcast cannot cross the subnet boundaries, so how can clients reach the server with their requests?

DHCP relay provides the solution (see Figure 6-8). A router or Layer 3 switch, generally the clients' default gateway, receives the DHCPDISCOVER broadcasts and routes them to the DHCP server as unicasts.

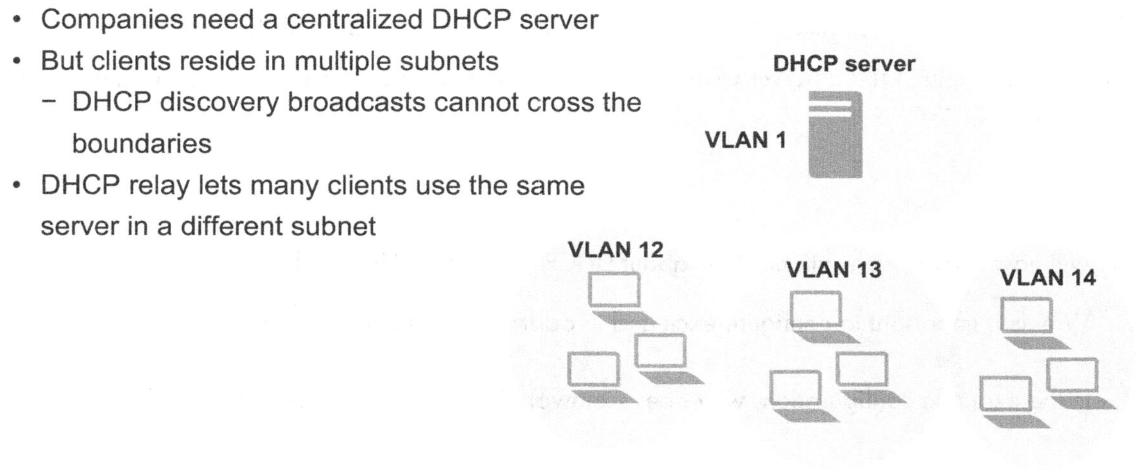


Figure 6-8: DHCP relay use case

Now the company has a centralized solution for its entire network.

In a typical environment, you will configure DHCP relay on the Layer 3 switches that support your clients. You will not need to configure a DHCP server on your switches because network servers generally fulfill that role. (Using your switch as a DHCP server is a less common use case; the switch might fulfill this role, for example, for guest clients in a guest authentication scenario.)

## DHCP relay process

DHCP relay follows this process, which is outlined in Figure 6-9:

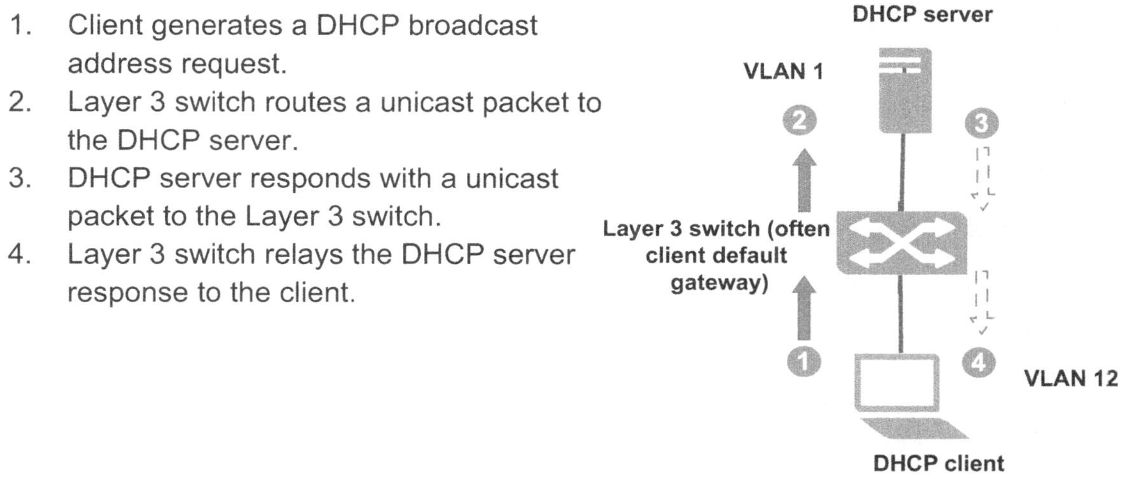


Figure 6-9: DHCP relay process

1. The client generates a DHCPDISCOVER broadcast.
2. The DHCP-relay enabled device, often the Layer 3 switch that is the client's default gateway, hears the broadcast. It creates a unicast DHCPDISCOVER message based on the client's original DHCPDISCOVER broadcast, which it sends to the server.
3. The server sends a DHCPOFFER back to the Layer 3 switch's IP address.
4. The switch sends the DHCPOFFER to the client, targeting it to its MAC address.

A similar process continues as the client sends its DHCPREQUEST and the server responds with a DHCPACK (or a DHCPNACK if the server cannot fulfill the request).

In Figure 6-10, you can see that the general process is similar to the basic DHCP process. The client sends or receives four messages:

1. The client sends a DHCP Discover message
2. The server sends a DHCP Offer.
3. The client sends a DHCP Request.
4. The server sends a DHCP ACK.

However, notice the source IP address of the DHCP Offer and DHCP ACK. As you, see the client is receiving the server's messages from 10.1.12.1, the DHCP relay. If you ran a capture on the server, you would similarly see that the client messages arrive from the relay and that the server directs responses to the relay.

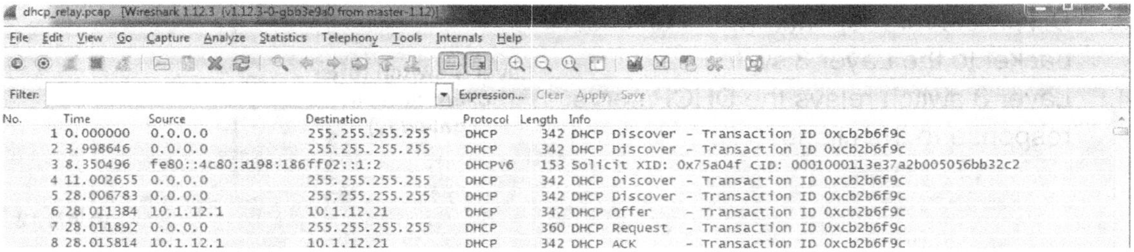


Figure 6-10: DHCP relay process as shown in a packet capture

You can also see the relay address in the details, shown in Figure 6-11.

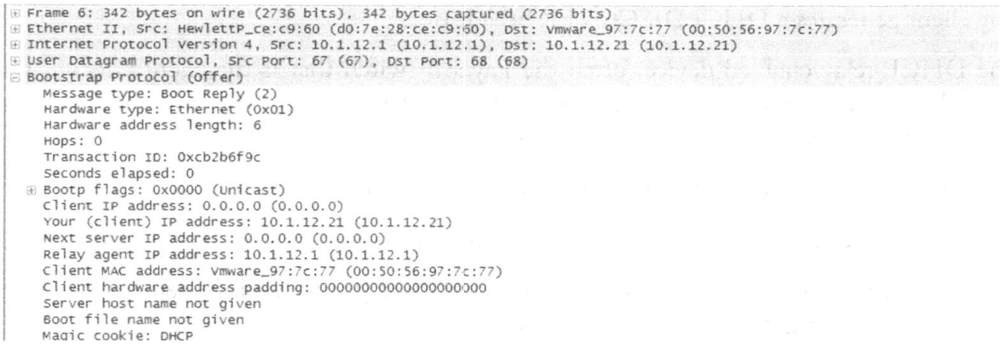


Figure 6-11: Relay agent IP address in DHCP Offer

## Implement DHCP relay on HP switches

Setting up DHCP relay is simple on both ProVision and Comware switches.

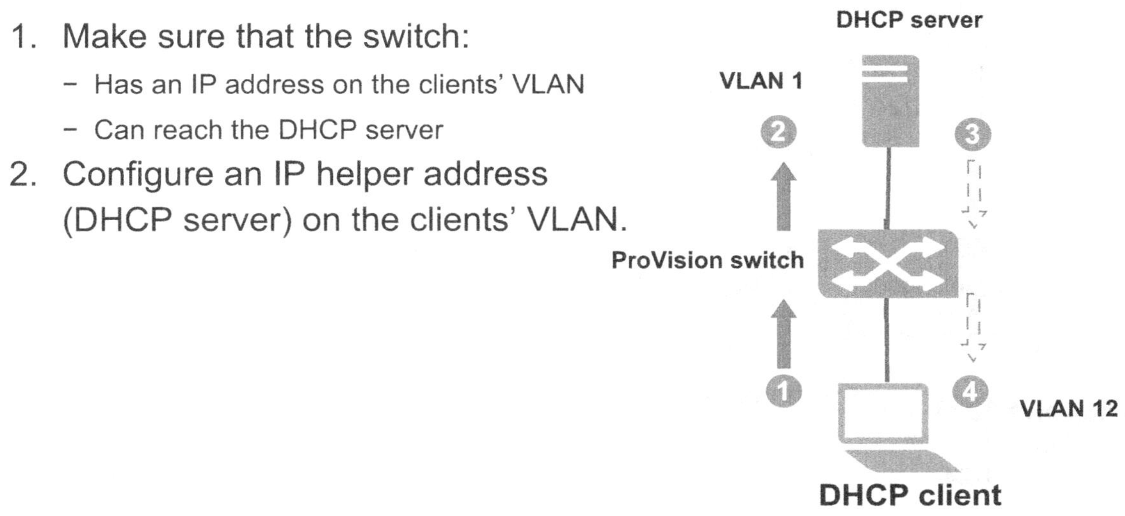


Figure 6-12: Implement DHCP relay on HP switches

## HP ProVision: Configuring DHCP relay

The steps for configuring DHCP relay on ProVision switches are outlined in Figure 6-12.

The ProVision switch needs to meet the requirements for a DHCP relay:

- It has an IP address on the DHCP clients' VLAN.
- It has IP connectivity to the DHCP server (the proper routing is in place).

In the example setup show in Figure 6-12, you need to add an IP address to the ProVision switch. It already has connectivity to the DHCP server.

```
ProVision(config)# vlan 12 ip address 10.1.12.3/24
```

Configure the DHCP server's IP address as a helper address on the DHCP clients' VLAN.

```
ProVision(config)# vlan 12 ip helper-address 10.1.1.2
```

Those simple steps complete the configuration. You do not need to enable DHCP relay, which is always enabled on ProVision switches.

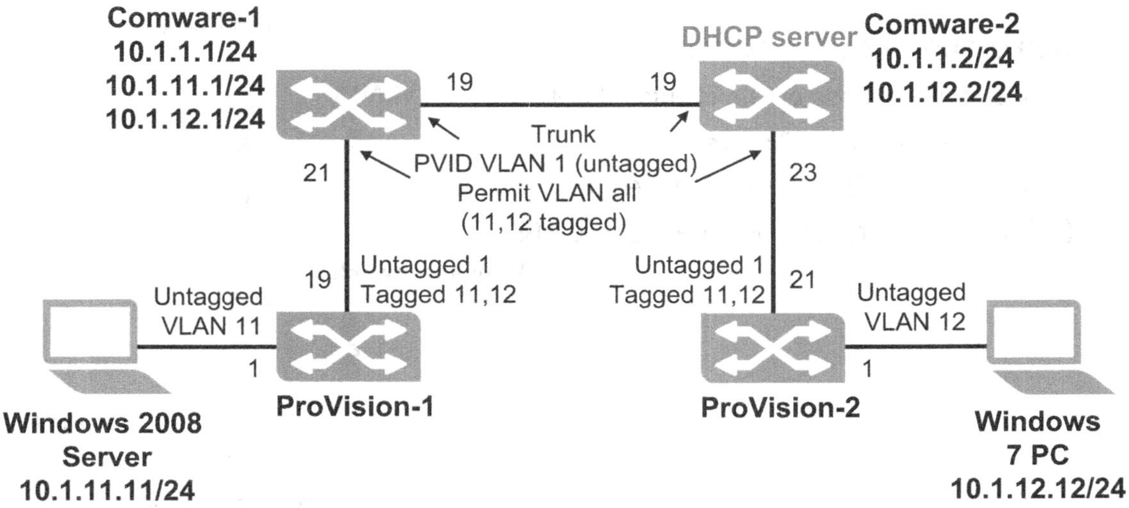


Figure 6-13: Example network topology

## HP Comware: Configuring DHCP relay

On Comware switches, you just need to configure the DHCP server's IP address, enable DHCP relay on the client's VLAN, and enable DHCP globally. The Comware switch does not need to have an IP address on the same VLAN as the clients, but it still must support the VLAN. To verify this is the case, use the `display interface brief` command. The switch also required connectivity to the same subnet even though it no longer has an IP address in that subnet. It should use its default route. You can verify this by using the `ping` command.

In the example network shown in Figure 6-13, Comware-1 is the default router for VLAN 12. You want to set it up as a DHCP relay, forwarding DHCP requests in VLAN 12 to the Comware-2 DHCP server residing in VLAN 1.

From the system view on the Comware switch, you enable DHCP services.

```
[Comware-1] dhcp enable

 DHCP is enabled successfully!
```

You then create a DHCP server group that references the DHCP server address, which is 10.1.1.2 in this example.

```
[Comware-1] dhcp relay server-group 1 ip 10.1.1.2
```

You enable DHCP relay on the VLAN interface—VLAN 12 in this example.

```
[Comware-1] interface vlan 12
```

```
[Comware-1-Vlan-interface12] dhcp select relay
```

You add the server group that you configured globally to the VLAN interface settings.

```
[Comware-1-Vlan-interface12] dhcp relay server-select 1
[Comware-1-Vlan-interface12] quit
```

To verify the status of your DHCP relay configuration, enter.

```
[Comware-1] display dhcp relay all
    Interface name                                              Server-group
    Vlan-interface12                                                  1
[Comware-1] display dhcp relay server-group 1
    No.             Group IP
    1               10.1.1.2
```

You can also view the DHCP relay statistics:

```
[Comware-1] display dhcp relay statistics
     Bad packets received:                      0
DHCP packets received from clients:             3
         DHCPDISCOVER packets received:         1
         DHCPREQUEST packets received:          1
         DHCPINFORM packets received:           1
         DHCPRELEASE packets received:          0
         DHCPDECLINE packets received:          0
         BOOTPREQUEST packets received:         0
DHCP packets received from servers:             3
         DHCPOFFER packets received:            1
         DHCPACK packets received:              2
         DHCPNAK packets received:              0
         BOOTPREPLY packets received:           0
     DHCP packets relayed to servers:           3
         DHCPDISCOVER packets relayed:          1
```

```
        DHCPREQUEST packets relayed:         1
        DHCPINFORM packets relayed:          1
        DHCPRELEASE packets relayed:         0
        DHCPDECLINE packets relayed:         0
        BOOTPREQUEST packets relayed:        0
DHCP packets relayed to clients:             3
        DHCPOFFER packets relayed:           1
        DHCPACK packets relayed:             2
        DHCPNAK packets relayed:             0
        BOOTPREPLY packets relayed:          0
<-output omitted>
```

## Summary

You have learned about the importance of DHCP services for IP networks. You learned how to configure a Comware switch as a DHCP server, which clients can contact to receive their IP settings.

You also learned that typical corporate networks use a network DHCP server. You learned how to set up DHCP relay on HP switches, allowing clients in many different subnets to obtain IP settings from that server.

## Learning check 2

You will now answer several questions about what you have learned in this chapter.

1. Why did Comware-2 only need its 10.1.1.2 IP address for this solution?

2. Why did you configure Comware-1 with the DHCP relay settings?

3. Which settings did you configure globally on Comware-1 and which ones did you configure on a VLAN interface?

4. Which ProVision command enables DHCP relay for a VLAN?

   a. ip relay
   b. ip helper-address
   c. dhcp-relay
   d. relay-address

## Answers to learning check 2

1. Why did Comware-2 only need its 10.1.1.2 IP address for this solution?

   Comware-1 acts as a DHCP relay, so the DHCP server (Comware-2) does not need to have an IP address on DHCP clients' subnet.

2. Why did you configure Comware-1 with the DHCP relay settings?

   The DHCP relay needs an IP address on the clients' VLAN. It also needs IP connectivity to the DHCP server. Comware-1, which is the default router for VLAN 12, meets these criteria.

3. Which settings did you configure globally on Comware-1 and which ones did you configure on a VLAN interface?

   You must enable DHCP globally. You must also configure a global DHCP relay server group, which specifies the DHCP server's IP address.

   On the VLAN interface, you choose relay as the DHCP mode and specify the ID of the DHCP relay server group.

4. Which ProVision command enables DHCP relay for a VLAN?

   ☑ **B** ip helper-address

   ☒ **A**, **C**, and **D** are incorrect

# 7 Spanning Tree

## EXAM OBJECTIVES

✓ Understand and configure Rapid Spanning Tree Protocol (RSTP)

✓ Understand how Multiple Spanning Tree Protocol (MSTP) provides load-sharing and implement MSTP to do so

## INTRODUCTION

This chapter introduces spanning tree protocol standards, which help switches to manage redundant links without introducing loops and broadcast storms. It focuses on RSTP and MSTP.

## Assumed knowledge

Before reading this chapter, you should have a basic understanding of:

- Virtual LANs (VLANs)
- Broadcast domains
- IP addressing

## Issues adding redundant links to the topology

Networks deliver critical services to users. Failure of a network link may make the network unavailable to users, resulting in lost time or revenue. To protect a network against these failures, you can install redundant links. Redundant links help to ensure that a path continues to exist across the network even if one link or even one switch fails. For example, in the topology illustrated in Figure 7-1, ProVision-1 could continue to reach Comware-1 if its direct link to Comware-1 failed. It could also continue to reach the core if Comware-1 failed entirely. (Of course, default router services would also have to be made redundant, but that is beyond the scope of this chapter, which focuses on link redundancy.)

HP ExpertOne If you want to view a video about Spanning Tree Protocol (STP), access the HP Aurasma application on your smartphone and use the application to focus on Figure 7-1. This action will launch the video.

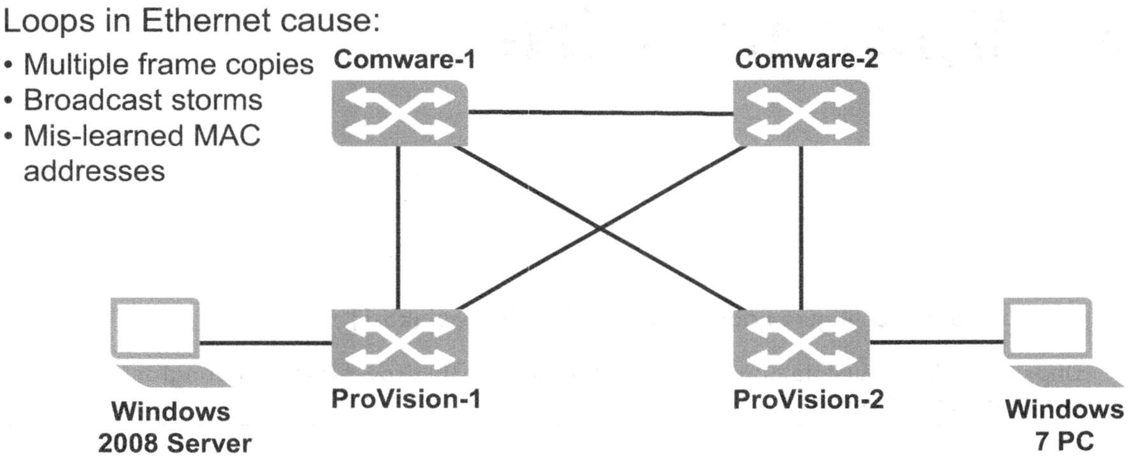


Figure 7-1: Issues adding redundant links to the topology

However, simply adding redundant physical links does not ensure that the switches can use those links correctly.

Adding redundant Layer 2 links without a protocol to manage the links results in network loops. These loops, in turn, create broadcast storms and mislearned MAC addresses, making the network inaccessible. To function properly, an Ethernet network must have only *one* active pathway between any two devices.

The next few sections describe the problems that occur when you add redundant Ethernet links, but there is no protocol to manage them.

## Multiple frame copies

An Ethernet switch floods three kinds of frames: broadcast, multicast, and unknown destination unicast frames.

While flooding is a necessary part of how Ethernet works, it can create serious performance issues in a Layer 2 network that has loops. For example, suppose you cleared the Address Resolution Protocol (ARP) cache on the PC in the small test network, as shown in Figure 7-2, and then attempted to ping the server. To complete the ping, the PC would broadcast an ARP request for its default gateway at 10.1.12.1. The broadcast would flood to both Comware-1 and Comware-2, which then floods the broadcast again to Comware-1 and ProVision-1. ProVision-1 floods the frame to Comware-1 and Comware-2, and so on.

Although the PC only generated one ARP request, Comware-1 (the default gateway) receives many copies of this request and needs to reply many times.

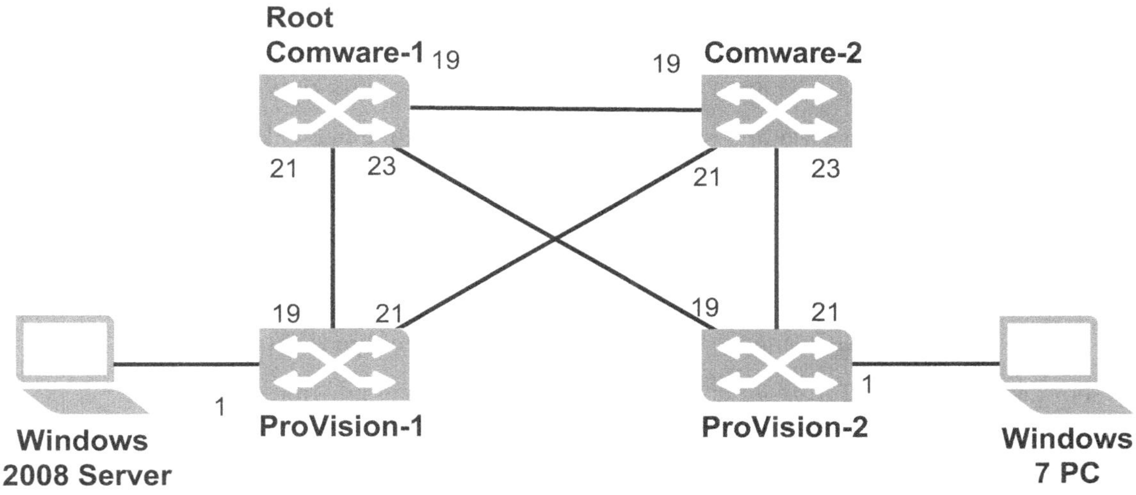


Figure 7-2: Example network

The same issue holds true for other types of broadcasts and multicasts. The application that receives the duplicate frames might see the duplication as an error and reset its connection to the source of the transmissions, creating connectivity issues.

## Broadcast storms

An even greater issue occurs. As mentioned above, ProVision-2 floods the broadcast to Comware-1 and Comware-2. Comware-1 floods the broadcast to Comware-2 and ProVision-1, and Comware-2 floods the broadcast to Comware-1 and ProVision-1. ProVision-1 floods both broadcasts to switches that flood the broadcast again. As you see, the broadcasts continue to be flooded between the switches, multiplying again and again, and affecting all devices in the broadcast domain, or VLAN.

The devices are affected because their network interface cards (NICs) must process all of these broadcasts. Eventually, the network will crash because devices will run out of CPU cycles to process the broadcasts.

Note that NICs only process multicasts for which they have registered (typically because an application running on the endpoint uses that address). Therefore, a flood of multicasts does not affect the CPU cycles of all devices to as great a degree. However, floods of multicasts still affect everyone's bandwidth, and a high-speed video stream, for example, could quickly consume all the available bandwidth.

## Mislearned MAC addresses

Layer 2 loops can also cause the switches to mislearn the interface associated with a MAC address.

Remember: Ethernet switches examine the source MAC address for incoming frames and associate that MAC address with the ingress interface for the frame. In a network that contains a loop, such as the one described in the example scenario, a switch might receive an ARP request sent by the same PC on several different interfaces. The switch then learns the PC's MAC address on interface 19, then 23, then 19, and so on. This happens again and again as the broadcasts loop.

The mislearning can cause a number of problems. In the example network shown in Figure 7-2, Comware-1 might currently have learned the PC's MAC address on port 19. It forwards an ARP reply frame destined to the PC to Comware-2. But Comware-2, due to the flip-flopping of learned interfaces, might currently have learned the PC's MAC address on its port 19, connected to Comware-1. Because switches do not forward traffic back out the port on which they receive it, the switch drops the frame. The PC does not receive its reply, which causes connectivity issues. It might even prompt the PC to send another ARP request, which adds to the number of broadcasts.

In short, a looped Layer 2 network quickly becomes congested and unusable.

---

**Important**

A common first indication of a Layer 2 loop is that the computing devices (PCs, servers, printers, switches, routers, and so on) experience very slow performance. Once a large number of broadcasts are built up within the loop, these computing devices must use their CPU (not ASICs) to process broadcasts. If you look at the CPU utilization on a device, you will see that it is running very high.

Once you notice a high CPU utilization on switches, look at the MAC address table on those switches. If you see flapping MAC addresses (the address constantly changes between two or more interfaces), then you are almost guaranteed to have a loop.

---

## Spanning tree solution

STP provides the traditional solution for adding redundant links to a Layer 2 network without causing loops. In this study guide you will begin by learning how to set up an IEEE 802.1w RSTP solution.

RSTP features a single root bridge. (Bridge is the spanning tree terminology for switch; this chapter will use the term "switch.") The spanning tree switches elect the root by exchanging Bridge Protocol Data Units (BPDUs) on their links. You will learn more details about the election process later in this chapter. For now, you should understand that you can affect the election by configuring certain spanning tree settings. In the example network shown in Figure 7-3, Comware-1 will be configured as the primary root to make sure that it wins the election.

Every other spanning tree bridge has a lowest cost path to the root. The links in the lowest cost paths forward traffic. Other links are blocked. The network resembles a tree, rooted at the root bridge, branching to other switches, and finally branching to the endpoints on switch edge ports. Because links that are not part of this tree are blocked, the network has no loops.

HP ExpertOne

If you want to view a video about more advanced STP concepts, access the HP Aurasma application on your smartphone and use the application to focus on Figure 7-3. This action will launch the video.

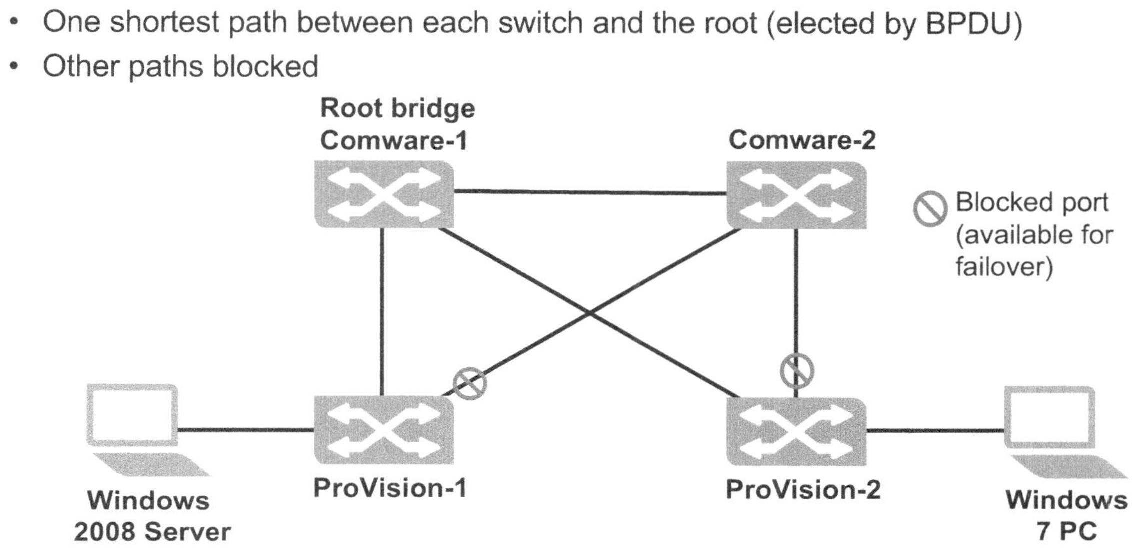


Figure 7-3: Spanning tree solution

The blocked links remain available for use if a change occurs in the topology—for example, a forwarding switch-to-switch link fails. You will learn more about failover later.

## Overview of STP

Several STP versions exist. The original standard, STP, is defined in 802.1D. This standard defined the basic behaviors still used in spanning tree, such as electing a root bridge, calculating the best path to that bridge, and blocking alternate paths. However, STP was designed in a time when a minute or so for convergence was acceptable. It requires ports to move through several port states, each with a relatively long (10–20 seconds) wait time, before the ports can start forwarding. Any change in the topology can force ports to progress slowly through these states again.

As network connectivity became more crucial, companies could no longer tolerate such long convergence times when any topology change occurred. RSTP, 802.1w, was developed in 1998 to speed convergence. It also defined port costs that better fit with modern port speeds. In 2004, RSTP was merged into the 802.1D, superseding STP. Therefore, this chapter focuses on how RSTP functions over STP. As you work through this study guide, you will see how RSTP handles convergence in under a second.

You will also learn about MSTP, which operates in the same basic way as RSTP but supports multiple instances. Each instance has its own topology, and, as you will see, this behavior allows for load sharing of traffic on redundant paths.

HP switches, both Comware and ProVision, use MSTP for their default spanning tree mode. However, the switches do not initially have any MSTP region settings defined. Until you define consistent MSTP region settings, the switches operate as if they were running RSTP. Therefore, you will first learn how to set up RSTP.

**Important**

Although MSTP is the default spanning tree mode, spanning tree as a whole is *disabled* on HP switches by default. Do *not* simply start establishing redundant links before configuring a spanning tree protocol, or you will cause a broadcast storm.

## Spanning tree port roles and states

RSTP switches do not have a complete view of the topology. They maintain the tree by exchanging BPDUs, which indicate how close they are to the root—in other words, the cost of their shortest path to the root, or their root path cost. These BPDUs help switches to calculate the correct port role for each of their ports. The port roles are summarized in Table 7-1.

Table 7-1: RSTP port roles

| Port role | Description | Potential port state |
|---|---|---|
| Designated | Port is closer to the root than any other port on this link | Learning (when the port first comes up)<br>Discarding (when certain events occur)<br>Forwarding (final state during normal operation) |
| Root | ■ Port is not the closest port to the root on this link.<br>■ This link offers the switch its best path to the root. | Forwarding |
| Alternate | ■ Port is not the closest port to the root on this link.<br>■ This link does not offer the switch its best path to the root. It offers a second (or third, and so forth) best path.<br>■ Port is not connected to the same switch. | Discarding |
| Backup | ■ Port is not the closest port to the root on this link.<br>■ This link does not offer the switch its best path to the root.<br>■ Port is connected to the same switch. The switch has a looped connection to itself. Or it has a loop through downstream devices that do not support spanning tree. | Discarding |

Look first at the links that will become part of the active spanning tree, or, in other words, part of switches' set of lowest cost paths to the root. For each of these links:

- The port on one side is Designated. This is the port that is closer to the root (or is, in fact, on the root).
- The port on the other side of the link is the root port.

In other words, the root port is the port that gives a nonroot switch its best path to the root. A switch can only have one root port because it can have only one best path.

The port role affects the states that the port is allowed to pass through. In normal operation, designated and root ports transition to forwarding state. In this state, ports learn MAC addresses and forward traffic. Because both the designated port and root port are in a forwarding state, the link is active.

Now look at the links that are *not* part of the spanning tree. For each of these links:

- The port on one side is Designated. Again, this is the port that is closer to the root.
- The port on the other side of the link is either:
    - Alternate (it is connected to a different switch)
    - Backup (it is connected to the same switch)

Typically, you will see alternate ports. A backup port occurs if a switch port is accidently looped to a port on the same switch. It also occurs if a switch port is connected to downstream switches that do not support spanning tree and that introduce a loop. From the point of view of spanning tree, those switches are transparent, so the spanning tree switch appears to be connected to itself.

Alternate and backup ports remain in the discarding state. The discarding state is sometimes called the blocking state, and you might see this in switch CLI output. In this state, ports drop incoming traffic, do not learn MAC addresses from incoming traffic, and do not forward traffic. They continue to receive and process BPDUs from the designated port; these BPDUs communicate information about and maintain the spanning tree.

In short, links that are not part of the spanning tree have a designated port on one side and an alternate or a backup port on the other. These links are effectively blocked because the downstream side of the link is in a discarding state.

---

**Note**

These rules properly apply to full-duplex Ethernet links, which are between two switches. The rules are a bit different for half-duplex links, which might connect several switches through a hub. In the latter case, *one* switch port on the link would be the designated port and all other ports would be alternate, backup, or root ports, depending on their lowest cost path to the root. However, this study guide focuses on the full-duplex behavior.

---

Note that ports that do not connect to spanning tree devices—such as the ProVision switches' connections to the server and PC—are still part of the spanning tree. These switch ports have a Designated role so that they can forward traffic to the endpoints.

## Configuration example: RSTP

In this section, you will learn how to set up RSTP on HP switches. The instructions provided are based on the example network, which is using the settings shown in Figure 7-4.

This example configuration was developed in a lab environment, which provides permanent connections between switches. As a result, the interfaces that are not in use are shut down or disabled. In Figure 7-4, the number next to each link indicates the port used. For example, on Comware-1, port 19 connects to Comware-2 and port 21 connects to ProVison-1.

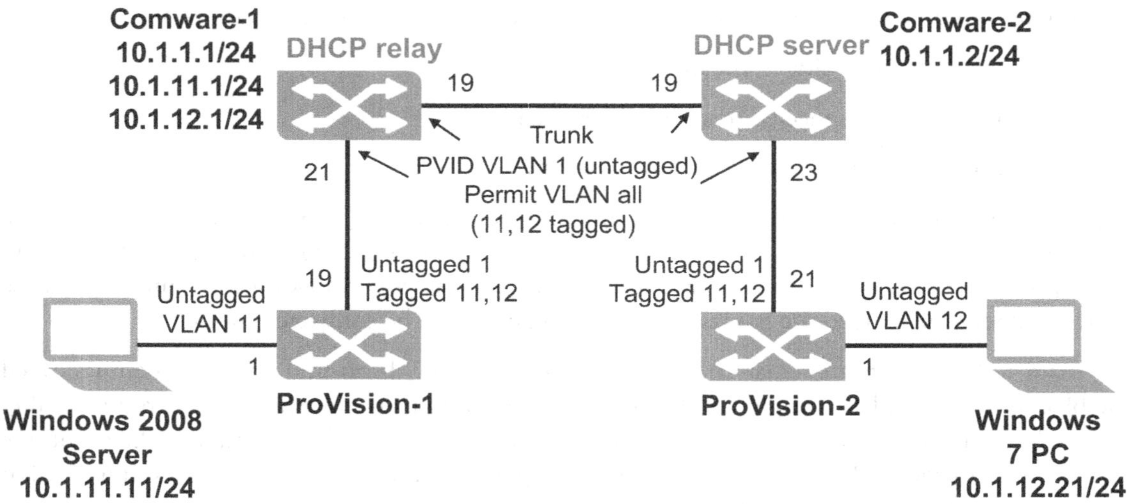


Figure 7-4: Starting network configuration

You will configure RSTP on each of the switches in the example network, changing the network topology as shown in Figure 7-5.

HP ExpertOne If you want to watch a video that shows the steps outlined in this section, access the HP Aurasma application on your smartphone and use the application to focus on Figure 7-5. This action will launch the video.

You will use the switch output provided to determine the port roles and states for each port on the switches, as shown in Figure 7-6.

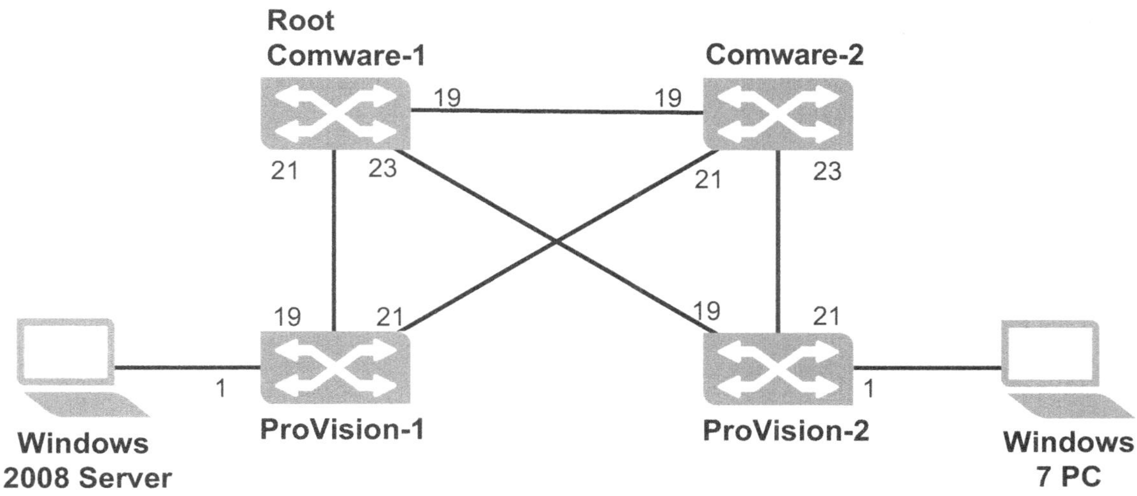


Figure 7-5: Add redundant links

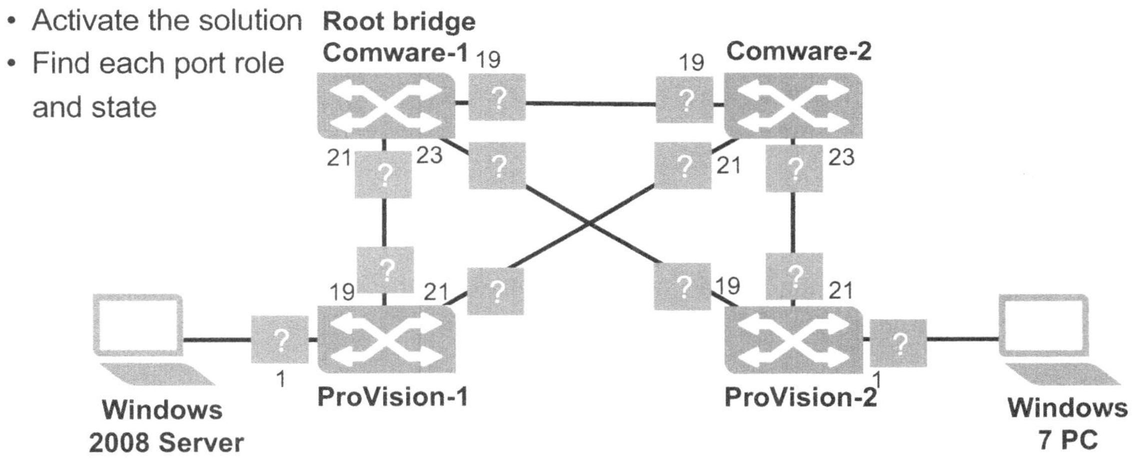


Figure 7-6: Configuration example: RSTP

## Configure Comware-1 as the root

You learned that RSTP sets up the spanning tree based on the root bridge. You will first configure Comware-1 as the primary root bridge, which ensures that it wins the root bridge election.

1. Access the Comware-1 CLI and move to system view.
2. Configure Comware-1 as the primary root bridge.

```
[Comware-1] stp root primary
```

## Enable spanning tree on each switch

You will enable spanning tree on the HP switches. Recall that STP is disabled by default on both the Comware and ProVision switches.

1. Enable spanning tree on Comware-1.

```
[Comware-1]stp enable
```

2. Access the Comware-2 CLI and move to system view.
3. Enable spanning tree.

```
[Comware-2] stp enable
```

4. Access the ProVision-1 CLI and move to global configuration mode.
5. Enable spanning tree.

```
ProVision-1(config)# spanning-tree
```

6. Access the ProVision-2 CLI and move to global configuration mode.
7. Enable spanning tree.

```
ProVision-2(config)# spanning-tree
```

## Verify the root bridge

You will now verify that Comware-1 has been elected the root bridge, as desired.

1. Enter the following command on the Comware-1 switch. Find the bridge ID, which is listed next to CIST Bridge.

```
[Comware-1]display stp

-------[CIST Global Info][Mode MSTP]-------

CIST Bridge         :0.d07e-28ce-c94f

Bridge Times        :Hello 2s MaxAge 20s FwDly 15s MaxHop 20

CIST Root/ERPC      :0.d07e-28ce-c94f / 0

CIST RegRoot/IRPC   :0.d07e-28ce-c94f / 0

CIST RootPortId     :0.0

<-output omitted->
```

2. Record Comware-1's bridge ID.
3. Now view the root bridge ID on Comware-1 and verify that it is this switch's bridge ID.

```
[Comware-1]display stp root
 MSTID Root Bridge ID   ExtPathCost IntPathCost Root Port
   0   0.d07e-28ce-c94f 0           0
```

4. Enter the same command on Comware-2. The root bridge ID should match Comware-1's ID.

```
[Comware-2] display stp root
 MSTID Root Bridge ID   ExtPathCost IntPathCost Root Port
   0   0.d07e-28ce-c94f 20          0           GigabitEthernet1/0/19
```

5. Enter the following command on ProVision-1 and verify that the root is Comware-1.

**Note**
The Common Spanning Tree (CST) includes all spanning tree bridges.

```
ProVision-1(config)# show spanning-tree root-history cst
 Status and Counters - CST Root Changes History
  MST Instance ID          : 0
  Root Changes Counter     : 2
  Current Root Bridge ID : 0:d07e28-cec94f
  Root Bridge ID      Date     Time
  ------------------- -------- --------
```

6. Access the ProVision-2 CLI. Use the same command to verify that Comware-1 is this switch's root bridge.

```
ProVision-2(config)# show spanning-tree root-history cst
 Status and Counters - CST Root Changes History
  MST Instance ID          : 0
  Root Changes Counter     : 2
  Current Root Bridge ID : 0:d07e28-cec94f
  Root Bridge ID      Date     Time
  ------------------- -------- --------
```

## Add redundant links

You will now add the redundant links, as shown in Figure 7-5 on page 187. The links should be configured to carry traffic on all VLANs supported on the switches.

1. Access Comware-1 and set up the physical port that connects to ProVision-2.

```
[Comware-1] interface g1/0/23
[Comware-1-GigabitEthernet1/0/23] port link-type trunk
[Comware-1-GigabitEthernet1/0/23] port trunk permit vlan all
[Comware-1-GigabitEthernet1/0/23] undo shutdown
[Comware-1-GigabitEthernet1/0/23] quit
```

2. Access Comware-2 and set up the physical port that connects to ProVision-1.

```
[Comware-2] interface g1/0/21
[Comware-2-GigabitEthernet1/0/21] port link-type trunk
[Comware-2-GigabitEthernet1/0/21] port trunk permit vlan all
[Comware-2-GigabitEthernet1/0/21] undo shutdown
[Comware-2-GigabitEthernet1/0/21] quit
```

3. Access ProVision-1 and set up the physical port that connects to Comware-2.

```
ProVision-1(config)# vlan 11 tagged 21
ProVision-1(config)# vlan 12 tagged 21
```

4. Enable the port.

```
ProVision-1(config)# interface 21 enable
```

5. If necessary, establish the physical link between ProVision-1 and Comware-2. (Remember that this example configuration was developed in a lab environment, which provides permanent connections between switches. To eliminate potential issues, the interfaces that are not in use were shut down.)

6. Verify that the link has come up.

```
ProVision-1(config)# show interface 21
 Status and Counters - Port Counters for port 21
  Name  :
  MAC Address      : e4115b-cccc53
  Link Status      : Up
<-output omitted->
```

7. Access ProVision-2 and set up the physical port that connects to Comware-1.

```
ProVision-2(config)# vlan 11 tagged 19
ProVision-2(config)# vlan 12 tagged 19
```

8. Enable the port.

```
ProVision-2(config)# interface 19 enable
```

9. If necessary in your environment, establish the physical link between ProVision-2 and Comware-2.
10. Verify that the link is up.

```
ProVision-2(config)# show interface 19
 Status and Counters - Port Counters for port 19
  Name  :
  MAC Address      : e4115b-ccd215
  Link Status      : Up
<-output omitted->
```

11. Access the Comware-1 CLI.
12. Quickly validate that the port connections are correct by examining the LLDP neighbor table.

```
[Comware-1] display lldp neighbor-information list
System Name  Local Interface  Chassis ID      Port ID
Comware-2    GE1/0/19         d07e-28ce-d31e  GigabitEthernet1/0/19
ProVision-1  GE1/0/21         d4c9-ef85-1680  19
ProVision-2  GE1/0/23         d4c9-ef84-fe80  19
```

**Note**
If you do not see ProVision-2, check the interface status for g1/0/23.

13. Access the terminal session with Comware-1.
14. Quickly validate that the port connections are correct by examining the LLDP neighbor table.

```
[Comware-2] display lldp neighbor-information list
System Name  Local Interface  Chassis ID      Port ID
Comware-1    GE1/0/19         d07e-28ce-c94f  GigabitEthernet1/0/19
```

```
ProVision-1  GE1/0/21         d4c9-ef85-1680  21
ProVision-2  GE1/0/23         d4c9-ef84-fe80  21
```

**Note**

If you do not see ProVision-1, check the interface status for g1/0/21.

## Check CPU

1. Verify that the CPU utilization on *each* switch is normal. A high CPU utilization could indicate a loop, but RSTP should have removed the loops.

```
[Comware-1] display cpu-usage
Slot 1 CPU usage:
       5% in last 5 seconds
       5% in last 1 minute
       5% in last 5 minutes
[Comware-2] display cpu-usage
Slot 1 CPU usage:
       5% in last 5 seconds
       5% in last 1 minute
       5% in last 5 minutes
ProVision-1(config)# show cpu
5 percent busy, from 300 sec ago
1 sec ave: 1 percent busy
5 sec ave: 1 percent busy
1 min ave: 1 percent busy
ProVision-2(config)# show cpu
4 percent busy, from 300 sec ago
1 sec ave: 4 percent busy
5 sec ave: 1 percent busy
1 min ave: 1 percent busy
```

## Verify the topology

You will now use `show` and `display` commands to explore the topology created by RSTP.

> **Note**
> Some of the command output depends on your switch hardware. For example, the system MAC address is unique to your equipment.

Using Figure 7-7, you will fill in the port role and indicate blocking ports.

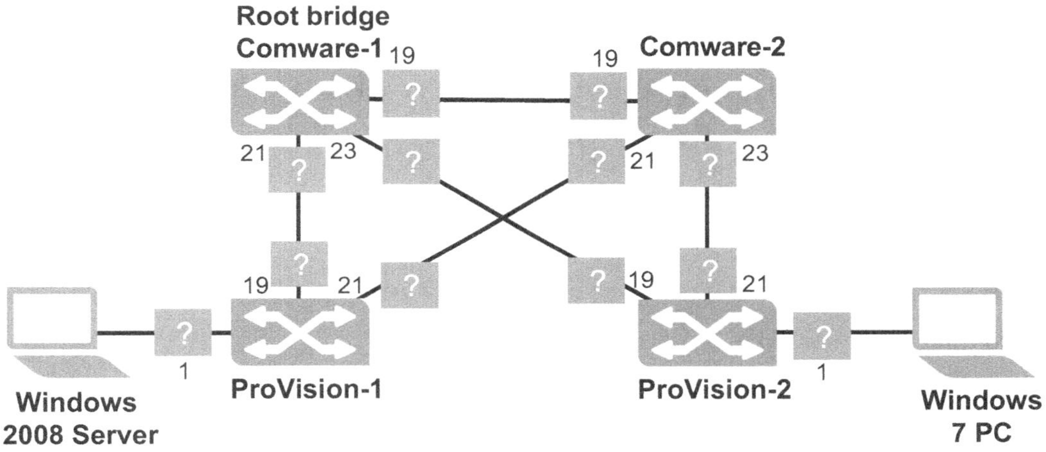


Figure 7-7: RSTP port roles and states in the example network topology

1. Access the Comware-1 CLI.
2. Use this command to see a summary of the port role and state for each STP-enabled port.

```
[Comware-1] display stp brief
 MSTID  Port                        Role  STP State   Protection
   0    GigabitEthernet1/0/19       DESI  FORWARDING  NONE
   0    GigabitEthernet1/0/21       DESI  FORWARDING  NONE
   0    GigabitEthernet1/0/23       DESI  FORWARDING  NONE
```

3. Fill in the port roles in Figure 7-7. You can abbreviate (such as DP for designated port.) If the state is blocking, draw an X to indicate the blocked link. If the state is forwarding, you do not need to do anything else.

4. Use the same command on Comware-2 and fill in the figure.

```
[Comware-2] display stp brief
MSTID  Port                     Role  STP State   Protection
  0    GigabitEthernet1/0/19    ROOT  FORWARDING  NONE
  0    GigabitEthernet1/0/21    DESI  FORWARDING  NONE
  0    GigabitEthernet1/0/23    DESI  FORWARDING  NONE
```

5. Also find the cost on the Comware-2 root port. Enter this command and look for the value in the Port Cost row.

```
[Comware-2] display stp int g1/0/19
----[CIST][Port19(GigabitEthernet1/0/19)][FORWARDING]----
 Port Protocol          :enabled
 Port Role              :CIST Designated Port
 Port Priority          :128
 Port Cost(Legacy)      :Config=auto / Active=20
```

6. Record the cost.
7. Access the ProVision-1 CLI.
8. Use the following command to view the port state on desired interfaces. Scroll to the bottom of the output to find the port state information.

```
ProVision-1(config)# show spanning-tree 1,19,21
<-output omitted->
                 |       Prio             | Designated    Hello
Port Type        | Cost  rity State       | Bridge        Time PtP Edge
---- ---------  + ----- ---- ---------- + ------------- ---- --- --
1    100/1000T | 20000 128  Forwarding | d4c9ef-851680 2    Yes
19   100/1000T | 20000 128  Forwarding | d07e28-cec94f 2    Yes No
21   100/1000T | 20000 128  Blocking   | d07e28-ced31e 2    Yes No
```

**Note**

The interface IDs are not required, but they limit the output so that it is easier for you to read.

9. If you want to see both the port role and state, you must use a slightly different command.

```
ProVision-1(config)# show spanning-tree 1,19,21 instance ist

<-output omitted->                                    Designated

Port  Type      Cost  Priority Role       State      Bridge

----- --------- ----- -------- ---------- ---------- --------------

1     100/1000T 20000 128      Designated Forwarding d4c9ef-851680

19    100/1000T 20000 128      Root       Forwarding d07e28-cec94f

21    100/1000T 20000 128      Alternate  Blocking   d07e28-ced31e
```

**Note**

In this command, IST refers to internal spanning tree, a concept for MSTP, which you will learn about a bit later. For now, you just need to understand that when a ProVision switch is using RSTP to interoperate with other switches, the IST is used.

10. You can now fill in the figure with the ProVision-1 port states. Remember to add an X to any blocking link.
11. Look in the Cost column for the root port in the current command output. What is the cost on this port?
12. Access the ProVision-2 CLI.
13. Use this command to fill in the Figure 7-7. Remember to add an X to any blocking link.

```
ProVision-2(config)# show spanning-tree 1,19,21 instance ist

<-output omitted->

                                                      Designated

Port  Type      Cost  Priority Role       State      Bridge

----  --------- ----- -------- ---------- ---------- --------------

1     100/1000T 20000 128      Designated Forwarding d4c9ef-84fe80

19    100/1000T 20000 128      Root       Forwarding d07e28-cec94f

21    100/1000T 20000 128      Alternate  Blocking   d07e28-ced31e
```

Save the current configuration on each of the four switches. For ProVision, use the `write memory` command. For Comware, use the `save [force]` command.

## Learning check 1

Answer the following questions to test your understanding of RSTP.

1. Which commands did you use to discover this information:
   - The root switch:
   - The port role and port state for each interface

2. The topology created in the example configuration is shown in Figure 7-8.
   a. How was this topology established?
   b. Which switch has only designated ports?
   c. How many root ports does each switch have?

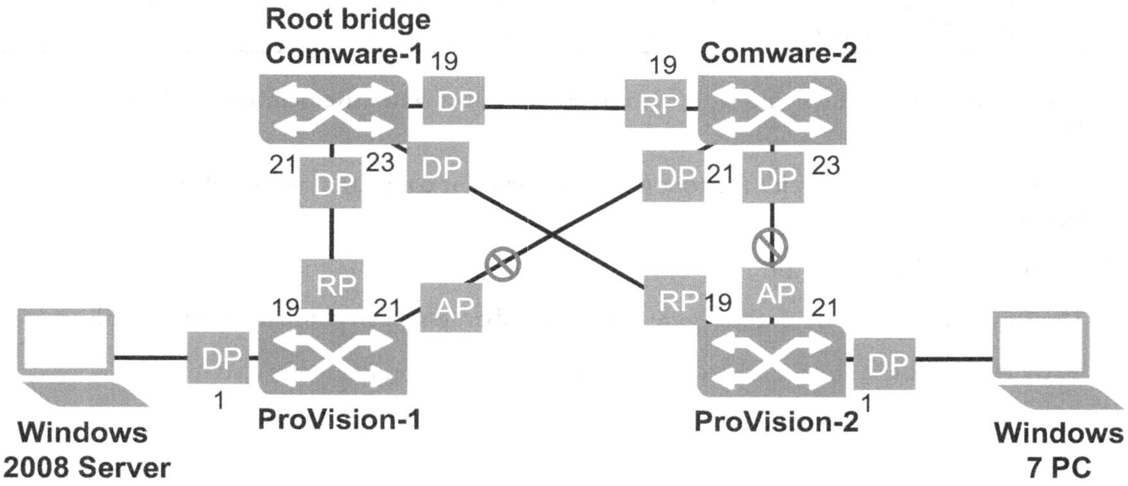


Figure 7-8: RSTP topology with the port states filled in

3. The `display spanning-tree brief` and `show spanning-tree` commands also showed the port cost on each port.
   a. What was the port cost on Comware-2's link to Comware-1?
   b. What was the port cost on ProVision-1's link Comware-1?
   d. Considering these costs, why did Comware-2's port on the link to ProVision-1 become the designated port?

## Learning check 1 answers

1. Which commands did you use to discover this information:

   - The root switch:

   Comware: `display stp root`

   ProVision: `show spanning-tree root-history cst`

   - The port role and port state for each interface:

   Comware: `display stp brief`

   ProVision: `show spanning-tree [interface ID list] instance ist`

2. The topology created in the example configuration is shown in Figure 7-8.

   a. How was this topology established?

   Comware-1 was elected the root. Switches sent BPDUs that indicated how close they were to the root. ProVision-1 and ProVision-2 both have two uplinks that offer paths to the root. In this topology, each switch's root port connects directly to Comware-1. Redundant links between ProVision-1 and Comware-2 and between ProVision-2 and Comware-2 were blocked. As you see, Comware-2 has the designated port on these links, and the ProVision switches have alternate ports.

   b. Which switch has only designated ports?

   The root, which is Comware-1, has only designated ports. Ports on the root are always closer to the root than connected ports, by definition. Therefore, the root bridge has only designated ports and no root port. The one exception is if the root had a looped connection to itself (directly or through non-STP switches). In that case, it could also have a backup port.

   c. How many root ports does each switch have?

   Each switch has *one* root port except Comware-1, which is the root, and has no root ports.

3. The `display spanning-tree brief` and `show spanning-tree` commands also showed the port cost on each port.

   a. What was the port cost on Comware-2's link to Comware-1?

   20

   b. What was the port cost on ProVision-1's link Comware-1?

   20,000

c. Considering these costs, why did Comware-2's port on the link to ProVision-1 become the designated port?

Comware-2 offers a lower root path cost to ProVision-1 (20) than ProVision-1 offers to Comware-2 (20,000). As you learned, the designated port is the port on a link that is closest to the root (has the lowest root path cost).

## Root election

You will now examine in more detail how a root bridge is elected. In the example configuration, the `stp root primary` command was used on Comware-1 to ensure that it won the election. (See Figure 7-9.)

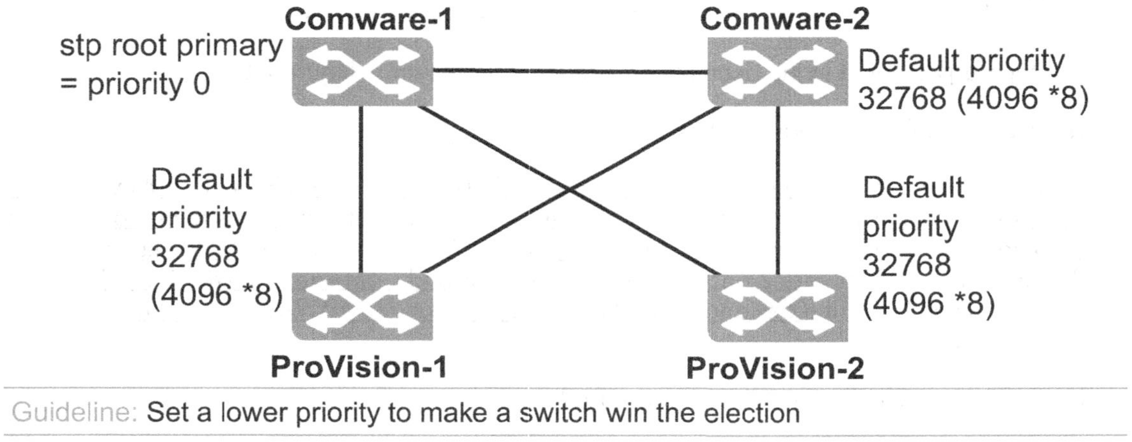


Figure 7-9: Root election

What did this command actually do? It set the switch's spanning tree priority value to 0. A *lower* priority value gives a *higher* priority for being elected root, so Comware-1 was elected root.

Consider what happens if multiple switches have the same lowest priority value, creating a tie, as you see in Figure 7-10. This situation occurs if you do not set a lower priority on any switches (nor use the Comware `stp root` command). In this case, all switches use the default priority value, which is 32,768 or 8 (priorities are set in steps of 4096; 8 × 4096 = 32,768).

In the example network shown in Figure 7-10, this situation could occur if Comware-1 failed, leaving only active switches with the default priority.

It is actually difficult to predict which switch becomes root in this case.

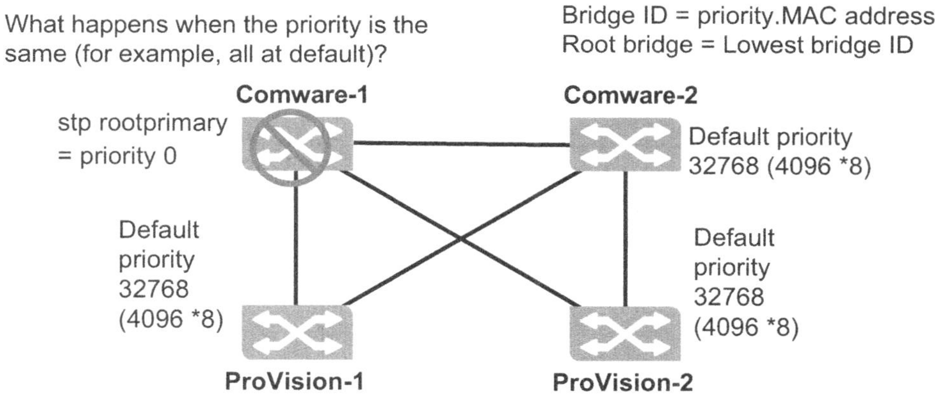


Figure 7-10: Situation in which all active switches are using the default priority

To understand why, you need to understand that switches do not actually use the priority alone to elect the root. They actually use their bridge ID for the election. The bridge ID is:

Priority + system ID (MAC address)

The priority value is more significant, so it affects the election most. However, when multiple switches have the same priority, the switch with the lowest MAC address becomes root. See Table 7-2 for a priority value list.

To prevent an unexpected root, you should always manually configure a priority on the desired root *and* the desired backup (or secondary) root bridge, as indicated in Figure 7-11.

Table 7-2: Priority

| Priority value | Step | Comware root command |
|---|---|---|
| 0 | 0 | root primary |
| 4096 | 1 | root secondary |
| 8092 | 2 | |
| 12288 | 3 | |
| 16384 | 4 | |
| 20480 | 5 | |
| 24576 | 6 | |
| 28672 | 7 | |
| 32768 | 8 | |
| 36864 | 9 | |
| 40960 | 10 | |
| 45056 | 11 | |
| 49152 | 12 | |
| 53248 | 13 | |
| 57344 | 14 | |
| 61440 | 15 | |

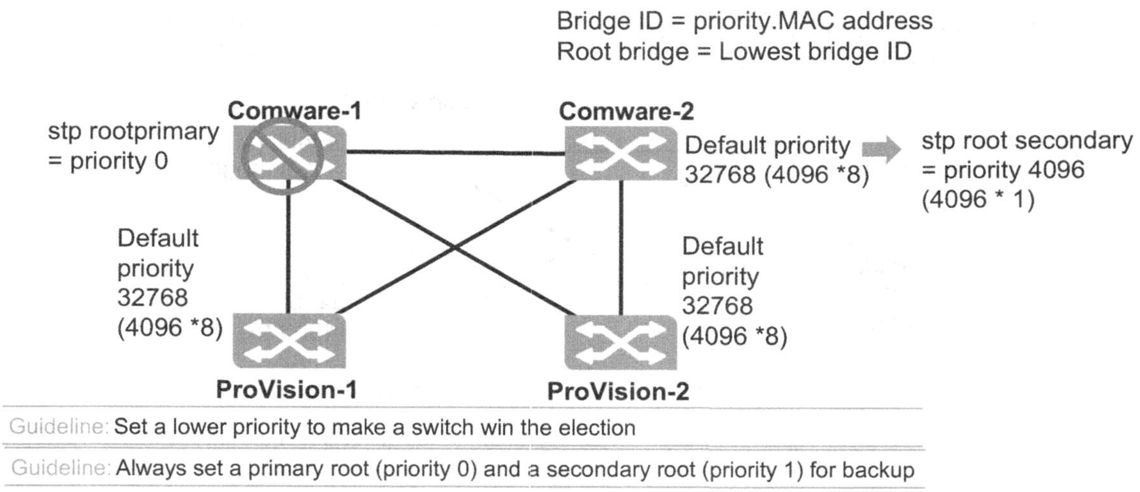


Figure 7-11: Setting a secondary root (priority 1 or 4096) for backup

## Port costs

Up to this point, you have been looking at the topology as if every link has the same cost, and you can simply count links to determine the best path to the root. This is not always true, however. Sometimes links have different bandwidth, so the ports have different costs. Sometimes switches are using different ways to set the port cost.

For example, look at both Comware-2's and ProVision-1's link to the root. As Figure 7-12 shows, Comware-2 has a port cost of 20, and ProVision-1 has a cost of 20,000—even though both switches

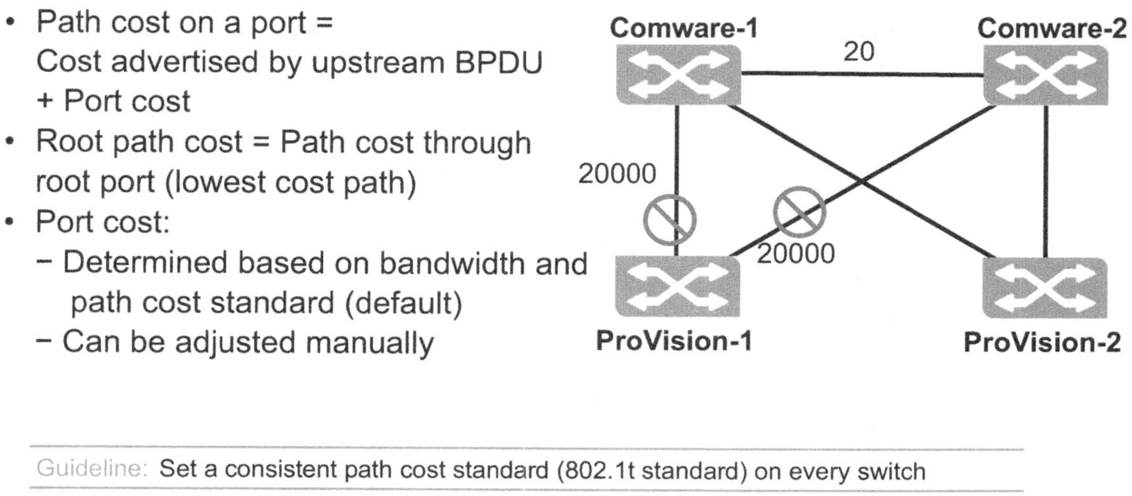


Figure 7-12: Port costs

have a GbE link. The difference occurs because the two switches are using different path cost standards.

You will now learn a bit more about how the root path cost is calculated so that you can follow best practices and understand how topologies are formed. Table 7-3 shows a list of default port costs.

The cost for reaching the root on any port is:

The cost advertised in the BPDU + The port cost

The port cost derives from the bandwidth on the port, as defined by the spanning tree standard. Comware switches, however, by default use a proprietary port cost stetting.

Table 7-3: Default port costs

| Connection Type | RSTP/MSTP | STP (802.1D pre-2004) | Comware proprietary (default setting) |
|---|---|---|---|
| 10 Gbps | 2000 | - | 2 |
| 1 Gbps | 20,000 | 4 | 20 |
| 100 Mbps | 200,000 | 10 | 200 |
| 10 Mbps | 2,000,000 | 100 | 2000 |

Look at the cost for reaching the root on Comware-2's root port. The cost advertised in the BPDU is 0. Comware-1 is the root, so it does not have a cost for reaching the root. The port cost is 20. So the path cost is:

0 + 20 = 20

This is Comware-2's best path to the root, so this becomes its *root path cost.* Comware-2 advertises this cost in BPDUs that it sends on any designated port.

Next examine ProVision-1's cost for reaching the root on its root port. Again Comware-1 advertises a root path cost of 0 because it is root. The port cost is 20,000, which is the RSTP standard cost for a GbE link. So the path cost to the root on this ProVision switch is:

0 + 20,000 = 20,000

Now look at the path cost on ProVision-1's alternate port. Comware-2, as you saw earlier, advertises a root path cost of 20. The port cost is 20,000. Therefore, the path cost to the root is:

20 + 20,000 = 20,020

This cost is higher than the cost on the root port, which is exactly why the first port is the root port and the second one is the alternate port. ProVision-1 has a root path cost of 20,000, which is what it would advertise to any downstream devices.

In the example network topology, you did not experience any negative effects from the discrepancy between the port costs on the Comware and ProVision switches. However, in a larger topology with more links of varying bandwidths, the discrepancy could cause issues. As a best practice, you should set consistent costs on all switches—generally, you should use the standard RSTP costs. ProVision switches use these costs by default. The Comware switches use this command: `stp pathcost dot1t`.

Figure 7-13 shows the port costs after you have set a consistent standard. Note that Comware-2 and ProVision-1 now have equal root path costs. How do the switches decide which has the designated port on the link between them? They look at other criteria to break the tie. The criteria are:

- Bridge ID (lower value wins; remember: the bridge ID is the priority and MAC address)
- If Bridge ID is the same, port priority (lower value wins)
- If port priority is the same, port ID (lower ID wins)

In this example, Comware-2 has a lower priority value, so it has a lower Bridge ID, and its port is designated.

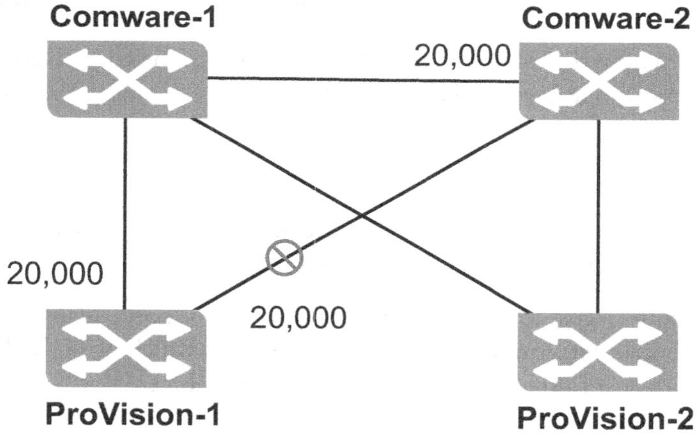


Figure 7-13: Link costs after all switches have been configured with a consistent standard

The consistent standard will help to ensure that the best path is chosen if bandwidths change. For example, consider a situation in which ProVision-1 has a GbE link to Comware-1 and a 10GbE link to Comware-2. Comware-2 also has a 10GbE link to Comware-1. Figure 7-14 shows the new port costs in this situation.

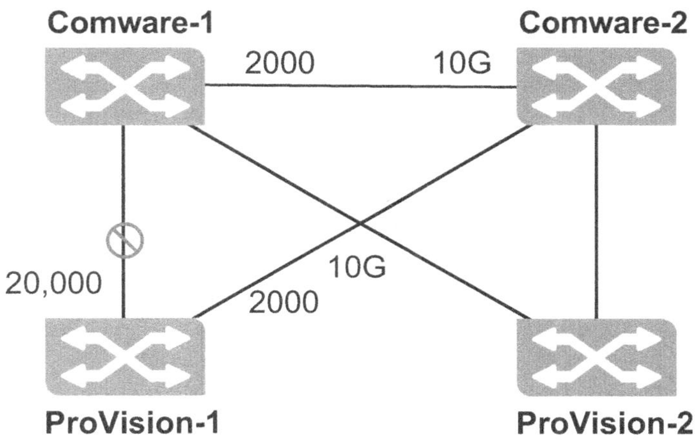


Figure 7-14: Topology in a network with GbE and 10GbE links

You see that, in this case, ProVision-1's link to Comware-1 has a root path cost of:

0 + 20,000 = 20,000

ProVision-1's link to Comware-2 has a root path cost of:

2000 + 2000 = 4000

Therefore, the direct GbE link to Comware-1 actually becomes the alternate port, and the 10GbE link to Comware-2 becomes the root port.

## Failover

You will now learn how failover works in an RSTP environment, as well as how other topology changes are handled.

### Failing over from a root to an alternate port

First you will examine a scenario in which ProVision-1 loses its active link to Comware-1 and must failover to its alternate link through Comware-1, as illustrated in Figure 7-15.

As you see in Figure 7-16, as soon as ProVision-1 detects that its root port is down, it changes its alternate port to a root port, which is immediately set to the forwarding state. Other ports on ProVision-1 and on other switches are unaffected. The convergence takes under a second.

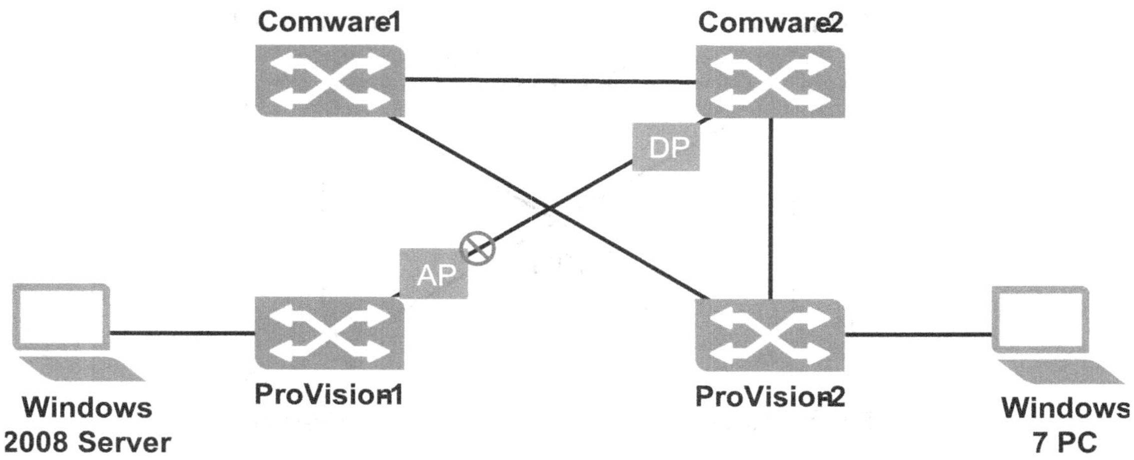


Figure 7-15: Root port lost on a switch with an alternate port

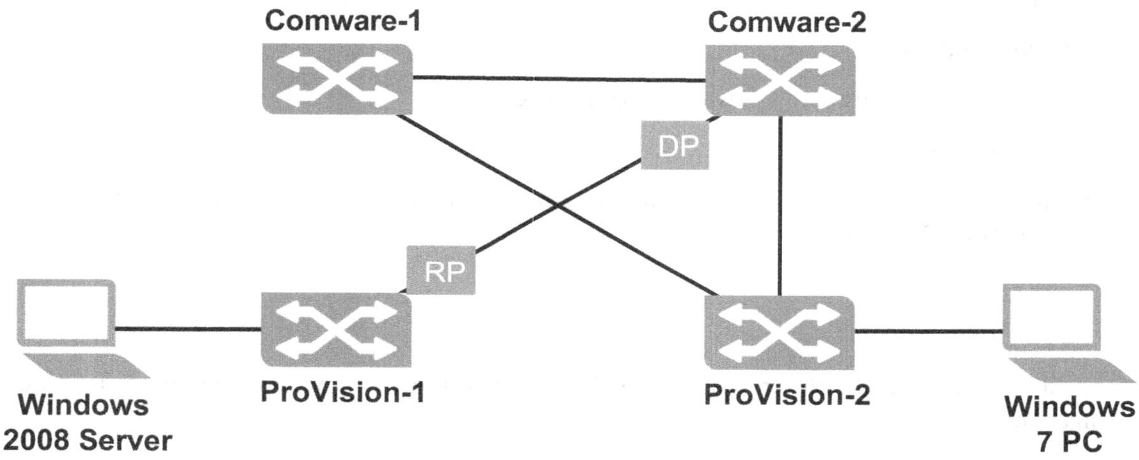


Figure 7-16: Alternate port becomes root port

Event messages on Comware-1 and Comware-2 also indicate that ProVision-1 sends a topology change message to update the root about the change.

## Failing over from a root to a designated port

Next you will examine a scenario in which Comware-2 loses its active link to Comware-1, as shown in Figure 7-17.

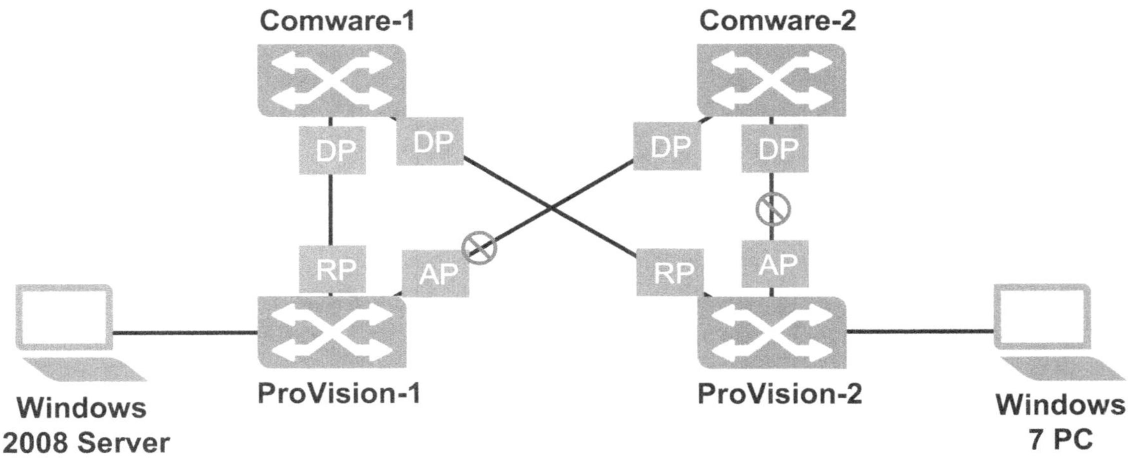


Figure 7-17: Root port lost on a switch without an alternate port

The convergence in this case is a bit more complicated. Comware-2 must transition one of its designated ports (which used to offer the connected switch a better path to the root) to a root port (which now receives a better path to the root from the connected switch). It must do so without introducing temporary loops due to the convergence process.

Several events occur when Comware-2 detects that it has lost its root port:

1. It sends a BPDU with a topology change indication on each of its designated ports.
   a. This BPDU indicates that Comware-2 is now the root. (It has lost contact with the root, and it believes that it has the best priority.)
   b. In legacy STP, the BPDU from Comware-2—called an inferior BPDU because Comware-2 has an inferior switch ID to Comware-1—would have caused the ProVision switch ports to undergo a lengthy reconvergence process. But with RSTP, the process is much faster. The ProVision switches simply rapidly propagate a topology change message along the tree and toward the root, as you can see in Figure 7-18.
   c. Comware-1 sends messages to the ProVision switches, indicating that it is still up, as illustrated in Figure 7-19.
2. Each ProVision switch knows that it has a better path to the root than Comware-2 is advertising in its inferior BPDU. It undergoes a quick exchange with the connected Comware-2 port.
   a. The ProVision port asserts itself as now offering a better connection to the root. The port becomes a designated port, but it is in discarding state to prevent temporary loops during convergence, as shown in Figure 7-20.

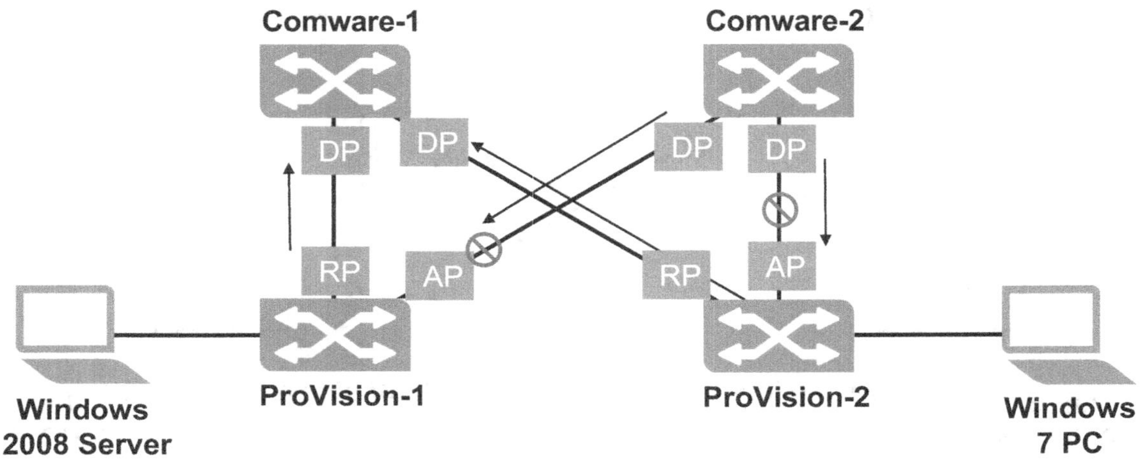


Figure 7-18: Step 1a and b in convergence

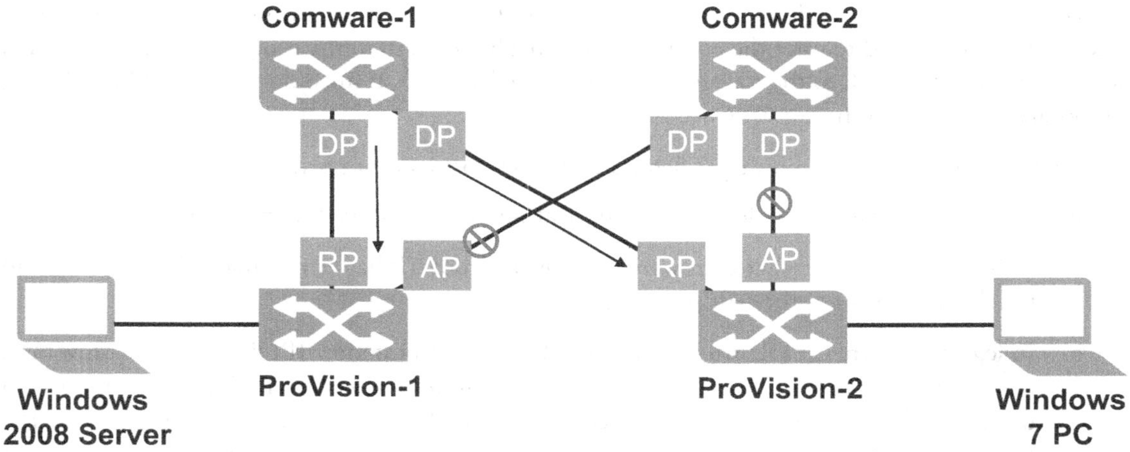


Figure 7-19: Step 1c in convergence

b. Comware-2 chooses the best BPDU—the lowest cost path—and sets the root port. This port can immediately transition to forwarding. The other port is an alternate port, which is discarding, illustrated in Figure 7-21.

c. Comware-2 informs each connected port that it has set a single root port and other ports are discarding (as shown in Figure 7-22), ensuring that it will not cause loops. The ProVision switches then let their designated ports transition to forwarding.

Although this process involves more steps than the previous failover scenario, it occurs in about a second or less.

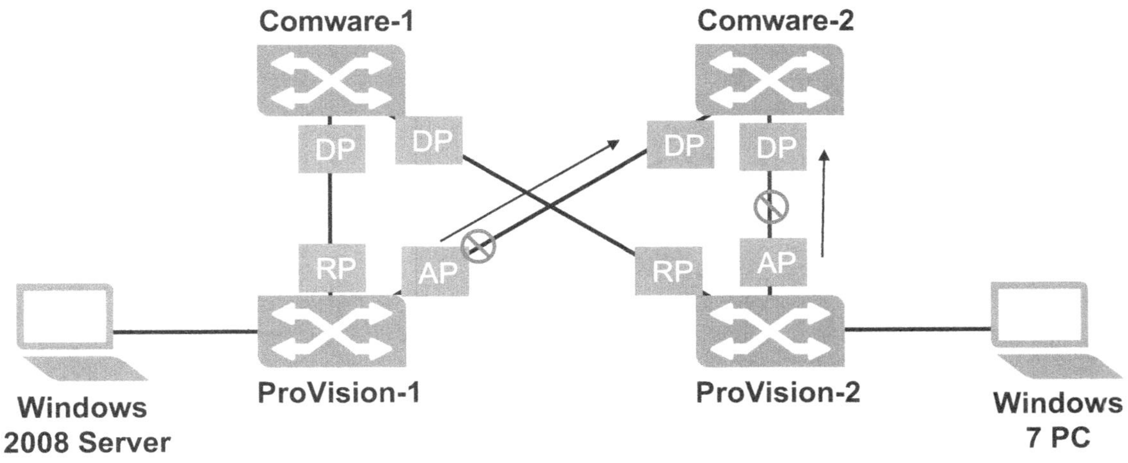


Figure 7-20: Step 2a in convergence

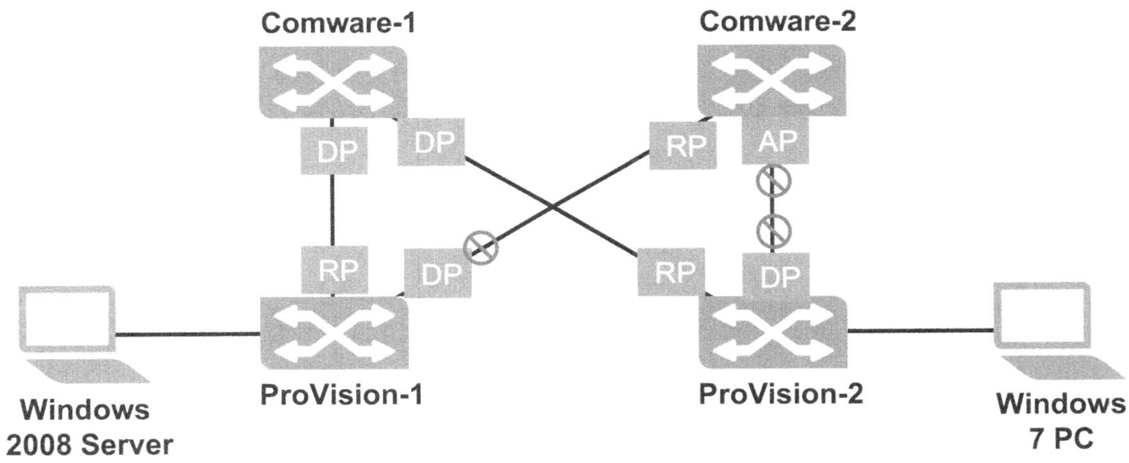


Figure 7-21: Step 2b in convergence

These are the messages and command outputs you would see on the Comware-2 switch:

```
<Comware-2>display stp brief
MSTID Port                     Role  STP State   Protection
  0   GigabitEthernet1/0/19    ROOT  FORWARDING  NONE
  0   GigabitEthernet1/0/21    DESI  FORWARDING  NONE
  0   GigabitEthernet1/0/23    DESI  FORWARDING  NONE
```

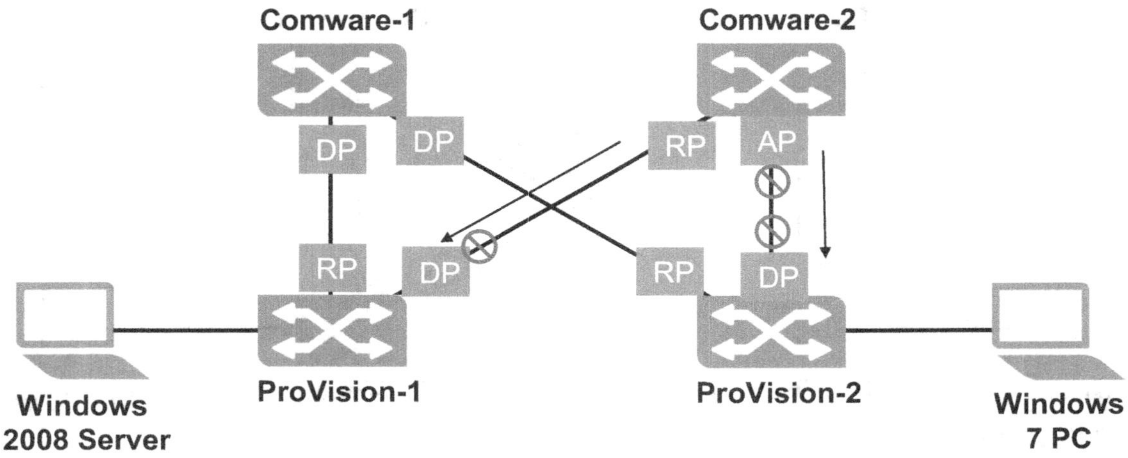


Figure 7-22: Step 2c in convergence

```
<Comware-2>

#Apr 27 04:13:59:596 2000 Comware-2 MSTP/1/PDISC: hwPortMstiStateDiscarding:
Instance 0's Port 0.9437204 has been set to discarding state!

%Apr 27 04:13:59:767 2000 Comware-2 MSTP/6/MSTP_DISCARDING: Instance
0's GigabitEthernet1/0/21 has been set to discarding state.

%Apr 27 04:13:59:927 2000 Comware-2 MSTP/6/MSTP_NOTIFIED_TC: Instance
0's GigabitEthernet1/0/23 was notified a topology change.

%Apr 27 04:14:00:088 2000 Comware-2 IFNET/3/LINK_UPDOWN: GigabitEthernet
1/0/19 link status is DOWN.

%Apr 27 04:14:01:160 2000 Comware-2 MSTP/6/MSTP_NOTIFIED_TC: Instance
0's GigabitEthernet1/0/23 was notified a topology change.

<Comware-2>display stp brief

MSTID  Port                     Role  STP State   Protection

  0    GigabitEthernet1/0/21    ALTE  DISCARDING   NONE

  0    GigabitEthernet1/0/23    ROOT  FORWARDING   NONE
```

## Reconvergence when a better path is added

You will now examine what happens when the link between Comware-2 and Comware-1 is repaired so that Comware-2 now has a better path to the root. As you see in Figure 7-23, the link comes up blocking to prevent loops.

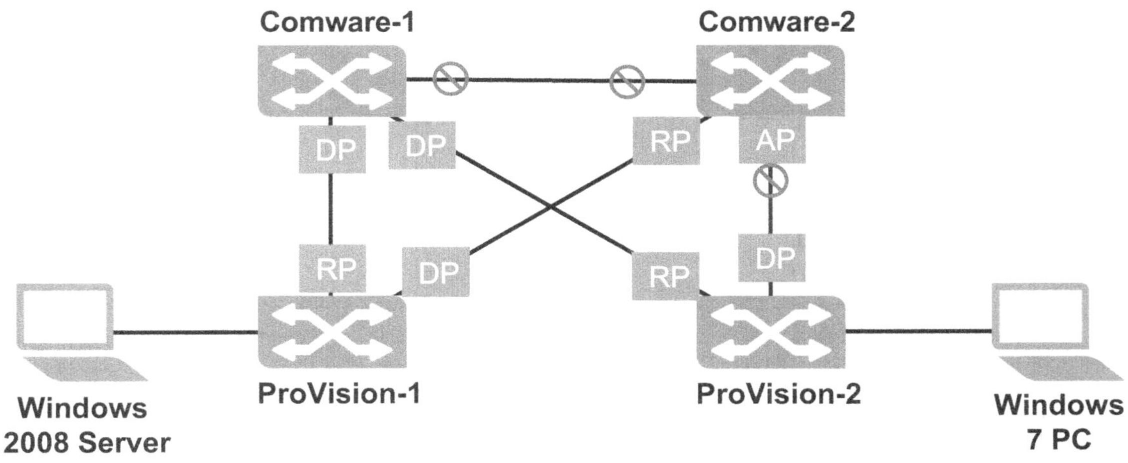


Figure 7-23: Better path to the root restored

Several events now occur:

1. Comware-1 asserts that it has a superior path to the root, and Comware-2 sees that this is the case.
   a. The new port on Comware-2 becomes its root port, which transitions immediately to forwarding. As Figure 7-24 shows, the link is still blocked on the Comware-1 side, so loops do not occur.

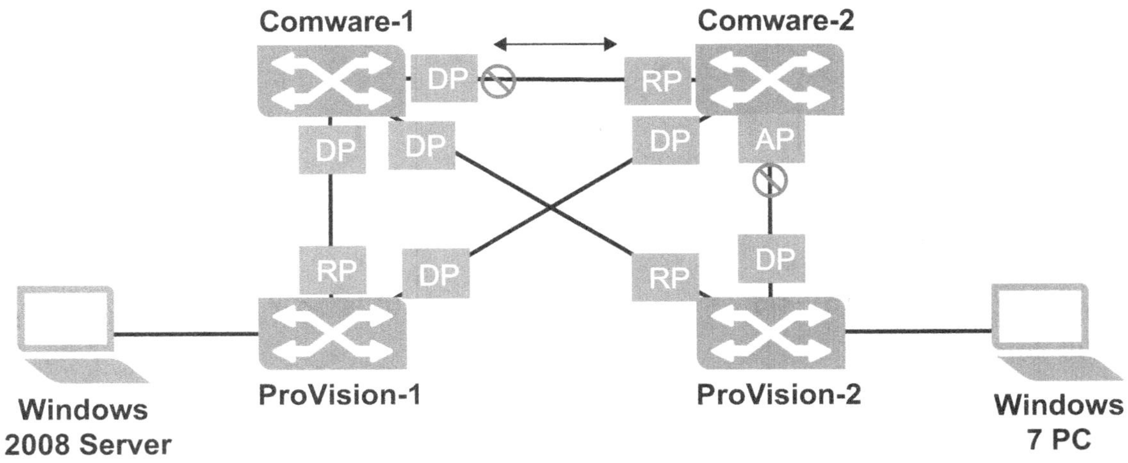


Figure 7-24: Convergence step 1a

   b. At the same time, Comware-2 sets any other port that is currently forwarding to discarding, as shown in Figure 7-25. This action guarantees that loops will not occur during convergence. (Comware-2 now believes that it has a better path to the root than each connected ProVision switch does, so the ports are designated.)

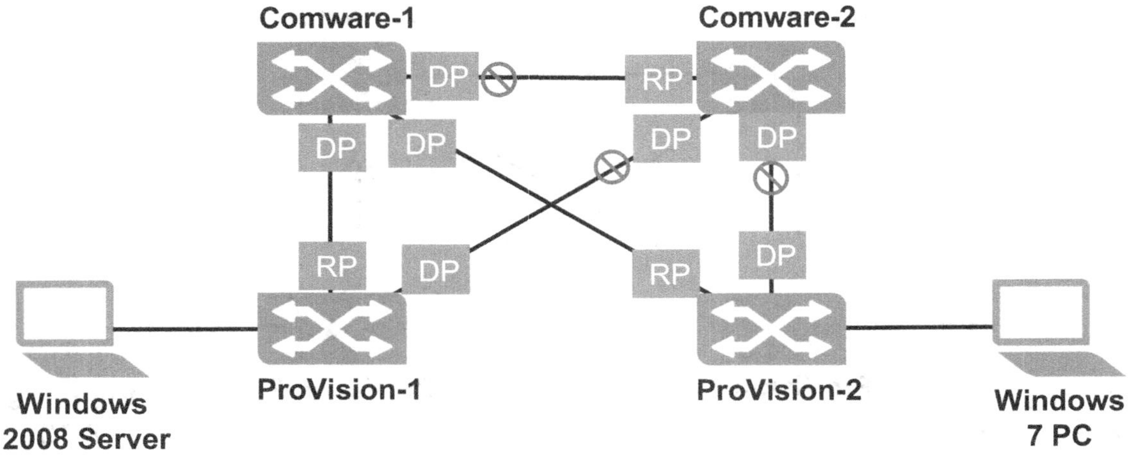


Figure 7-25: Convergence step 1b

c. Comware-2 informs Comware-1 that it has a single root port and other ports are discarding, as you see in Figure 7-26. With this assurance, Comware-1 sets its designated port to forwarding.

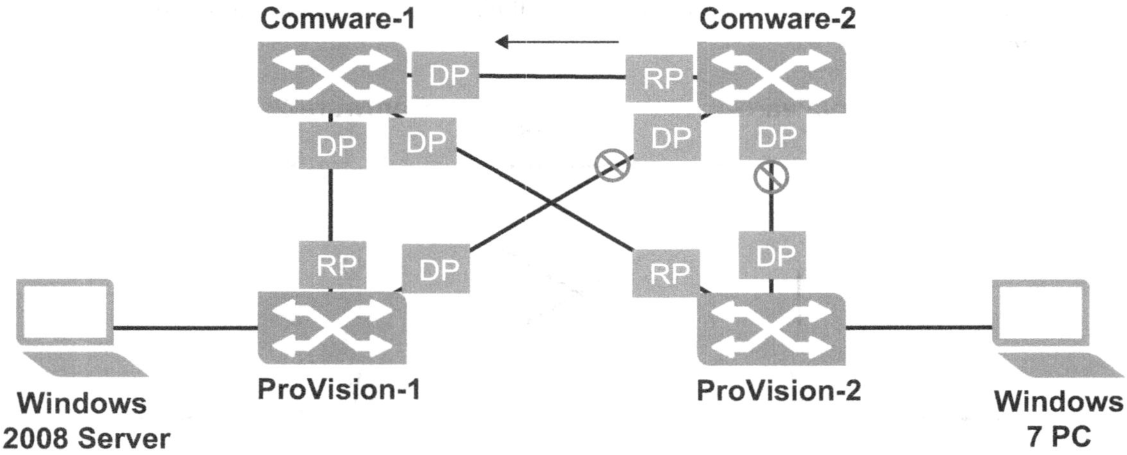


Figure 7-26: Convergence step 1c

2. The new link is now up, but Comware-2 needs to move any discarding designated ports to a forwarding state. To do this, it completes an agreement proposal handshake mechanism with the ProVision switches similar to the one that it completed with Comware-1.

   a. In this exchange, Comware-2 takes the role that Comware-1 took because Comware-2 believes that it offers a better path to the root than the correct switch, as shown in Figure 7-27.

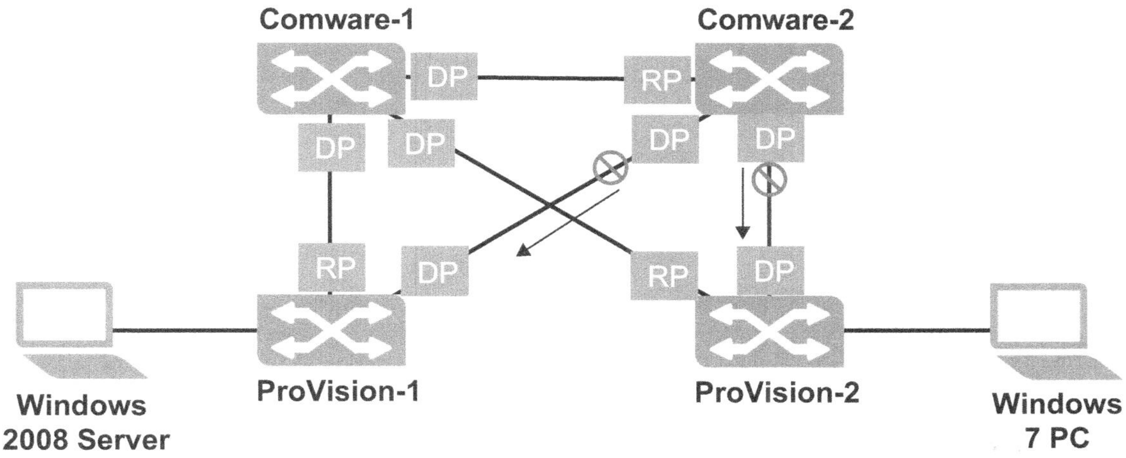


Figure 7-27: Convergence step 2a

b. Each ProVision switch recognizes that Comware-2 is using the same root as it is, but Comware-2 does now offer a better root path (superior BPDU) than it can. So each ProVision switch changes the port role accordingly, as illustrated in Figure 7-28.

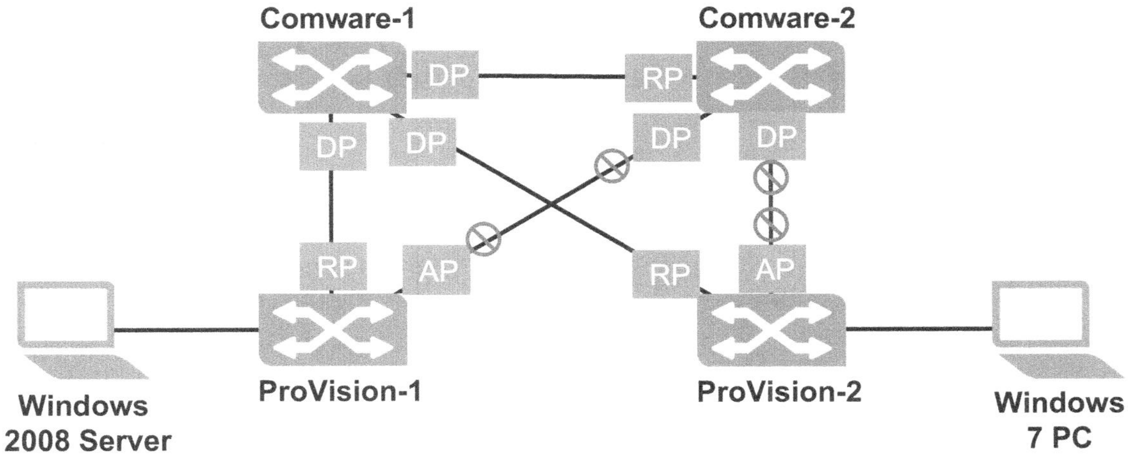


Figure 7-28: Convergence step 2b

c. Each ProVision switch informs the connected Comware-2 port that it has one root port and other ports are blocking, as you can see in Figure 7-29. (If the switch had other designated ports, it would need to block them, and the convergence would ripple downstream.) Therefore, Comware-1 sets its port to forwarding.

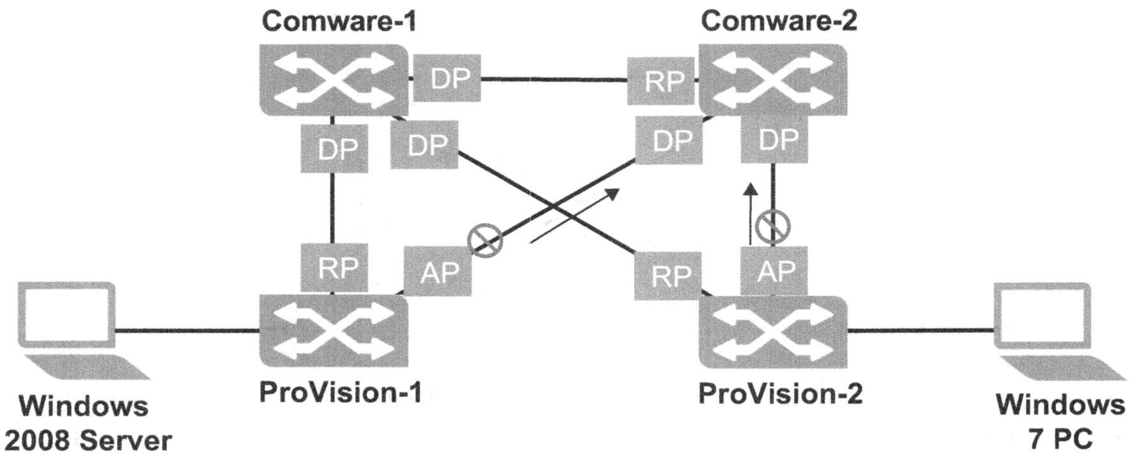


Figure 7-29: Convergence step 2c

These are the messages and command outputs you would see on the Comware-2 switch:

```
<Comware-2>display stp brief
 MSTID  Port                     Role  STP State   Protection
   0    GigabitEthernet1/0/21    ALTE  DISCARDING  NONE
   0    GigabitEthernet1/0/23    ROOT  FORWARDING  NONE
<Comware-2>
%Apr 27 04:17:08:934 2000 Comware-2 IFNET/3/LINK_UPDOWN: GigabitEthernet1/0/19
link status is UP.
%Apr 27 04:17:09:221 2000 Comware-2 LLDP/6/LLDP_CREATE_NEIGHBOUR:
New neighbor created on Port GigabitEthernet1/0/19 (IfIndex 9437202),
Chassis ID is d07e-28ce-c94f, Port ID is GigabitEthernet1/0/19.
#Apr 27 04:17:10:222 2000 Comware-2 MSTP/1/PDISC: hwPortMstiStateDiscarding:
Instance 0's Port 0.9437206 has been set to discarding state!
#Apr 27 04:17:10:392 2000 Comware-2 MSTP/1/PFWD: hwPortMstiStateForwarding:
Instance 0's Port 0.9437202 has been set to forwarding state!
%Apr 27 04:17:10:563 2000 Comware-2 MSTP/6/MSTP_DISCARDING: Instance
0's GigabitEthernet1/0/23 has been set to discarding state.
%Apr 27 04:17:10:723 2000 Comware-2 MSTP/6/MSTP_FORWARDING: Instance
0's GigabitEthernet1/0/19 has been set to forwarding state.
%Apr 27 04:17:10:884 2000 Comware-2 MSTP/6/MSTP_DETECTED_TC: Instance
0's GigabitEthernet1/0/19 detected a topology change.
```

```
%Apr 27 04:17:11:034 2000 Comware-2 MSTP/6/MSTP_NOTIFIED_TC: Instance
0's GigabitEthernet1/0/19 was notified a topology change.

%Apr 27 04:17:11:194 2000 Comware-2 MSTP/6/MSTP_FORWARDING: Instance
0's GigabitEthernet1/0/21 has been set to forwarding state.

%Apr 27 04:17:11:365 2000 Comware-2 MSTP/6/MSTP_DETECTED_TC: Instance
0's GigabitEthernet1/0/21 detected a topology change.

#Apr 27 04:17:11:515 2000 Comware-2 MSTP/1/PFWD: hwPortMstiStateForwarding:
Instance 0's Port 0.9437204 has been set to forwarding state!

#Apr 27 04:17:11:757 2000 Comware-2 MSTP/1/PFWD: hwPortMstiStateForwarding:
Instance 0's Port 0.9437206 has been set to forwarding state!

%Apr 27 04:17:11:927 2000 Comware-2 MSTP/6/MSTP_FORWARDING: Instance
0's GigabitEthernet1/0/23 has been set to forwarding state.

%Apr 27 04:17:12:220 2000 Comware-2 MSTP/6/MSTP_NOTIFIED_TC: Instance
0's GigabitEthernet1/0/19 was notified a topology change.
```

## Spanning tree edge ports

The example network shown in Figure 7-30 has two ports that connect to endpoints, one on ProVision-1 and one on ProVision-2. Of course, a real network would have many more such ports. Because these ports connect to endpoints, with a single link to the spanning tree topology, they can never introduce loops.

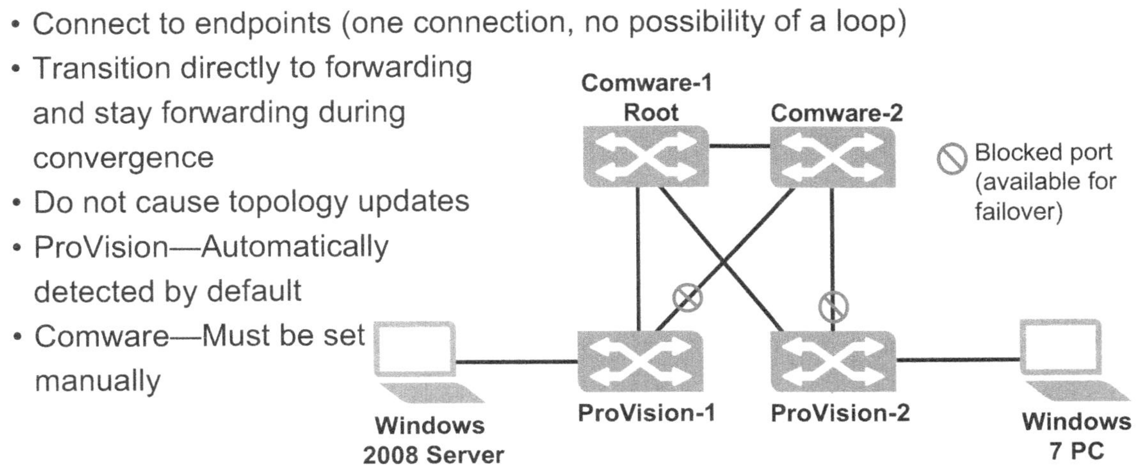


Figure 7-30: Spanning tree edge ports

RSTP lets you define such ports as "edge ports." The edge ports have the designated role, and they are allowed to transition to forwarding immediately.

On the previous pages, you learned about various failover situations. Several of these situations cause a switch to temporarily set all of its designated ports to a discarding state to prevent loops during convergence. If one of those designated ports connects to an endpoint, that endpoint would lose connectivity while the switch waits to receive a BPDU or for a timeout. The temporary loss of connectivity could be damaging, particularly for a server. However, a switch keeps edge ports forwarding during these situations—because it knows that the edge ports cannot introduce loops—helping endpoints and servers to maintain connectivity.

In addition, changes in edge port status do not affect the RSTP topology. If an edge port goes down, the switch does not send a topology update to the root. It does not need to recalculate its port roles, and neither do other switches. In this way, properly defined edge ports make the overall RSTP topology more stable.

HP ProVision switch ports are enabled for auto edge port detection by default. If the port does not receive a BPDU within 3 seconds, the switch defines the port as an edge port. If it later receives a BPDU, the port is reclassified as non edge. You can enable and disable this feature with this command:

```
ProVision(config)# [no] spanning-tree <int-id-list> auto-edge-port
```

You can also configure a port as an edge port manually, which makes the port always act as an edge port immediately. However, if the switch detects a BPDU on the port, it disables the edge port function.

```
ProVision(config)# [no] spanning-tree <int-id-list> admin-edge-port
```

Comware ports do not support the auto detection feature. You must configure the port as an edge port manually. Make sure that the port actually connects to an endpoint that cannot introduce loops and enter this command from the interface view:

```
[Comware-int-id] stp edged-port enable
```

You can disable the edge port feature with this command:

```
[Comware-int-id] stp edged-port disable
```

## Issues with RSTP

RSTP guarantees a loop-free network, but it does not guarantee optimization.

For example, in the network shown in Figure 7-31, ProVision-1 and ProVision-2 each have two uplinks. However, only one uplink is active, so half of the uplink bandwidth is essentially blocked.

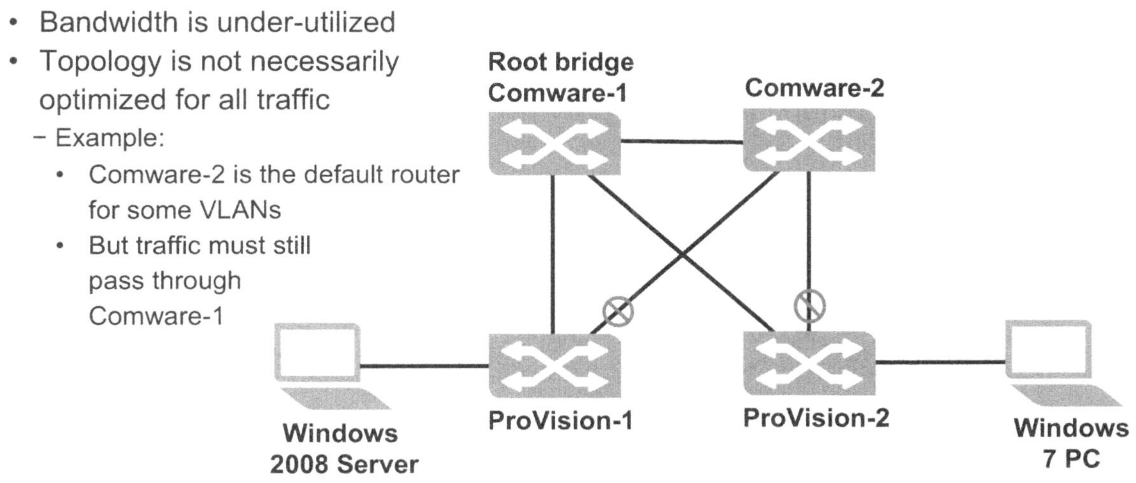


Figure 7-31: Issues with RSTP

In addition, the spanning tree always sets up shortest paths to the root, which might not be the best path for specific traffic. You set Comware-1 as the root because it is the default gateway, so most traffic is directed to it. However, what if you wanted to set up Comware-2 as the default gateway for some VLANs? Traffic would still need to pass through Comware-1 because Comware-1 is the root.

## MSTP solution

MSTP, IEEE standard 802.1s, provided an extension to STP and RSTP in order to deal with the issues outlined in this chapter. Later 802.1s was merged into 802.1Q-2005, so MSTP is now part of the Ethernet standard.

As Figure 7-32 shows, MSTP allows switches to set up multiple spanning trees, called spanning tree instances. Each spanning tree instance is associated with a different group of VLANs. Different spanning tree instances can have different root bridges, leading to different topologies that use different links. Traffic for some VLANs can use some links while traffic for other VLANs use other links, which significantly improves network resource utilization while maintaining a loop-free environment in each VLAN broadcast domain.

Figure 7-32 shows two MSTP instances. In addition, each network also requires the default instance 0, which has some special roles.

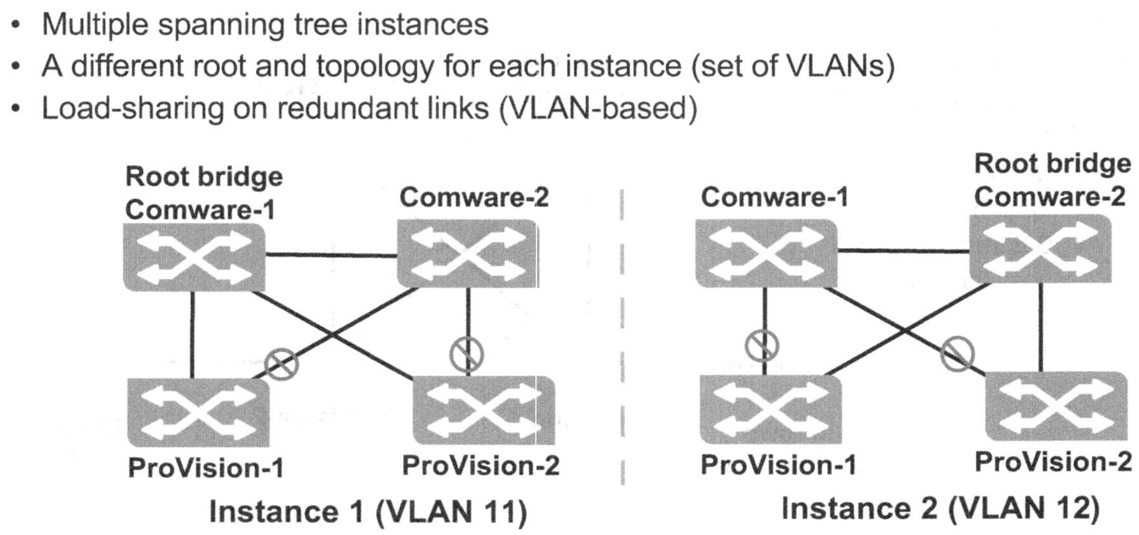


Figure 7-32: MSTP solution

Table 7-4 indicates the root bridge for each instance in the example network shown in Figure 7-32, as well as the VLANs that are associated with each instance.

As you see in Figure 7-32, this plan creates two spanning tree topologies, one used by instance 1 (and 0) and one used by instance 2. The first topology uses the uplinks between the ProVision switches and Comware-1, and the second topology uses the uplinks between the ProVision switches and Comware-2.

Table 7-4: MSTP instances

| Instance | VLANs | Root | Secondary root |
|---|---|---|---|
| 0 | 1 (and all unused) | Comware-1 | Comware-2 |
| 1 | 11 | Comware-1 | Comware-2 |
| 2 | 12 | Comware-2 | Comware-1 |

## MSTP region

Before configuring MSTP, you should understand how to set up a region, as illustrated in Figure 7-33.

The MSTP region lets MSTP switches guarantee that they are using the same VLAN-to-instance mappings. It consists of these settings:

- Region name
- Revision number
- Instance-to-VLAN mapping

All of the settings must match for switches to become part of the same MSTP region.

- Switches in an MSTP region must match region settings exactly
- Ports inside a region can use all instances

| Parameter | Setting for example network |
|---|---|
| Name | hp |
| Revision number | 1 |
| Instance-to-VLAN mapping | 0 = VLAN 1 and all unused VLANs<br>1 = VLAN 11<br>2 = VLAN 12 |

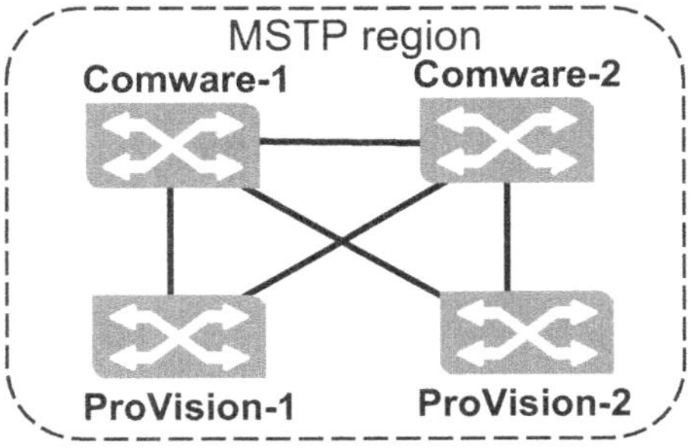


Figure 7-33: MSTP region

## MSTP region incompatibility

When two connected switches run MSTP but do not have the same settings, they have an MSTP region incompatibility. Switches in different MSTP regions cannot use their instances because they do not know that they map the same VLANs to these instances. Therefore, the switches use RSTP between each other. The ports between the regions, which are called boundary ports, have a forwarding or blocking state that applies to *all* instances configured on the switch. As far as each switch is concerned, the different MSTP region is an RSTP switch.

"Configuration example: RSTP" on page 186 actually shows this setup. The HP switches operate in MSTP mode, so you cannot truly configure RSTP on them. Instead, when you want to use RSTP, you simply do not establish any MSTP region settings. Because the switches have incompatible regions, each switch is in its own MSTP region and looks like an RSTP switch to its neighbors.

As Figure 7-34 shows, all of the switches across the different regions, as well as any actual RSTP switches, establish a common internal spanning tree (CIST). The CIST elects a single root. MSTP switches use their instance 0 settings in the CIST election. (The CIST is sometimes called just the common spanning tree [CST] when the connections between regions are emphasized.)

Instance 0 is also called the internal spanning tree (IST). The IST connects the MSTP region to the rest of the CIST. The region selects one boundary port—the one closest to the CIST root—as the region master port. In switch command outputs, you will see this port designated as a root port in the IST and CIST.

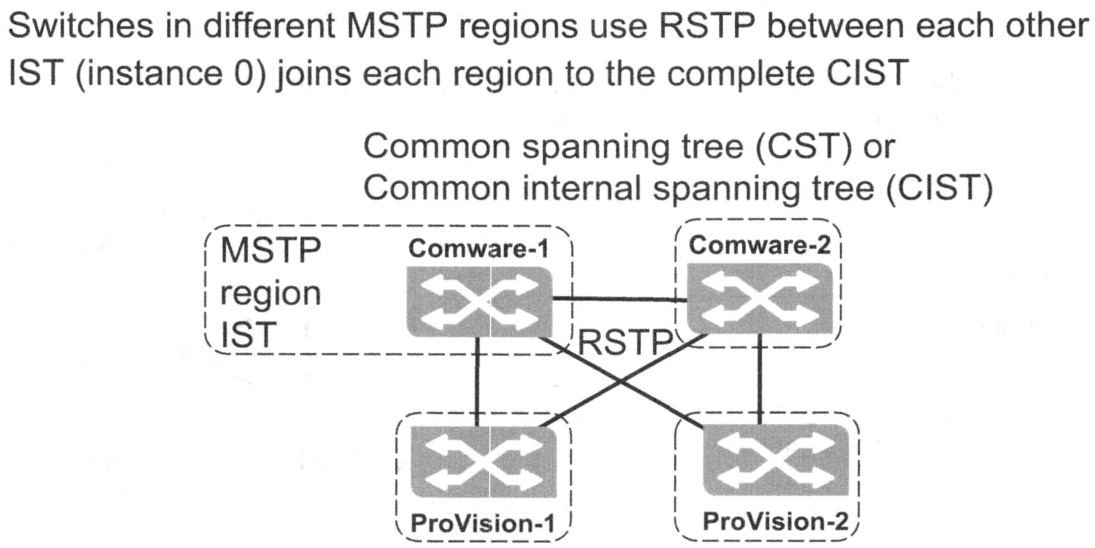


Figure 7-34: MSTP region incompatibility

If the region has other boundary ports—such as ProVision-1's second uplink in the figure—those ports become designated, alternate, or backup ports in the CIST. You can determine the selected role by thinking of each MSTP region as a single RSTP switch and using the rules that you learned for RSTP. Remember: boundary ports function like RSTP ports, so the boundary port role is the *single* role used by *all* VLANs on the port, regardless of MSTP instance settings.

Note that, for HP switches, any settings that you configure globally for spanning tree apply to instance 0. When you configured Comware-1 with the `stp root primary` command, you actually configured it with a priority of 0 in MSTP instance 0. Other switches had the default priority in their instance 0, so Comware-1 won the CST election.

## Example configuration: MSTP

In this section, you will learn how to configure MSTP on HP switches. This section builds on the network topology established in "Configuration example: RSTP" on page 186.

HP ExpertOne If you want to watch a video that shows the steps outlined in this section, access the HP Aurasma application on your smartphone and use the application to focus on Figure 7-35. This action will launch the video.

Specifically, you will learn how to configure a consistent MSTP region on all of the switches. Figure 7-35 shows the root bridge for the MSTP instances, and Table 7-5 lists the region settings for the example network.

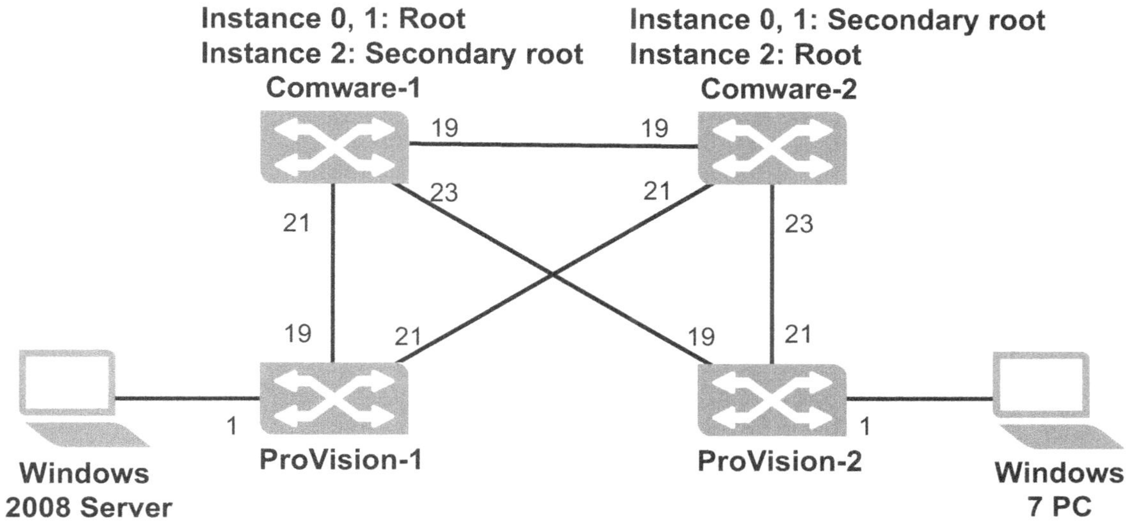


Figure 7-35: Root and secondary root bridges for the MSTP regions

Table 7-5: MSTP region settings

| Parameter | Value |
|---|---|
| Name | hp |
| Revision | 1 |
| Instance-to-VLAN mapping | 0: 1-10,13-4094<br>1: 11<br>2: 12 |

## Configure MSTP settings on Comware-1

1. Access the Comware-1 CLI and move to the system view.
2. Access the MSTP region view.

```
[Comware-1] stp region-configuration
```

3. Establish the settings indicated in the table.

```
[Comware-1-mst-region] region-name hp

[Comware-1-mst-region] revision-level 1

[Comware-1-mst-region] instance 1 vlan 11

[Comware-1-mst-region] instance 2 vlan 12
```

4. You do not need to configure the instance 0 mapping; by default, instance 0 includes all VLANs that have not been mapped to a different instance. However, if you did need to move a VLAN back to instance 0, use this command:

```
[Comware-1-mst-region] instance 0 vlan 1
```

5. If you want, verify the configuration.

```
[Comware-1-mst-region] check region-configuration
 Admin configuration
   Format selector     :0
   Region name         :hp
   Revision level      :1
Configuration digest :0xbe0284d20f4d46a8da89c5d9b3b4f78a
   Instance   Vlans Mapped
   0          1 to 10, 13 to 4094
   1          11
   2          12
```

**Important**

You must activate the MSTP region configuration on Comware switches. Otherwise, the MSTP region settings do not take effect, even though they appear in the configuration.

6. Activate the configuration.

```
[Comware-1-mst-region] active region-configuration
[Comware-1-mst-region] quit
```

## Configure MSTP settings on Comware-2

1. Access the Comware-2 CLI and move to system view.
2. Follow the same steps to establish the MSTP region.

```
[Comware-2] stp region-configuration
[Comware-2-mst-region] region-name hp
[Comware-2-mst-region] revision-level 1
[Comware-2-mst-region] instance 1 vlan 11
[Comware-2-mst-region] instance 2 vlan 12
```

3. Verify the configuration. Make sure that the configuration digest matches the digest on Comware-2. If the digest matches, then all settings are compatible.

```
[Comware-2-mst-region] check region-configuration
 Admin configuration
   Format selector    :0
   Region name        :hp
   Revision level     :1
Configuration digest :0xbe0284d20f4d46a8da89c5d9b3b4f78a
   Instance   Vlans Mapped
      0       1 to 10, 13 to 4094
      1       11
      2       12
```

4. Remember to activate the region.

```
[Comware-2-mst-region] active region-configuration
[Comware-2-mst-region] quit
```

## Configure MSTP settings on ProVision-1

You will now configure the MSTP region on the ProVision switches.

1. Access the terminal session with ProVision-1 and move to the global configuration context.
2. Configure identical MSTP region settings as on the Comware switches.

```
ProVision-1(config)# spanning-tree config-name hp
ProVision-1(config)# spanning-tree config-revision 1
ProVision-1(config)# spanning-tree instance 1 vlan 11
ProVision-1(config)# spanning-tree instance 2 vlan 12
```

**Note**

Again, you do not need to configure the instance 0 mapping; by default, instance 0 includes all VLANs that have not been mapped to a different instance. If you did need to move a VLAN back to instance 0, you must remove the VLAN from its current instance: no spanningtree instance <ID> vlan <vid>

3. Verify the settings. (The settings are automatically activated on ProVision switches.) Once again, check the configuration digest against that on the other switches.

```
ProVision-1(config)# show spanning-tree mst-config

 MST Configuration Identifier Information

  MST Configuration Name : hp

  MST Configuration Revision : 1

  MST Configuration Digest : 0xBE0284D20F4D46A8DA89C5D9B3B4F78A

IST Mapped VLANs : 1-10,13-4094

  Instance ID Mapped VLANs

  ----------  ----------------------------------------------

  1            11

  2            12
```

## Configure MSTP settings on ProVision-2

1. Access the terminal session with the ProVision-2 switch and move to the global configuration context.
2. Complete the same steps to configure and verify the MSTP region.

```
ProVision-2(config)# spanning-tree config-name hp

ProVision-2(config)# spanning-tree config-revision 1

ProVision-2(config)# spanning-tree instance 1 vlan 11

ProVision-2(config)# spanning-tree instance 2 vlan 12

ProVision-2(config)# show spanning-tree mst-config

 MST Configuration Identifier Information

  MST Configuration Name : hp

  MST Configuration Revision : 1

  MST Configuration Digest : 0xBE0284D20F4D46A8DA89C5D9B3B4F78A
```

```
IST Mapped VLANs : 1-10,13-4094
  Instance ID Mapped VLANs
  ----------- --------------------------------------------------
  1           11
  2           12
```

## Configure the instance root settings and set standard costs for Comware switches

Now that the switches are in the same MSTP region, they will elect a root for each instance. It is best practice to select different roots for different instances so as to load-balance traffic. Table 7-6 shows the settings for this example configuration.

Table 7-6: MSTP instances

| Instance | VLANs | Root | Secondary root |
|---|---|---|---|
| 0 | 1 (and all unused) | Comware-1 | Comware-2 |
| 1 | 11 | Comware-1 | Comware-2 |
| 2 | 12 | Comware-2 | Comware-1 |

You will also learn how to configure the Comware switches to use 802.1t for their path cost standard. Setting a standard is best practice for RSTP and MSTP.

## Configure root settings and cost on Comware-1

1. Access the Comware-1 terminal session and make sure that you are in the system view.
2. To configure the switch as a primary or secondary root, enter:

```
[Comware-1] stp instance 0 root primary
[Comware-1] stp instance 1 root primary
[Comware-1] stp instance 2 root secondary
```

3. Configure the switch to use the 802.1t port cost standard, making the standard consistent with ProVision switches. *Make sure to confirm the change.*

```
[Comware-1] stp pathcost dot1t
Cost configuration of every port will be reset and auto-calculation is
available after changing current pathcost standard. Continue?[Y/N]:y
```

## Configure root settings and cost on Comware-2

4. Access the Comware-2 switch terminal session.

5. Use similar commands to configure this switch as the secondary and primary root of the correct instances. (See Table 7-6.)

```
[Comware-2] stp instance 0 root secondary

[Comware-2] stp instance 1 root secondary

[Comware-2] stp instance 2 root primary
```

6. Configure the switch to use the 802.1t port cost standard.

```
[Comware-2] stp pathcost dot1t

Cost configuration of every port will be reset and auto-calculation is
available after changing current pathcost standard. Continue?[Y/N]:y
```

## Verify the configuration

If all of the switches are operating in the same MSTP region, they should have elected the same switches as root in each instance:

- Comware-1 in instance 0 and 1
- Comware-2 in instance 2

You will now verify that this is the case.

1. Access the Comware-1 CLI. Verify that it is the root for instances 0 and 1. (Even if you do not remember Comware-1's bridge ID, you can see that it is root because it has no root port.)

```
[Comware-1] display stp root

MSTID Root Bridge ID     ExtPathCost IntPathCost Root Port

  0   0.d07e-28ce-c94f   0           0

  1   0.d07e-28ce-c94f   0           0

  2   0.d07e-28ce-d31e   0           20000       GigabitEthernet1/0/19
```

2. Record the instances 0 and 1 root bridge ID (or at least the last four characters).

3. Access the terminal session for Comware-2. Verify that it is the root for instance 2.

```
[Comware-2] display stp root

MSTID Root Bridge ID   ExtPathCost IntPathCost Root Port

  0   0.d07e-28ce-c94f 0           20000       GigabitEthernet1/0/19
```

```
  1    0.d07e-28ce-c94f 0            20000        GigabitEthernet1/0/19
  2    0.d07e-28ce-d31e 0            0
```

4. Record the instance 2 root bridge ID (or at least the last four characters).
5. Check the command output to verify that Comware-2 is using Comware-1 for the instance 0 and 1 root bridge (ID recorded in 2).

```
[Comware-2] display stp root
MSTID Root Bridge ID   ExtPathCost IntPathCost Root Port
  0    0.d07e-28ce-c94f 0            20000        GigabitEthernet1/0/19
  1    0.d07e-28ce-c94f 0            20000        GigabitEthernet1/0/19
  2    0.d07e-28ce-d31e 0            0
```

6. Check back in the command output on Comware-1. Verify that it is using Comware-2 for the instance 2 root bridge (ID recorded in 4).

```
[Comware-1] display stp root
MSTID Root Bridge ID   ExtPathCost IntPathCost Root Port
   0   0.d07e-28ce-c94f 0            0
   1   0.d07e-28ce-c94f 0            0
   2   0.d07e-28ce-d31e 0            20000        GigabitEthernet1/0/19
```

7. Access the ProVision-1 CLI and verify that the switch's root for each instance matches the root on the Comware switches. You must enter the `show spanning-tree root-history` command for each instance separately.

---

**Note**

You can also use the `show spanning-tree` command; however, this command produces more output. The `show spanning-tree root-history` command makes it easier for you to focus on the root ID. It also helps you to troubleshoot because you can see changes to the root.

---

```
ProVision-1(config)# show spanning-tree root-history ist
 Status and Counters - IST Regional Root Changes History
  MST Instance ID         : 0
  Root Changes Counter    : 2
  Current Root Bridge ID : 0:d07e28-cec94f
<-output omitted->
```

```
ProVision-1(config)# show spanning-tree root-history msti 1

 Status and Counters - MST Instance Regional Root Changes History

  MST Instance ID          : 1

  Root Changes Counter     : 3

  Current Root Bridge ID : 0:d07e28-cec94f

<-output omitted->

ProVision-1(config)# show spanning-tree root-history msti 2

 Status and Counters - MST Instance Regional Root Changes History

  MST Instance ID          : 2

  Root Changes Counter     : 4

  Current Root Bridge ID : 0:d07e28-ced31e

<-output omitted->
```

8. Complete the same step on ProVision-2. The output should be identical.

```
ProVision-2(config)# show spanning-tree root-history ist

 Status and Counters - IST Regional Root Changes History

  MST Instance ID          : 0

  Root Changes Counter     : 2

  Current Root Bridge ID : 0:d07e28-cec94f

<-output omitted->

ProVision-2(config)# show spanning-tree root-history msti 1

 Status and Counters - MST Instance Regional Root Changes History

  MST Instance ID          : 1

  Root Changes Counter     : 3

  Current Root Bridge ID : 0:d07e28-cec94f

<-output omitted->

ProVision-2(config)# show spanning-tree root-history msti 2

 Status and Counters - MST Instance Regional Root Changes History

  MST Instance ID          : 2

  Root Changes Counter     : 4

  Current Root Bridge ID : 0:d07e28-ced31e

<-output omitted->
```

## Map the topology

Based on what you have learned in this chapter, you should be able to predict the port roles for each switch-to-switch port in each instance. Use Figures 7-36, 7-37, and 7-38 to record your predictions.

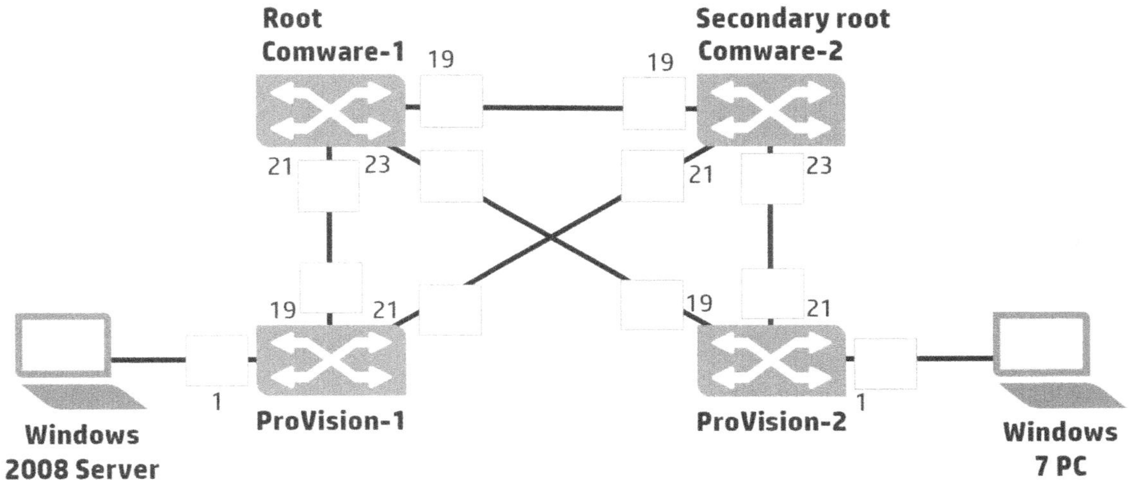


Figure 7-36: Instance 0 topology

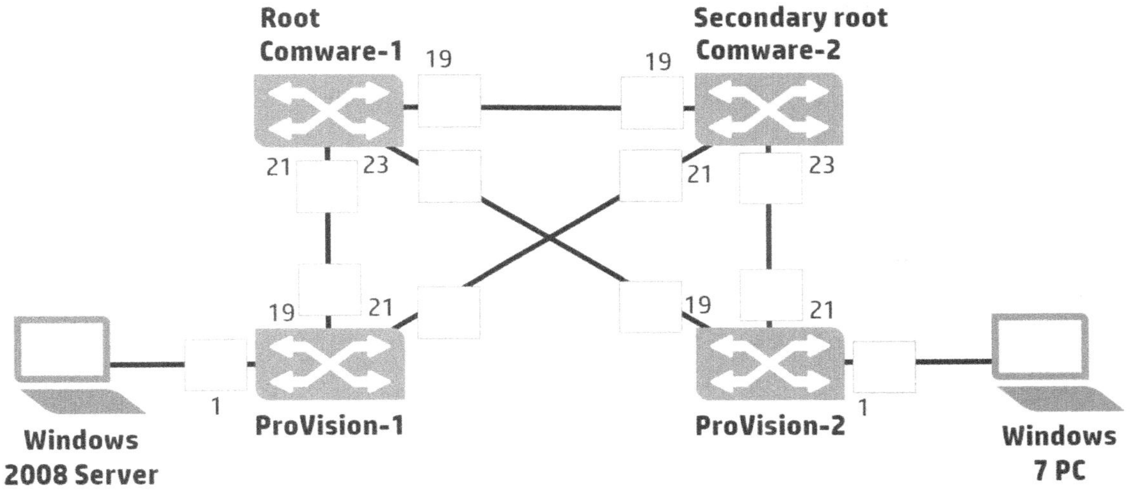


Figure 7-37: Instance 1 topology

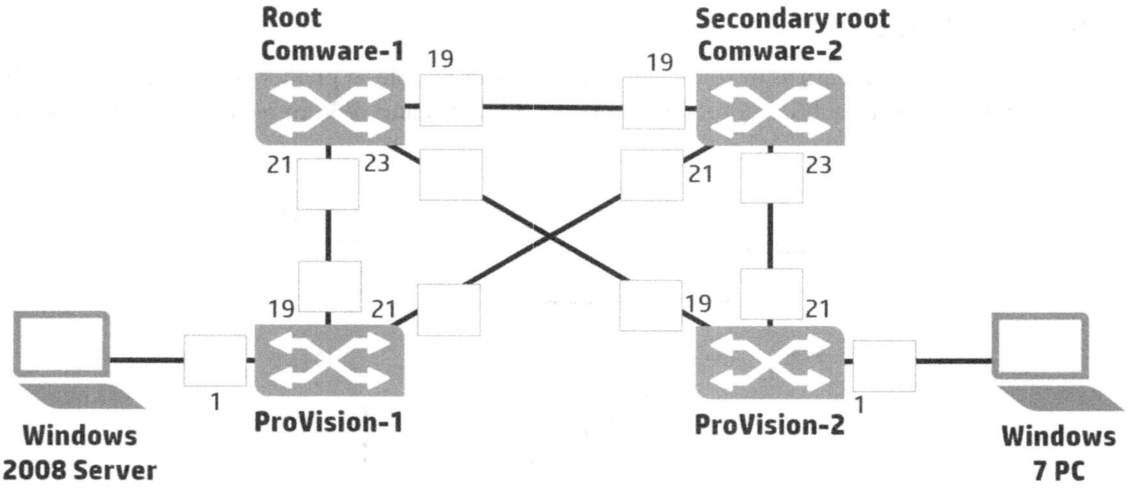


Figure 7-38: Instance 2 topology

You will now use switch output to verify your selections.

## Map instance 0

1. Access the terminal session with Comware-1. View the port roles for instance 0. Check them against your predictions and adjust if necessary.

```
[Comware-1] display stp instance 0 brief
 MSTID  Port                     Role  STP State   Protection
   0    GigabitEthernet1/0/19    DESI  FORWARDING  NONE
   0    GigabitEthernet1/0/21    DESI  FORWARDING  NONE
   0    GigabitEthernet1/0/23    DESI  FORWARDING  NONE
```

2. Access the terminal session with Comware-2. View the port roles for instance 0. Check them against your predictions and adjust if necessary.

```
[Comware-2] display stp instance 0 brief
 MSTID  Port                     Role  STP State   Protection
   0    GigabitEthernet1/0/19    ROOT  FORWARDING  NONE
   0    GigabitEthernet1/0/21    DESI  FORWARDING  NONE
   0    GigabitEthernet1/0/23    DESI  FORWARDING  NONE
```

3. Access the terminal session with ProVision-1. View the port roles for instance 0 (also called the IST). Check them against your predictions and adjust if necessary. Remember that you will find the port roles and states at the bottom of the output.

```
ProVision-1(config)# show spanning-tree 1,19,21 instance ist

<-output omitted->

                                                      Designated
Port Type      Cost  Priority Role       State      Bridge
---- --------- ----- -------- ---------- ----------  -------------
1    100/1000T Auto  128      Designated Forwarding d07e28-cec94f
19   100/1000T 20000 128      Root       Forwarding d07e28-cec94f
21   100/1000T 20000 128      Alternate  Blocking   d07e28-cec94f
```

4. Access the terminal session with ProVision-2. View the port roles for instance 0. Check them against your predictions and adjust if necessary.

```
ProVision-2(config)# show spanning-tree 1,19,21 instance ist

<-output omitted->

Designated
Port Type      Cost  Priority Role       State      Bridge
---- --------- ----- -------- ---------- ---------- -----------
1    100/1000T Auto  128      Designated Forwarding d07e28-cec94f
19   100/1000T 20000 128      Root       Forwarding d07e28-cec94f
21   100/1000T 20000 128      Alternate  Blocking   d07e28-cec94f
```

## Map the topology in instance 1

Follow the same steps to check your predictions for instance 1.

```
[Comware-1] display stp instance 1 brief

MSTID  Port                   Role  STP State   Protection
  0    GigabitEthernet1/0/19  DESI  FORWARDING  NONE
  0    GigabitEthernet1/0/21  DESI  FORWARDING  NONE
  0    GigabitEthernet1/0/23  DESI  FORWARDING  NONE
```

```
[Comware-2] display stp instance 1 brief
MSTID  Port                       Role  STP State   Protection
 1     GigabitEthernet1/0/19      ROOT  FORWARDING  NONE
 1     GigabitEthernet1/0/21      DESI  FORWARDING  NONE
 1     GigabitEthernet1/0/23      DESI  FORWARDING  NONE
ProVision-1(config)# show spanning-tree 1,19,21 instance 1
<-output omitted->
                                                      Designated
Port Type      Cost   Priority Role       State      Bridge
---- --------- ------ -------- ---------- ---------- -------------
1    100/1000T 20000  128      Designated Forwarding d4c9ef-851680
19   100/1000T 20000  128      Root       Forwarding d07e28-cec94f
21   100/1000T 20000  128      Alternate  Blocking   d07e28-ced31e
ProVision-2(config)# show spanning-tree 1,19,21 instance 1
<-output omitted->
                                                      Designated
Port Type      Cost   Priority Role       State      Bridge
---- --------- ------ -------- ---------- ---------- -------------
1    100/1000T 20000  128      Designated Forwarding d4c9ef-84fe80
19   100/1000T 20000  128      Root       Forwarding d07e28-cec94f
21   100/1000T 20000  128      Alternate  Blocking   d07e28-ced31e
```

## Map the topology in instance 2

Follow the same steps to check your predictions for instance 2.

```
[Comware-1] display stp instance 2 brief
MSTID  Port                       Role  STP State   Protection
 2     GigabitEthernet1/0/19      ROOT  FORWARDING  NONE
 2     GigabitEthernet1/0/21      DESI  FORWARDING  NONE
 2     GigabitEthernet1/0/23      DESI  FORWARDING  NONE
```

```
[Comware-2] display stp instance 2 brief
MSTID Port                     Role  STP State   Protection
 2     GigabitEthernet1/0/19   DESI  FORWARDING  NONE
 2     GigabitEthernet1/0/21   DESI  FORWARDING  NONE
 2     GigabitEthernet1/0/23   DESI  FORWARDING  NONE
ProVision-1(config)# show spanning-tree 1,19,21 instance 2
<-output omitted->
                                                       Designated
Port Type      Cost  Priority Role       State      Bridge
---- --------- ----- -------- ---------- ---------- -------------
1    100/1000T 20000 128      Designated Forwarding d4c9ef-851680
19   100/1000T 20000 128      Alternate  Blocking   d07e28-cec94f
21   100/1000T 20000 128      Root       Forwarding d07e28-ced31e
ProVision-2(config)# show spanning-tree 1,19,21 instance 2
<-output omitted->
                                                       Designated
Port Type      Cost  Priority Role       State      Bridge
---- --------- ----- -------- ---------- ---------- -------------
1    100/1000T 20000 128      Designated Forwarding d4c9ef-84fe80
19   100/1000T 20000 128      Alternate  Blocking   d07e28-cec94f
21   100/1000T 20000 128      Root       Forwarding d07e28-ced31e
```

## Save the configurations

Save the current configuration on each of the four switches. For ProVision, use the `write memory` command. For Comware, use the `save [force]` command.

## Additional exploration of MSTP: Add a VLAN

Consider what happens if you add a VLAN to a switch that is running MSTP.

1. Access the Comware-1 CLI and move to the system view.

2. Add VLAN 13.

```
[Comware-1] vlan 13
[Comware-1-vlan13] quit
```

3. Follow the same steps on Comware-2.

```
[Comware-2] vlan 13
[Comware-2-vlan13] quit
```

4. Check the spanning tree settings on ProVision-1. You should not see any differences.

```
ProVision-1# show spanning-tree
```

5. On Comware-1, add VLAN 13 to instance 1.

```
[Comware-1] stp region-configuration
[Comware-1-mst-region] instance 1 vlan 13
```

6. Complete the same steps on Comware-2.

```
[Comware-2] stp region-configuration
[Comware-2-mst-region] instance 1 vlan 13
```

7. Activate the MSTP region configuration on both Comware-1 and Comware-2.

```
[Comware-1-mst-region] active region-configuration
[Comware-2-mst-region] active region-configuration
```

The switches will exchange a number of reconvergence log messages as soon as the configuration is activated. Although you have not made any changes to the ProVision switches, you will see that this configuration has actually affected them.

8. On ProVision-1, check the switch-to-switch interfaces' port roles for *each* MSTP instance.

```
ProVision-1(config)# show spanning-tree 19,21 instance ist
<-output omitted->
Port Type      Cost  Priority Role       State      Bridge
---- --------- ----- -------- ---------- ---------- -------------
19   100/1000T 20000 128      Root       Forwarding d07e28-cec94f
21   100/1000T 20000 128      Alternate  Blocking   d07e28-cec94f
```

```
ProVision-1(config)# show spanning-tree 19,21 instance 1
<-output omitted->
Port Type      Cost  Priority Role       State      Bridge
---- --------- ----- -------- ---------- ---------- -------------
19   100/1000T 20000 128      Master     Forwarding d4c9ef-851680
21   100/1000T 20000 128      Alternate  Blocking   d4c9ef-851680
ProVision-1(config)# show spanning-tree 19,21 instance 2
<-output omitted->
                                                    Designated
Port Type      Cost   Priority Role       State      Bridge
---- --------- ------ -------- ---------- ---------- -------------
19   100/1000T 20000  128      Master     Forwarding d4c9ef-851680
21   100/1000T 20000  128      Alternate  Blocking   d4c9ef-851680
```

9. As you see, ProVision-1 now is forwarding only on the link to Comware-1 no matter what the instance or VLAN.
10. Compare the MSTP region settings on Comware-1, Comware-2, and ProVision-1 for a clue as to what has happened.

```
ProVision-1# show spanning-tree mst-config
MST Configuration Identifier Information
 MST Configuration Name : hp
 MST Configuration Revision : 1
MST Configuration Digest : 0xBE0284D20F4D46A8DA89C5D9B3B4F78A
 IST Mapped VLANs : 1-10,13-4094
 Instance ID Mapped VLANs
 ------------------------------------------------------------
 1           11
 2           12
```

```
[Comware-1-mst-region] display stp region-configuration
Oper configuration
   Format selector        :0
   Region name            :hp
   Revision level         :1
Configuration digest :0xf843355b493955bcd42bee4c4e2ffb00
   Instance    Vlans Mapped
      0        1 to 10, 14 to 4094
      1        11, 13
      2        12
[Comware-2-mst-region] display stp region-configuration
Oper configuration
   Format selector        :0
   Region name            :hp
   Revision level         :1
Configuration digest :0xf843355b493955bcd42bee4c4e2ffb00
   Instance    Vlans Mapped
      0        1 to 10, 14 to 4094
      1        11, 13
      2        12
```

11. As you see, ProVision-1 has different MST region settings. This discrepancy explains why ProVision-1 is interacting with Comware-1 and Comware-2 as if it was running RSTP. It is using the CIST rather than its individual MSTP instances.

12. Why are ProVision-1's MST region settings now different from the Comware switches' settings?

Any difference in VLAN-to-instance *mappings* will place switches in different regions even if some switches are not actively using the VLAN. As you learned earlier, when switches are in different regions, they interact using RSTP, and the boundary ports between regions are part of a CIST.

On an MSTP boundary port, all instances refer to the IST port role and state. As Figure 7-39 shows, ProVision-1's link to Comware-1 is its root port in the CIST. So instance 1 and instance 2 also forward on that port. (Master Port is another name for a region's root port in the CIST; this is the port that offers the MSTP region its link to the rest of the CST). ProVision-1's link to Comware-2 is its alternate port in the CIST, so the other instances also block on this port.

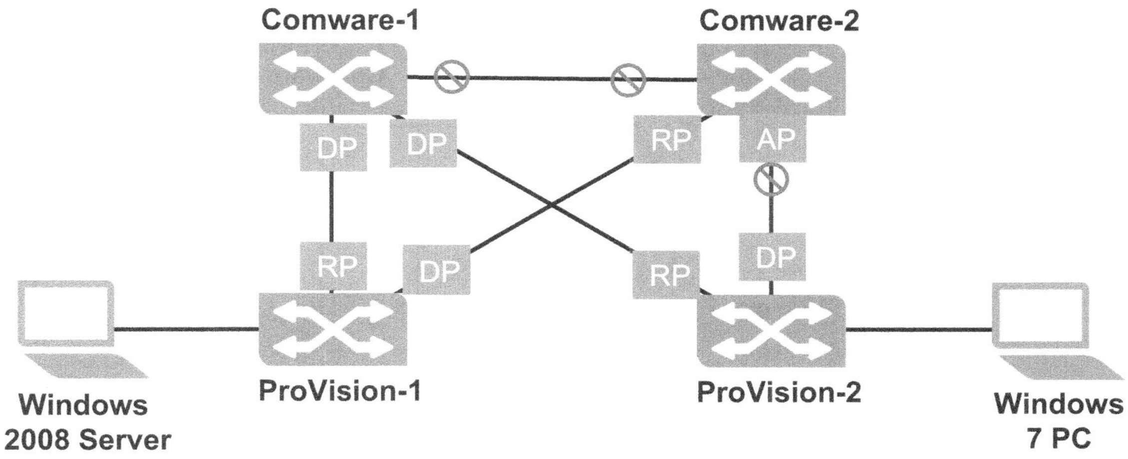


Figure 7-39: Topology with the MST region mismatch

In short, no load-balancing occurs on the MSTP boundary ports. Therefore, adding a VLAN and moving that VLAN from instance 0 to another instance in an active network can cause disruptions and non optimal link utilization.

**Note**

Plan in advance and place all VLANs that might be used in the future in the desired instance on all switches.

As you learned earlier, Comware-1 is elected as the CIST root because it has the lowest instance 0 priority.

Switches in other MSTP regions elect a regional root for the CIST based on which switch offers the best path to the CIST root. The regional root becomes the IST root.

As Figure 7-40 shows, ProVision-1 is the only switch in its region, so it is the regional root. Its boundary port with the lowest cost to the CIST root becomes the CIST root port (IST master port), and its other boundary port becomes an alternate port.

Note an important implication: an MSTP region could have an IST root that does not have the lowest priority value. This can occur when the region is part of a CIST and does not contain the CIST root. The switch with the best path to the CIST root might not be the switch with the lowest instance 0 priority value, but it still becomes the IST root.

It is important to set priorities for instance 0 even if you are not using that instance for carrying data traffic. You want to control which switch becomes CIST root in case of misconfigurations.

To clear the misconfigured settings in the example network, you should reboot the Comware-1 and Comware-2 switches. (Also reboot ProVision-1 if you made any changes to its configuration). Do not save the configuration.

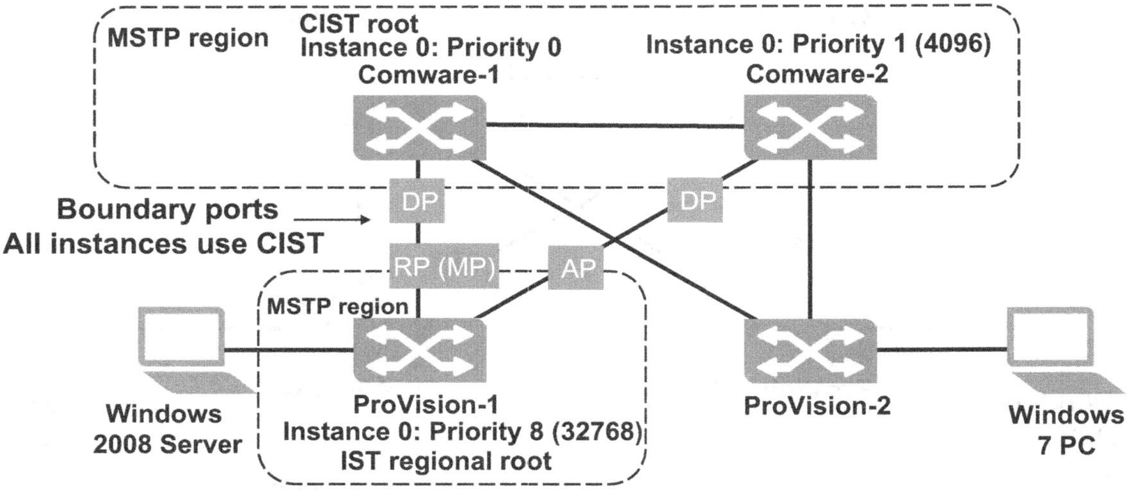


Figure 7-40: Topology with the MST region mismatch

## Additional exploration of MSTP: Remove a VLAN from a link

You will now consider what can happen if a VLAN is not supported on a particular switch-to-switch link.

1. Access the client desktop and open a command prompt.
2. Establish a continuous ping to the server.

```
ping 10.1.11.11 -t
```

3. Access the ProVision-1 CLI and move to the global configuration context.
4. Remove VLAN 12 from the links to Comware-1 and Comware-2.

```
ProVision-1(config)# no vlan 12 tagged 19,21
```

5. Assign ProVision-1 priority multiplier 4 (priority 16384) in each instance.

   The issue that you want to observe might or might not occur depending on which access layer switch is selected as being closer to the root, which often depends on MAC address in a real network. For this experiment, you want to make sure that ProVision-1 is selected so that you can see the potential issue.

```
ProVision-1(config)# spanning-tree priority 4
ProVision-1(config)# spanning-tree instance 1 priority 4
ProVision-1(config)# spanning-tree instance 2 priority 4
```

6. Access the terminal session with ProVision-2 and move to the global configuration mode.

7. Remove VLAN 11 from the links to Comware-1 and Comware-2.

```
ProVision-2(config)# no vlan 11 tagged 19,21
```

8. Check the ping.
9. As you see, at this point, the ping continues normally. During normal operation for this topology, ProVision-1 does not need to be part of VLAN 12 and ProVision-2 does not need to be part of VLAN 11.
10. Now observe what happens if the link between Comware-1 and Comware-2 fails. Access the Comware-1 CLI.
11. Shut down the port that connects to Comware-2.

```
[Comware-1] interface g1/0/19
[Comware-1-GigabitEthernet1/0/19] shutdown
```

12. View the ping on the client. The ping has failed. Why did the failover not occur properly? (If you do not know, do not worry. The steps below help you to discover the problem.)
13. The client needs to reach Comware-1, its default gateway in VLAN 12. VLAN 12 is associated with instance 2. Which links are active on Comware-1 in instance 2? View the port roles on Comware-1 in instance 2 to find out.
14. Access the terminal session with Comware-1 and enter this command:

```
[Comware-1] display stp instance 2 brief
MSTID  Port                   Role  STP State   Protection
  2    GigabitEthernet1/0/21  ROOT  FORWARDING  NONE
  2    GigabitEthernet1/0/23  ALTE  DISCARDING  NONE
```

15. The root port on Comware-1 connects to ProVision-1 port 19. View settings on ProVision-1 to see the issue. Access the Comware-1 CLI and enter the command below.

```
ProVision-1(config)# show spanning-tree 1,19,21 instance 2
<-output omitted->
                                                   Designated
Port Type      Cost  Priority Role       State      Bridge
---- --------- ----- -------- ---------- ---------- --------------
1    100/1000T 20000 128      Designated Forwarding d4c9ef-851680
19   100/1000T 20000 128      Designated Forwarding d4c9ef-851680
21   100/1000T 20000 128      Root       Forwarding d07e28-ced31e
```

16. Does the active link between Comware-1 and ProVision-1 carry VLAN 12? To remind yourself, view VLANs on port 19 on ProVision-1.

```
ProVision-1(config)# show vlan port 19

Status and Counters - VLAN Information - for ports 19

VLAN ID Name            | Status      Voice Jumbo

------- --------------- + ---------- ----- -----

 1      DEFAULT_VLAN    | Port-based No    No

 11     VLAN11          | Port-based No    No
```

17. As you see, the active link does not carry VLAN 12. Therefore, the client cannot reach its default gateway, causing a disruption in connectivity.
18. Re-enable the port on Comware-1.

```
[Comware-1-GigabitEthernet1/0/19] undo shutdown
```

19. Reboot the ProVision-1 and ProVision-2 switches. Do *not* save the configuration.
20. End the continuous ping on the PC with **<Ctrl+c>.**

## What happened?

MSTP uses a *single* BPDU, which is backward compatible with RSTP and also includes MSTP instance information within it. This approach has the advantage of causing minimal impact on the switch's CPU; only one BPDU must be processed. However, it also introduces some important implications, which you must understand.

An MSTP BPDU is carried on the Ethernet link regardless of the VLANs carried on that link, so the switches are essentially using a single physical Layer 2 topology for all instances. The switches then use the specific instance settings to calculate the port roles and states for each spanning tree instance. (See Figure 7-41 and Figure 7-42.) But, again, this process is not VLAN aware.

The VLAN mappings only come into play when the switch decides how to treat traffic in a specific VLAN on a specific port. For example, instance 2 is forwarding on a port. VLAN 12 is part of instance 2, so VLAN 12 traffic is allowed to be forwarded on that port.

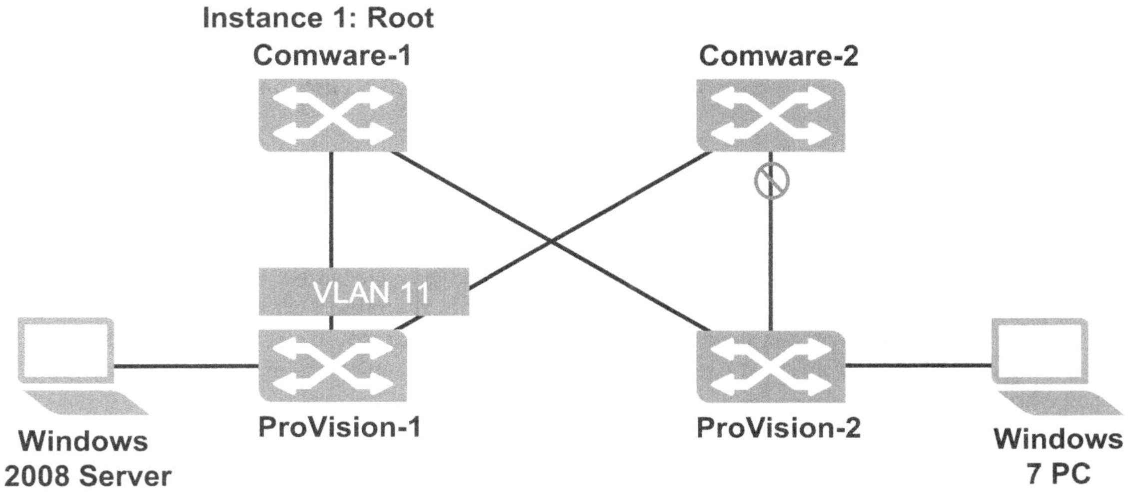


Figure 7-41: Instance 1 topology and VLAN 11 assignments

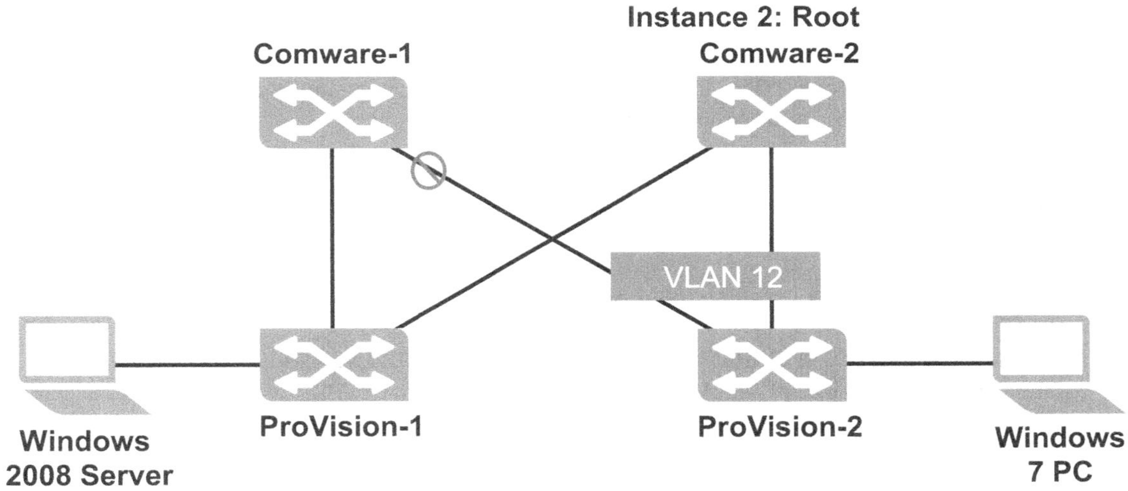


Figure 7-42: Instance 2 topology and VLAN 12 assignments

Because the spanning tree instances are set up without regard to which VLANs are actually carried on a link, it is possible that an instance could select a root port that does not carry one of the instance's VLANs and an alternate port that does—as occurred in this scenario (shown in Figure 7-42). In other words, even though *spanning tree* allows the forwarding of VLAN 12 on a port, VLAN 12 might not be allowed by the tagging on the port. And even though another port is available that *does* have the proper VLAN 12 tagging, that port is still blocked in instance 2.

You need to be aware of the VLAN topology on your switches and make sure that it works with the spanning tree instance topology *both in normal operation and failover situations.* To prevent misconfigurations, configure all switch-to-switch links to support all VLANs used in the network segment.

---

**Note**

This explanation helps you to understand more completely why MSTP region settings must match on switches. Because the spanning trees are set up without regard to VLAN topology, it is very important that every switch assigns the same VLANs to the same instance. Otherwise, one switch might use one topology to flood a broadcast in a VLAN while a different switches uses a different topology: loops could occur. If switches have different settings, they must use RSTP to prevent loops—which is, as you have learned, what they do.

---

## Learning check 2

1. Describe the differences between the topology in instance 1 and in instance 2.

2. You might have noticed that the Comware switches have you configure the MSTP region settings and then activate them. On the ProVision switches, you simply configure the settings. Discuss the differences between these approaches.

## Answers to learning check 2

1. Describe the differences between the topology in instance 1 and in instance 2.

   Comware-1 is the root in instance 1, so the ProVision switches use their direct link to Comware-1 and block their other link. Comware-2 is the root in instance 2, so the ProVision switches use their direct Comware-2 link and block their other link. In this way, each access layer switch has two links, one of which is used in each instance.

2. You might have noticed that the Comware switches have you configure the MSTP region settings and then activate them. On the ProVision switches, you simply configure the settings. Discuss the differences between these approaches.

   It can be useful to configure the region settings in advance without applying them. Then you can set up the new configuration on many switches in advance. When you are ready, you apply the changes to all the switches at once, minimizing the disruption. (Changing the region settings always causes a brief disruption as the topology reconverges.)

You can actually make the ProVision switches act more like the Comware switches. Instead of configuring the new settings directly, configure pending settings with these commands:

```
ProVision(config)# spanning-tree pending config-name <name>
ProVision(config)# spanning-tree pending config-revision <number>
ProVision(config)# spanning-tree pending instance <ID > vlan <vid range>
```

When you are ready, activate the pending region settings with this command:

```
ProVision(config)# spanning-tree pending apply
```

If, before you apply the pending settings, you decide that you want to revert them to match the currently active settings instead, use this command:

```
ProVision(config)# spanning-tree pending reset
```

## Plan instances for load-sharing

Figure 7-43 shows an expanded network with more VLANs. Consider the following questions.

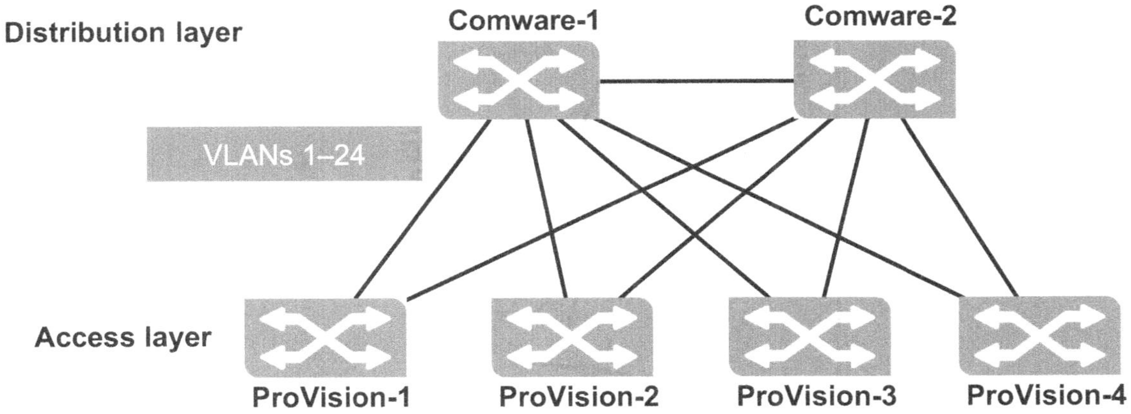


Figure 7-43: Plan load sharing on an MSTP network

1. How many MSTP instances, besides the IST (instance 0), do you need to have most effective load sharing?
2. Consider a network that has expanded even more, as shown in Figure 7-44.

   How many MSTP instances, besides the IST (instance 0), do you need to have most effective load sharing?

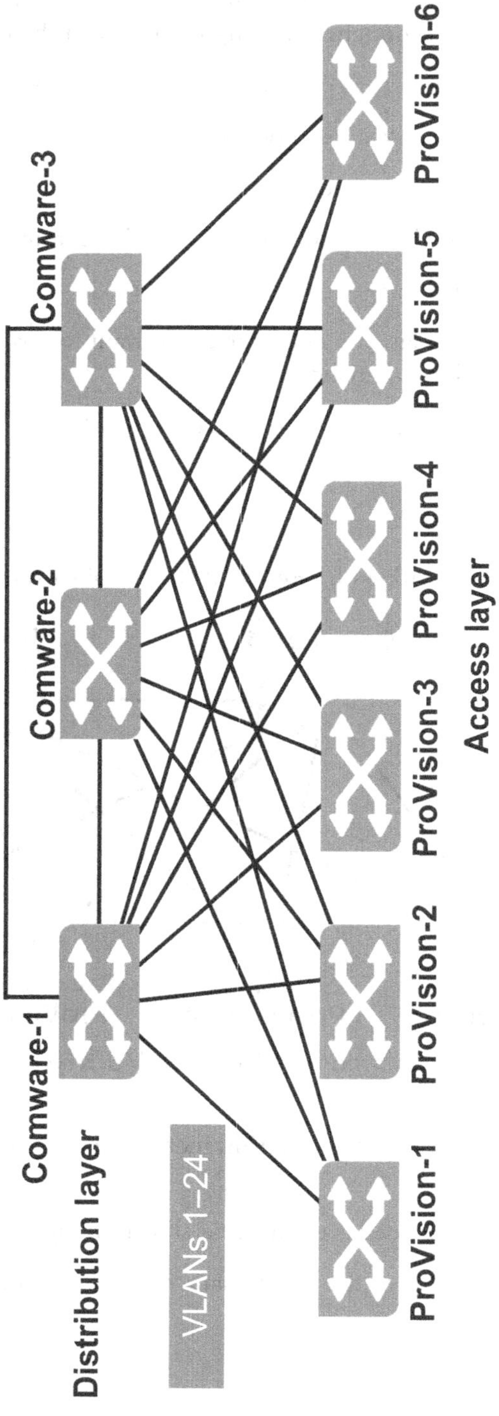


Figure 7-44: Plan load sharing on an MSTP network

## Answers

1. How many MSTP instances, besides the IST (instance 0), do you need to have most effective load sharing?

   To implement the most effective load sharing, you need a separate instance for each distribution/aggregation switch (Comware-1 and Comware-2). Then each of these switches can act as root in one instance and uplinks to that switch can be active in that instance. Therefore, in the first scenario, two instances (in addition to instance 0) work.

   The number of VLANs does not affect the number of instances that you need. You can assign half of the VLANs to one instance and half to the other. Or you might take a more nuanced approach to load balancing. For example, you can look at the typical traffic load for each VLAN and combine the VLANs such that each instance has about half the load (but not necessarily half the VLAN IDs).

   You also can choose not to load balance traffic evenly. For example, maybe you want to reserve a path for a critical VLAN, but you do not care that other VLANs have to share bandwidth. You would then give the critical VLAN its own instance and assign other VLANs to the other instance.

   The choice is yours and depends on the company's needs.

2. Consider a network that has expanded even more, as shown in Figure 7-44.

   How many MSTP instances, besides the IST (instance 0), do you need to have most effective load sharing?

   You now have three switches at the distribution layer, so you should use three instances. Again, you should distribute the VLANs among these instances evenly or evenly based on traffic load.

   As you can see from this second scenario, the more distribution layer switches you have, the more configuration you need to perform in order to utilize the bandwidth on the access layer switches' uplinks. The design becomes more complex, which is why many networks use a pair of distribution switches.

   Note finally that you do not have to load share traffic. You should think about why you do or do not want to. Maybe you can provision enough bandwidth so that an access layer switch can handle all traffic on a single uplink. If so, you can use RSTP and avoid the complexities of MSTP configuration. If the backup link has the same bandwidth as the main link, you also know that the backup link will have enough bandwidth to handle all traffic during failover.

## Summary

In this chapter, you learned about how to use RSTP and MSTP to add Layer 2 link redundancy without introducing dangerous loops and broadcast storms. You can set the appropriate priority values to select a primary and secondary root. You understand how an RSTP topology is created and the port roles and states used in the topology. You also understand how failover occurs.

You learned about how you can use MSTP to utilize your network bandwidth more efficiently. MSTP operates much like RSTP, but supports multiple instances each with its own root, topology, and port roles and states. You also explored some best practices for implementing MSTP and learned how to avoid potential issues.

## Learning check 3

1. The root switch in STP is elected using which of the following criteria?

   a. Lowest MAC address

   b. Highest MAC address

   c. Lowest priority

   d. Highest priority

   e. Lowest bridge ID

   f. Highest bridge ID

2. Which RSTP port role designates a port that offers a second (or third, and so forth) best path to the root? In other words, the port connects to a switch that

   - Offers a better path to the root than this switch does
   - But does not offer this switch its best path to the root

     a. Root port

     b. Backup port

     c. Designated port

     d. Alternate port

3. What ProVision command enables a spanning tree protocol? Also specify the default mode for the protocol.

4. What Comware command makes changes to the MSTP region effective?

## Answers to learning check 3

1. The root switch in STP is elected using which of the following criteria?
   a. Lowest MAC address
   d. Highest MAC address
   e. Lowest priority
   f. Highest priority
   g. **Lowest bridge ID**
   h. Highest bridge ID

2. Which RSTP port role designates a port that offers a second (or third, and so forth) best path to the root? In other words, the port connects to a switch that
   - Offers a better path to the root than this switch does
   - But does not offer this switch its best path to the root
     a. Root port
     b. Backup port
     c. Designated port
     d. **Alternate port**

3. What ProVision command enables a spanning tree protocol? Also specify the default mode for the protocol.

   The command is `spanning-tree`. The default mode is MSTP.

4. What Comware command makes changes to the MSTP region effective?

   `activate region-configuration`

# 8 Link Aggregation

## EXAM OBJECTIVES

✓ Differentiate between different types of link aggregation

✓ Describe the benefits of Link Aggregation Control Protocol (LACP)

✓ Configure and troubleshoot link aggregation on HP switches

## INTRODUCTION

Link aggregation helps you to add redundant links and more bandwidth between switches more easily and efficiently.

### Assumed knowledge

Before reading this chapter, you should have a basic understanding of:

- Virtual LANs(VLANs)
- Spanning Tree Protocol (STP)

### Add redundant links between the same two switches

In the previous chapter, you learned how to create a redundant Layer 2 topology using spanning tree. Now think about a situation in which you want to increase bandwidth between two switches—for example, Comware-1 and Comware-2 in the example network shown in Figure 8-1. You want to add another link between the switches.

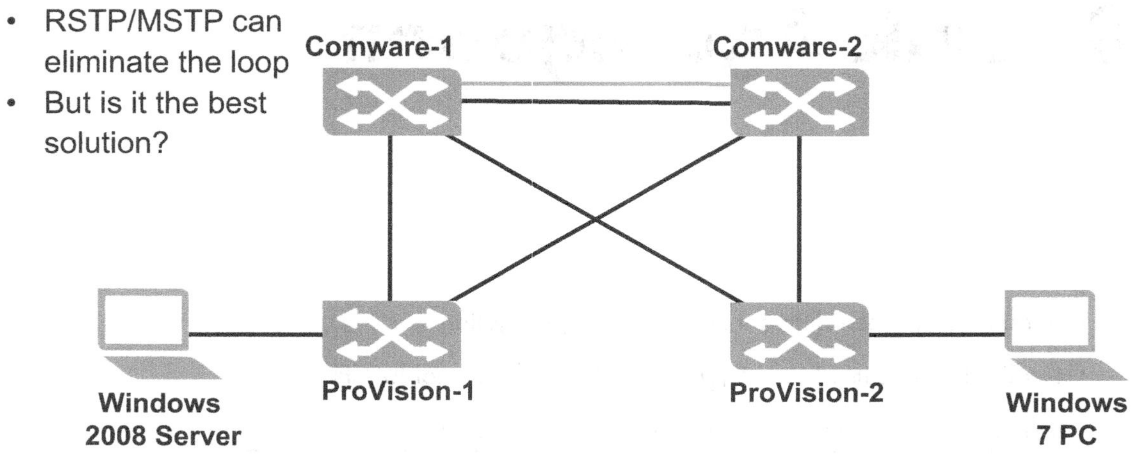


Figure 8-1: Add redundant links between the same two switches

If you use RSTP, how will this protocol deal with the new link? If you use Multiple Spanning Tree Protocol (MSTP), how could the protocol deal with this new link, and what would you need to configure to obtain the desired results? What are drawbacks of allowing RSTP/MSTP to handle the redundant links in this situation? You will learn the answers to these questions in the next section.

## Example configuration: Use MSTP for redundant links between two switches

In this section, a redundant link will be added between Comware-1 and Comware-2 in the example network shown in Figure 8-2. You will then consider how MSTP handles the new topology.

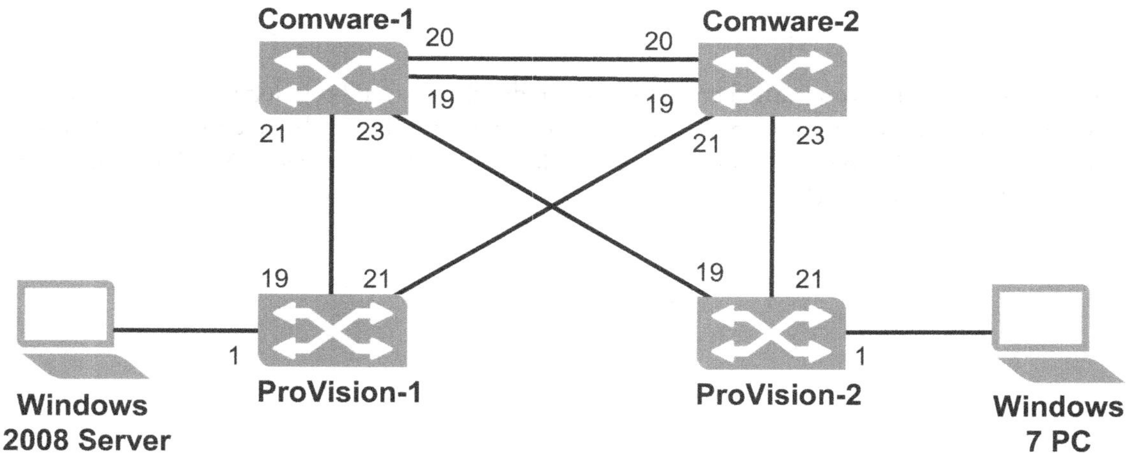


Figure 8-2: Using MSTP for redundant links between two switches

MSTP is already running on the example network, using the settings provided in "Example configuration: MSTP" on page 218 in Chapter 7. As you can see in Figure 8-3, Comware-1 is the root for instances 0 and 1, and the secondary root for instance 2. Comware-2 is the secondary root for instances 0 and 1, and the root for instance 2.

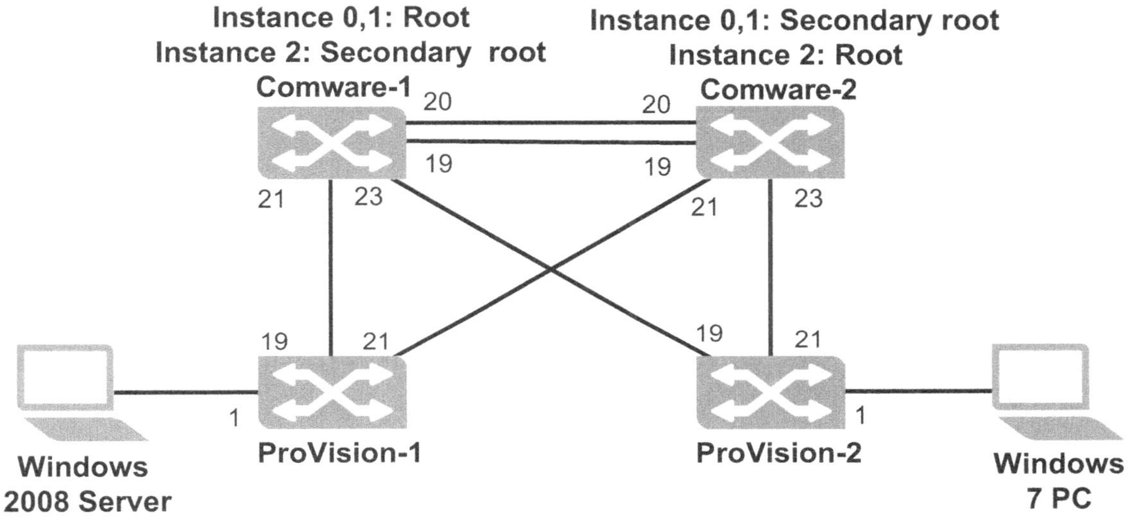


Figure 8-3: MSTP root settings for the example network

1. You can verify MSTP settings by accessing the Comware-1 CLI and viewing the MSTP root settings:

```
<Comware-1> display stp root
MSTID Root Bridge ID   ExtPathCost IntPathCost Root Port
  0   0.d07e-28ce-c94f 0           0
  1   0.d07e-28ce-c94f 0           0
  2   0.d07e-28ce-d31e 0           20000    GigabitEthernet1/0/19
```

2. Access the Comware-2 CLI and enter the same command:

```
<Comware-2> display stp root
MSTID Root Bridge ID   ExtPathCost IntPathCost Root Port
  0   0.d07e-28ce-c94f 0           20000     GigabitEthernet1/0/19
  1   0.d07e-28ce-c94f 0           20000     GigabitEthernet1/0/19
  2   0.d07e-28ce-d31e 0           0
```

## Add a redundant link

You will now add a redundant link between Comware-1 and Comware-2.

These instructions were developed for a remote lab environment that has permanent connections between switches, so you simply need to enable the second interface on each switch to provide a redundant link. If you are performing the instructions in a different test environment, you will need to add a second physical connection between Comware-1 and Comware-2.

1. On Comware-1, move to the second physical port that connects to Comware-2.

```
<Comware-1> system-view
[Comware-1] interface g1/0/20
```

2. Configure this interface as a trunk port that supports all VLANs.

```
[Comware-1-GigabitEthernet1/0/20] port link-type trunk
[Comware-1-GigabitEthernet1/0/20] port trunk permit vlan all
```

3. Enable the interface.

```
[Comware-1-GigabitEthernet1/0/20] undo shutdown
```

4. Access Comware-2. Follow similar steps to configure the other side of the link.

```
<Comware-2> system-view
[Comware-2] interface g1/0/20
[Comware-2-GigabitEthernet1/0/20] port link-type trunk
[Comware-2-GigabitEthernet1/0/20] port trunk permit vlan all
[Comware-2-GigabitEthernet1/0/20] undo shutdown
```

5. Verify that the new link is passing traffic in the correct VLANs.

```
[Comware-2] display port trunk
Interface  PVID  VLAN passing
GE1/0/19   1 1,  11-12,
GE1/0/20   1 1,  11-12,
GE1/0/21   1 1,  11-12,
GE1/0/23   1 1,  11-12,
```

6. Return to Comware-1. Also verify the VLANs on its side of the link.

```
[Comware-1] display port trunk
```

```
Interface  PVID  VLAN passing
GE1/0/19   1     1, 11-12,
GE1/0/20   1     1, 11-12,
GE1/0/21   1     1, 11-12,
GE1/0/23   1     1, 11-12,
```

7. Verify that Comware-1 detects Comware-2 as an LLDP neighbor on both interfaces.

```
[Comware-1] display lldp neighbor-information list
System Name  Local Interface  Chassis ID      Port ID
Comware-2    GE1/0/19         d07e-28ce-d31e  GigabitEthernet1/0/19
Comware-2    GE1/0/20         d07e-28ce-d31e  GigabitEthernet1/0/20
ProVision-1  GE1/0/21         d4c9-ef85-1680  19
ProVision-2  GE1/0/23         d4c9-ef84-fe80  19
```

## Observe MSTP with the new link

You will now examine how MSTP handles the two links between Comware-1 and Comware-2. You should be familiar with the HP `display` and `show` commands for STP from Chapter 7.

1. Closely examine the MSTP topology for Comware-1, looking at the status of the two links to Comware-2 in all instances.

```
[Comware-1] display stp brief
MSTID Port                   Role  STP State   Protection
 0    GigabitEthernet1/0/19  DESI  FORWARDING  NONE
 0    GigabitEthernet1/0/20  DESI  FORWARDING  NONE
 0    GigabitEthernet1/0/21  DESI  FORWARDING  NONE
 0    GigabitEthernet1/0/23  DESI  FORWARDING  NONE
 1    GigabitEthernet1/0/19  DESI  FORWARDING  NONE
 1    GigabitEthernet1/0/20  DESI  FORWARDING  NONE
 1    GigabitEthernet1/0/21  DESI  FORWARDING  NONE
 1    GigabitEthernet1/0/23  DESI  FORWARDING  NONE
 2    GigabitEthernet1/0/19  ROOT  FORWARDING  NONE
```

```
2     GigabitEthernet1/0/20  ALTE  DISCARDING  NONE
2     GigabitEthernet1/0/21  DESI  FORWARDING  NONE
2     GigabitEthernet1/0/23  DESI  FORWARDING  NONE
```

2. Closely examine the MSTP topology for Comware-2, looking at the status of the ports.

```
[Comware-2] display stp brief
 MSTID  Port                      Role  STP State   Protection
 0      GigabitEthernet1/0/19     ROOT  FORWARDING  NONE
 0      GigabitEthernet1/0/20     ALTE  DISCARDING  NONE
 0      GigabitEthernet1/0/21     DESI  FORWARDING  NONE
 0      GigabitEthernet1/0/23     DESI  FORWARDING  NONE
 1      GigabitEthernet1/0/19     ROOT  FORWARDING  NONE
 1      GigabitEthernet1/0/20     ALTE  DISCARDING  NONE
 1      GigabitEthernet1/0/21     DESI  FORWARDING  NONE
 1      GigabitEthernet1/0/23     DESI  FORWARDING  NONE
 2      GigabitEthernet1/0/19     DESI  FORWARDING  NONE
 2      GigabitEthernet1/0/20     DESI  FORWARDING  NONE
 2      GigabitEthernet1/0/21     DESI  FORWARDING  NONE
 2      GigabitEthernet1/0/23     DESI  FORWARDING  NONE
```

Table 8-1 lists the criteria used for selecting the root port.

Table 8-1: Criteria for selecting the root port

| Criteria |
|---|
| Port with the lowest cost path to the root (receives a BPDU with the lowest cost |
| If a tie, port that connects to the neighbor with the lowest switch ID (priority + MAC address) |
| If a tie, port that connects to the port with the lowest port priority value |
| If a tie, port with the lowest ID |

3. As an optional step, you can try adjusting spanning tree port priorities to achieve a degree of load balancing. Set the Comware-2 G1/0/20 port priority lower than the default in instance 2.

```
[Comware-2] interface g1/0/20
[Comware-2-GigabitEthernet1/0/20] stp instance 2 port priority 0
```

4. A lower value is preferred, so you should see that Comware-1 changes its root port to G1/0/20. Now the g1/0/20 link is forwarding in instance 2.

```
[Comware-1]dis stp brief

 MSTID  Port                     Role  STP State   Protection
   0    Bridge-Aggregation2      DESI  FORWARDING  NONE
   0    Bridge-Aggregation3      DESI  FORWARDING  NONE
   0    GigabitEthernet1/0/19 DESI  FORWARDING  NONE
   0    GigabitEthernet1/0/20 DESI  FORWARDING  NONE
   1    GigabitEthernet1/0/19 DESI  FORWARDING  NONE
   1    GigabitEthernet1/0/20 DESI  FORWARDING  NONE
   2    GigabitEthernet1/0/19 ALTE  DISCARDING  NONE
   2    GigabitEthernet1/0/20 ROOT  FORWARDING  NONE
```

5. Save the current configuration on each of the Comware switches: `save [force]`.

## Learning check 1

1. How did MSTP handle the two links between Comware-1 and Comware-2?

2. What are the implications? Does this new link add bandwidth to the connection between Comware-1 and Comware-2?

3. What would happen if you simply disabled spanning tree?

## Learning 1 check answers

1. How did MSTP handle the two links between Comware-1 and Comware-2?

   On Comware-1, both ports are designated ports and in a forwarding state in instances 1 and 0. In instance 2, for which Comware-2 is the root, interface g1/0/19 is the root port and interface g1/0/20 is the alternate port (placed in a discarding/blocking state). Interface g1/0/19 and interface g1/0/20 have the same root cost, the same bridge ID for the upstream (Comware-2), and the same port priority. Interface g1/0/19 was selected as the root port because it has a lower ID.

On Comware-2, both ports are alternate ports and in a blocking state for instances 1 and 0. They are designated in instance 2.

2. What are the implications? Does this new link add bandwidth to the connection between Comware-1 and Comware-2?

   Interface g1/0/20 is blocked in all instances. The new link is not adding any bandwidth to the connection. It adds some redundancy, but it does not increase the throughput between Comware-1 and Comware-2.

   You could try to manipulate the configuration to enable some load balancing across the links on a per instance basis. For example, you could assign a higher port priority to one port in instance 1 and a higher port priority to a different port in instance 2. However, the planning is more complex, and the load balancing is not granular. If you wanted to add more than two links, they would be blocked.

3. What would happen if you simply disabled spanning tree?

   You cannot simply disable spanning tree. That would cause a loop and broadcast storm.

## Link aggregation

As you saw in "Example configuration: Use MSTP for redundant links between two switches" on page 248, when you simply add a link between two switches, RSTP/MSTP sees it as another redundant link. It blocks the link as an alternate path to the root for Comware-2. The new link adds a bit of resiliency, but it does *not* add any bandwidth to the connection.

As Figure 8-4 shows, link aggregation solves this issue.

You define an aggregated link on the switch and add physical interfaces to that aggregated link. Now, from the point of view of the switch, the *aggregated link* is the logical link—not the individual physical interfaces assigned to it:

- RSTP/MSTP sees the aggregated link as an interface, and the aggregated link as a whole has a forwarding or blocking status.
- The switch learns MAC addresses on the aggregated link, not on an individual interface.

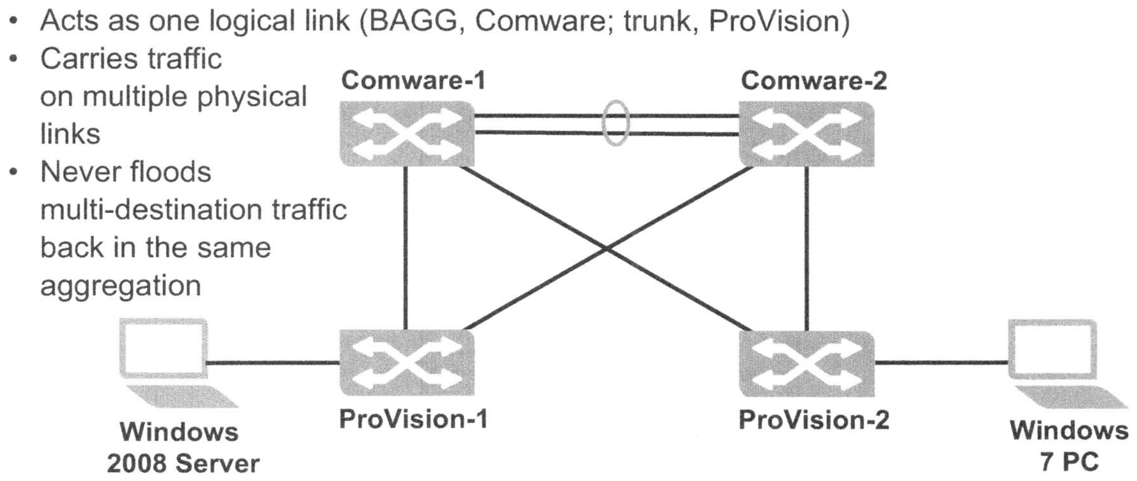


Figure 8-4: Link aggregation

The switch can forward traffic across any of the physical interfaces in the aggregated link. In effect, the aggregated link is a link with the combined bandwidth of the physical interfaces assigned to it. (At a lower level, of course, the switch is assigning specific traffic to specific physical interfaces within the aggregated link, and load balancing is not always perfect. You will learn about the load balancing mechanism in more depth a bit later.)

Aggregated links can handle unicast and multi destination traffic (multi destination traffic includes multicasts, broadcasts, and unknown unicasts).

When a switch receives multi destination traffic on an aggregated link, it never floods that traffic on another link in the same aggregated link. This behavior ensures that the aggregated link does not create a loop.

Table 8-2 lists the terminology used for aggregated links on ProVision and Comware switches.

Table 8-2: Terminology for aggregated links on ProVision and Comware switches

| Term | ProVision | Comware* |
|---|---|---|
| Link aggregation | trunk | bridge-aggregation (BAGG) interface |

*Some Comware devices also support route-aggregation (RAGG) interfaces, which are aggregated links that operate as Layer 3 interfaces.

## Example configuration: Set up a manual aggregated link

In this section, the two links between the Comware switches will be configured as an aggregated link. This section also provides the steps for configuring an aggregated link between Comware-1 and ProVision-1, as shown in Figure 8-5.

The example network is using the configuration set up in "Example configuration: Use MSTP for redundant links between two switches" on page 248.

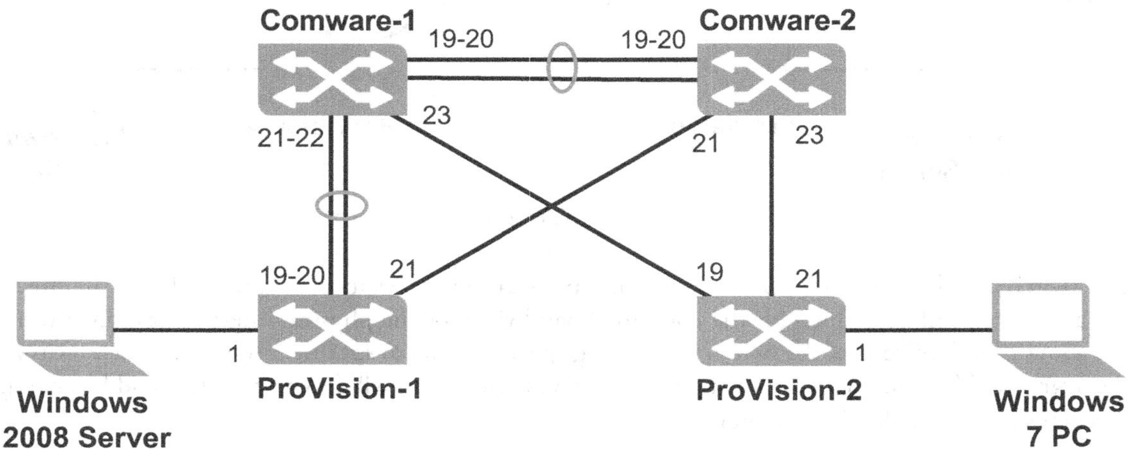


Figure 8-5: Example configuration: Set up a manual aggregated link

This section provides the steps for configuring manual aggregated links, sometimes called static aggregated links on Comware switches. These aggregated links do not use any protocol to establish the aggregated links. On each switch, you simply select the physical interfaces that you want for the aggregated link and assign them to the link aggregation interface.

### Configure an aggregated link between the Comware switches

An aggregated link is a logical interface, and on Comware switches, you create a bridge-aggregation interface as the logical interface for each aggregated link. You then assign physical interfaces to the logical interface.

1. Access the Comware-1 CLI and move to system view.
2. Create a bridge-aggregation interface with ID 1.

```
[Comware-1] interface Bridge-Aggregation 1

[Comware-1-Bridge-Aggregation1] quit
```

3. Assign the two ports that connect to Comware-2 to the bridge-aggregation interface.

```
[Comware-1] interface g1/0/19
[Comware-1-GigabitEthernet1/0/19] port link-aggregation group 1
[Comware-1-GigabitEthernet1/0/19] interface g1/0/20
[Comware-1-GigabitEthernet1/0/20] port link-aggregation group 1
[Comware-1-GigabitEthernet1/0/20] quit
```

4. Now you assign VLAN settings to the *bridge-aggregation* interface. Configure this interface as a trunk port and permit all VLANs. Note that when you enter these commands, the switch applies them to the physical interfaces.

```
[Comware-1] interface Bridge-Aggregation 1
[Comware-1-Bridge-Aggregation1] port link-type trunk
[Comware-1-Bridge-Aggregation1] port trunk permit vlan all
 Please wait................................. Done.
 Configuring GigabitEthernet1/0/19............ Done.
 Configuring GigabitEthernet1/0/20.............. Done.
[Comware-1-Bridge-Aggregation1] quit
```

5. Access the Comware-2 CLI and move to system view.

6. On Comware-2, follow similar steps to configure a manual aggregated link with the two interfaces that connect to Comware-1. Remember to configure the VLAN settings on the bridge-aggregation interface.

```
[Comware-2] interface Bridge-Aggregation 1
[Comware-2-Bridge-Aggregation1] quit
[Comware-2] interface g1/0/19
[Comware-2-GigabitEthernet1/0/19] port link-aggregation group 1
[Comware-2-GigabitEthernet1/0/19] interface g1/0/20
[Comware-2-GigabitEthernet1/0/20] port link-aggregation group 1
[Comware-2-GigabitEthernet1/0/20] quit
[Comware-2] interface Bridge-Aggregation 1
[Comware-2-Bridge-Aggregation1] port link-type trunk
```

```
[Comware-2-Bridge-Aggregation1] port trunk permit vlan all
[Comware-2-Bridge-Aggregation1] quit
```

## Observe the aggregated link

You will now verify that the aggregated link has come up and observe the effects on the MSTP topology.

1. On Comware-2, display the details of the aggregated link. Note that Comware refers to a manual aggregated link as **Static**. Also note that both interfaces have **S**, or Selected, status.

```
[Comware-2] display link-aggregation verbose Bridge-Aggregation 1
Loadsharing Type: Shar -- Loadsharing, NonS -- Non-Loadsharing
Port Status: S -- Selected, U -- Unselected
Flags:  A -- LACP_Activity, B -- LACP_Timeout, C -- Aggregation,
        D -- Synchronization, E -- Collecting, F -- Distributing,
        G -- Defaulted, H -- Expired
Aggregation Interface: Bridge-Aggregation1
Aggregation Mode: Static
Loadsharing Type: Shar
  Port      Status    Oper-Key
-----------------------------------------------
  GE1/0/1    S          1
  GE1/0/20   S          1
```

2. On Comware-2, examine the MSTP topology. Notice that you no longer see ports 19 and 20, but the bridge-aggregation interface instead.

```
[Comware-2] display stp brief
 MSTID  Port                  Role  STP State   Protection
   0    Bridge-Aggregation1   ROOT  FORWARDING  NONE
   0    GigabitEthernet1/0/21 DESI  FORWARDING  NONE
   0    GigabitEthernet1/0/23 DESI  FORWARDING  NONE
   1    Bridge-Aggregation1   ROOT  FORWARDING  NONE
```

```
1    GigabitEthernet1/0/21   DESI   FORWARDING   NONE
1    GigabitEthernet1/0/23   DESI   FORWARDING   NONE
2    Bridge-Aggregation1     DESI   FORWARDING   NONE
2    GigabitEthernet1/0/21   DESI   FORWARDING   NONE
2    GigabitEthernet1/0/23   DESI   FORWARDING   NONE
```

## Configure an aggregated link between a Comware and ProVision switch

You will now create a second manual aggregated link on Comware-1. This aggregated link connects to ProVision-1. You will also configure a manual aggregated link on the ProVision-1 side.

1. Access the Comware-1 CLI.
2. Create bridge-aggregation interface 2 and assign the link that connects to ProVision-1 to it.

```
[Comware-1] interface Bridge-Aggregation 2
[Comware-1-Bridge-Aggregation2] quit
[Comware-1] interface g1/0/21
[Comware-1-GigabitEthernet1/0/21] port link-aggregation group 2
[Comware-1] quit
```

3. View aggregated links on Comware-1. Observe that the interface has been added to the aggregated link but current has **U**, or Unselected, status.

```
[Comware-1] display link-aggregation verbose bridge 2
<-output omitted->
Aggregation Interface: Bridge-Aggregation2
Aggregation Mode: Static
Loadsharing Type: Shar
  Port       Status   Oper-Key
--------------------------------------------------
  GE1/0/21   U        2
```

4. Add the second port that connects to ProVision-1 to the aggregated link. Also enable the interface.

```
[Comware-] interface g1/0/22
[Comware-1-GigabitEthernet1/0/22] port link-aggregation group 2
[Comware-1-GigabitEthernet1/0/22] undo shutdown
[Comware-1-GigabitEthernet1/0/22] quit
```

5. Remember to configure the VLAN settings on the bridge-aggregation interface.

```
[Comware-1] interface Bridge-Aggregation 2
[Comware-1-Bridge-Aggregation2] port link-type trunk
[Comware-1-Bridge-Aggregation2] port trunk permit vlan all
[Comware-1] quit
```

6. View the aggregated link again.

```
[Comware-1-Bridge-Aggregation2] display link-aggregation verbose
bridge 2
Loadsharing Type: Shar -- Loadsharing, NonS -- Non-Loadsharing
Port Status: S -- Selected, U -- Unselected
Flags:  A -- LACP_Activity, B -- LACP_Timeout, C -- Aggregation,
        D -- Synchronization, E -- Collecting, F -- Distributing,
        G -- Defaulted, H -- Expired
Aggregation Interface: Bridge-Aggregation2
Aggregation Mode: Static
Loadsharing Type: Shar
  Port         Status    Oper-Key
------------------------------------------------------------
  GE1/0/21     S           2
  GE1/0/22     U           2
```

7. Note that interface g1/0/22 is unselected (U status) because the interface is down (you have not enabled the link on the ProVision-1 side). But observe that g1/0/21 is now selected (S status). Begin to think about what has changed since you checked the status previously. You will now configure the ProVision-1 side of the aggregated. On ProVision switches, you also define a logical interface for the aggregated link. The interfaces are called trunks (not to be confused with Comware trunk ports) and are labeled trk1, trk2, and so on. You define the trunk with the same command that you use to assign ports to it.

8. Access the ProVision-1 CLI and move to global configuration mode.
9. Assign the two interfaces that connect to Comware-1 to trk1. The `trunk` option indicates that this is a manual aggregated link.

```
ProVision-1(config)# trunk 19-20 trk1 trunk
```

10. Enable the second interface in the aggregated link.

```
ProVision-1(config)# interface 20 enable
```

11. On ProVision switches, you must define the VLAN settings on the trunk interface.

```
ProVision-1(config)# vlan 11 tag trk1

ProVision-1(config)# vlan 12 tag trk1
```

## Observe the second aggregated link

1. You will now view the aggregated link on the ProVision switch and verify that it has come up. On ProVision-1, enter:

```
ProVision-1(config)# show trunk

 Load Balancing Method: L3-based (Default), L2-based if non-IP traffic

  Port   | Name            Type       | Group  Type
  ------ + --------------- ---------- + ------ --------
  19     |                 100/1000T  | Trk1   Trunk
  20     |                 100/1000T  | Trk1   Trunk
```

2. Also verify the VLAN memberships. Notice that the *trk1* interface appears as a VLAN member instead of the physical interfaces of 19 and 20.

```
ProVision-1(config)# show vlan 11

 Status and Counters - VLAN Information - VLAN 11

  VLAN ID : 11
  Name : VLAN11
  Status : Port-based
  Voice : No
  Jumbo : No

  Port Information Mode     Unknown VLAN Status
```

```
  ---------------- -------- ------------ ----------
  1                Untagged Learn        Up
  21               Tagged   Learn        Up
  Trk1             Tagged   Learn        Up
ProVision-1(config)# show vlan 12
 Status and Counters - VLAN Information - VLAN 12
  VLAN ID : 12
  Name : VLAN12
  Status : Port-based
  Voice : No
  Jumbo : No
  Port Information Mode     Unknown VLAN Status
  ---------------- -------- ------------ ----------
  21               Tagged   Learn        Up
  Trk1             Tagged   Learn        Up
```

3. View the MSTP topology for instance 0. Again notice that trk1, as a single logical port, is forwarding.

```
ProVision-1(config)# show spanning-tree instance ist
 <-output omitted->
                                                          Designated
Port  Type      Cost   Priority Role       State      Bridge
----- --------- ------ -------- ---------- ---------- -----
1     100/1000T 20000  128      Designated Forwarding d07e28-cec94f
<-output omitted->
21    100/1000T 20000  128      Alternate  Blocking   d07e28-cec94f
<-output omitted->
Trk1            20000  64       Root       Forwarding d07e28-cec94f
```

### Comware-1

4. On Comware-1, view a summary of the aggregated links. Verify that the both ports are now selected for BAGG 2.

```
[Comware-1] display link-aggregation summary
Aggregation Interface Type:
BAGG -- Bridge-Aggregation, RAGG -- Route-Aggregation
Aggregation Mode: S -- Static, D -- Dynamic
Loadsharing Type: Shar -- Loadsharing, NonS -- Non-Loadsharing
Actor System ID: 0x8000, d07e-28ce-c94f
AGG          AGG      Partner ID     Select Unselect   Share
Interface    Mode                    Ports  Ports      Type
------------------------------------------------------------
BAGG1        S        none           2      0          Shar
BAGG2        S        none           2      0          Shar
```

Save the current configuration on each of the four switches. For ProVision, use the `write memory` command. For Comware, use the `save [force]` command.

## Learning check 2

Answer the following questions to test your understanding of how manual aggregated links are formed on ProVision and Comware switches.

1. What messages did you observe when you applied VLAN settings to the BAGG interface on a Comware switch?

2. You observed that after you first added interface g1/0/21 to BAGG 2, the interface was unselected. Later you observed that the g1/0/21 interface was selected. What changed in between these two occurrences?

3. What commands did you use to view the manual aggregated link on Comware switches and on ProVision switches?

## Learning check 2 answers

1. What messages did you observe when you applied VLAN settings to the BAGG interface on a Comware switch?

   The messages indicated that the switch was applying the configuration to the physical interfaces assigned to the BAGG interface.

2. You observed that after you first added interface g1/0/21 to BAGG 2, the interface was unselected. Later you observed that the g1/0/21 interface was selected. What changed in between these two occurrences?

   You added another interface to the link aggregation group, and you configured the VLAN settings on the BAGG interface.

   It is the second task that made the difference. Before you configured the VLAN settings on the BAGG, the VLAN settings on interface g1/0/21 (trunk port that permits all VLANs) did not match the settings on the BAGG 1 (access port in VLAN 1). After you configured the settings, the interface and BAGG settings matched. In a moment, you will learn about link compatibility issues that could prevent an interface from being selected (becoming active) in the aggregated link.

3. What commands did you use to view the manual aggregated link on Comware switches and on ProVision switches?

   The Comware commands are:

   - `display link-aggregation summary`
   - `display link-aggregation verbose bridge-interface <ID>`

   The ProVision command is `display trunks`.

   The Comware summary command shows all aggregated links on the switch, indicating the type of aggregated link and how many links are selected and unselected for the aggregation. The Comware verbose command provides more details about an individual aggregated link and the physical interfaces assigned to it.

   The ProVision command shows the interfaces that are assigned to trunk interfaces and the type of link aggregation.

## Requirements for links

Can you add any physical interface to an aggregated link, or does the physical interface have to meet any criteria?

You actually need to follow several guidelines. Links that are in the same aggregated link should have matching settings for:

- Duplex mode (full or half)
- Link speed and media (copper or fiber)
- VLAN assignment and other configurations such as QoS settings

## HP Comware: Behavior under incompatibility

Table 8-3 lists the parameters on Comware switches that must match. Note that the Comware switch is checking for compatibility on the ports *on its own side of the aggregation*. (A link typically negotiates its duplex mode and speed as it comes up, of course, but the Comware switch has no way of knowing the VLAN and other configurations on the other side of the link.)

Table 8-3: Configuration settings and link aggregation effects

| Configuration | Example Settings | Affects Aggregation State? |
|---|---|---|
| Port | Port rate Duplex Mode Link Status (up or down) | Yes |
| Class one | GVRP VLANs MSTP settings | No |
| Class two | Port type—trunk, access, or hybrid VLAN settings | Yes |

If you violate these rules—for example, you attempt to assign a 10 GbE interface and a GbE interface to the same aggregated link—incompatible links fail to become part of the aggregation. In Comware terminology, the links are unselected, unable to receive and forward traffic in the aggregation. Output for `display link-aggregation` commands indicates that the link is unselected with a U and selected with an S.

A Comware switch determines whether links are selected or unselected for an aggregation by this process:

1. The switch assigns the aggregated link a reference port. For a manual (static) aggregated link, it chooses the port as follows:
    - The port must meet this criteria to become a candidate port:
        - Be up
        - Match the class two settings on the BAGG interface
    - A reference port is selected from the candidate ports, which are sorted according to these criteria in this order:
        - Lower link aggregation priority value
        - Full-duplex higher speed
        - Full-duplex lower speed

- Half-duplex higher speed
- Half-duplex lower speed

A port with a lower link aggregation priority value is selected first. If ports have the same priority value, the full-duplex port that is higher speed will be selected, and so on.

2. When you add an interface to the aggregated link, the switch checks the interface's settings.
   - If any of the port or class two settings (see Table 8-2) do not match the settings on the reference port, the link becomes unselected.
   - If all of the settings that affect the aggregation state match, the link becomes selected.

### Guidelines for ensuring VLAN compatibility

When you configure VLAN settings (class 2 settings in Table 8-2) on a Comware BAGG interface, the settings apply to any physical interface that is assigned to it. You can see these settings on the physical interface in the configuration, and they persist even if you later remove the interface from the aggregated link.

These settings are applied to the physical interfaces at the moment that you configure them on the BAGG interface. If you later add another physical interface to the BAGG interface, the settings are not applied. If the physical interface's settings are incompatible, it will remain in the unselected state. It is recommended that you always configure these settings on the BAGG interface *after* you assign all physical interfaces to it.

## HP ProVision: Behavior under incompatibility

ProVision switches do not check for compatibility on links in a *manual* aggregated link. You could, theoretically, combine links of different speeds and duplex modes, and the links could become part of the aggregated link. However, you *should not* do so.

You must be very careful to check the compatibility yourself for:

- Duplex mode
- Link speed and media

The aggregated link will not work as well as expected if, for example, you combine links of different speeds. You could even disrupt network connectivity if you aggregate incompatible links between a ProVision switch and a Comware one. Then the ProVision switch will use the incompatible links, but the Comware switch will not accept traffic on its side because it considers the links unselected. Traffic will be dropped and connectivity disrupted.

### Guidelines for ensuring VLAN compatibility

For ProVision switches, you do not need to worry about VLAN incompatibilities. ProVision interfaces lose their individual VLAN settings when you add them to the aggregated link (trunk). You must then configure the VLAN settings on the trunk interface. The settings automatically apply to interfaces in the trunk, including ones that you add to the trunk at a later point. Therefore, you do not need to worry about making the settings match. You do, however, need to remember to configure the settings on the trunk interface.

### Requirements for maximum number of links

An aggregated link can only support a certain number of links per aggregated link. The number of links allowed in an aggregation depends on the switch model. For example, many switches have a maximum of eight links. Refer to the documentation for your switch for more information.

## Potential issue with manual aggregated links

You just considered the settings that you need to ensure match when you add interfaces to an aggregated link. You also need to be very careful that all interfaces actually connect to the same switch.

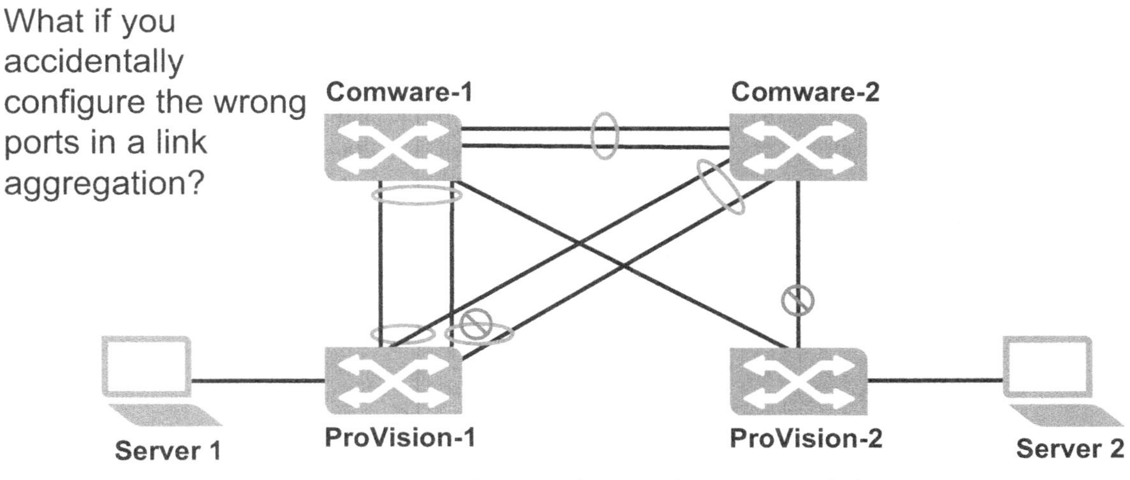


Figure 8-6: Potential issue with manual aggregated links

In Figure 8-6, you see a situation in which the administrator wanted to create two aggregated links on ProVision-1: one to Comware-1 and one to Comware-2. However, the admin accidentally misconfigured the two l aggregated links on ProVision-1. Each of the aggregated links has one link to Comware-1 and one link to Comware-2. The aggregated links are configured correctly on Comware-1 and Comware-2.

This mismatch can cause serious consequences.

Figure 8-7 shows two servers in VLAN 11, one connected to ProVision-1 and one connected to ProVision-2. What happens when server 2 tries to contact server 1?

The frame reaches Comware-1, which forwards the frame on the aggregated link to ProVision-1. The switch happens to select the link that is misconfigured on ProVision-1. This link is part of an aggregated link that is blocked by MSTP. The communication fails.

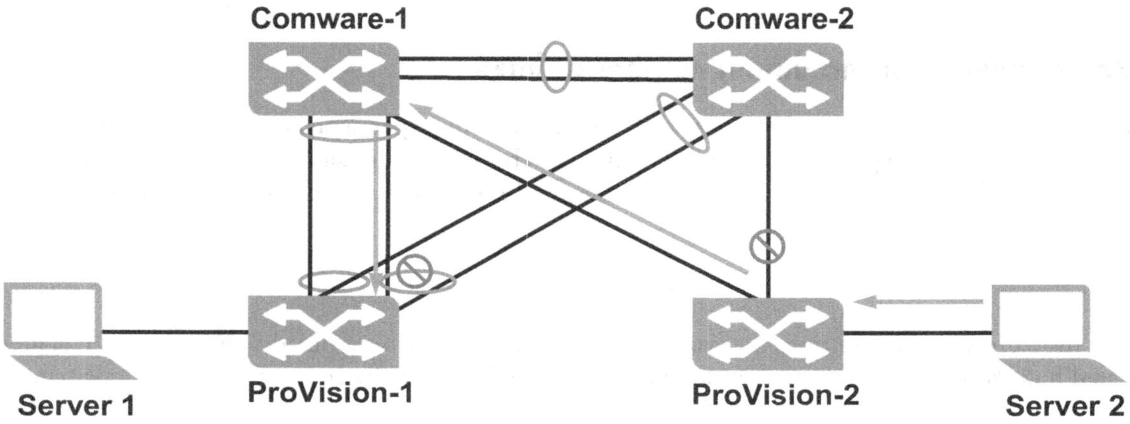


Figure 8-7: Traffic flow across the misconfigured aggregated links (1)

What can happen when server 1 sends a broadcast? ProVision-1 should flood the broadcast on its aggregated link to Comware-1 (its MSTP root port) but not on its aggregated link to Comware-2 (its MSTP alternate port). But, as shown in Figure 8-8, it happens to choose the misconfigured link and sends the broadcast to Comware-2. Comware-2 accepts the broadcast because it does not block the aggregated link; on its side, the aggregation is an MSTP designated port. Comware-2 floods the broadcast back to Comware-1, which will forward the broadcast to ProVision-2 *and* back to ProVision-1.

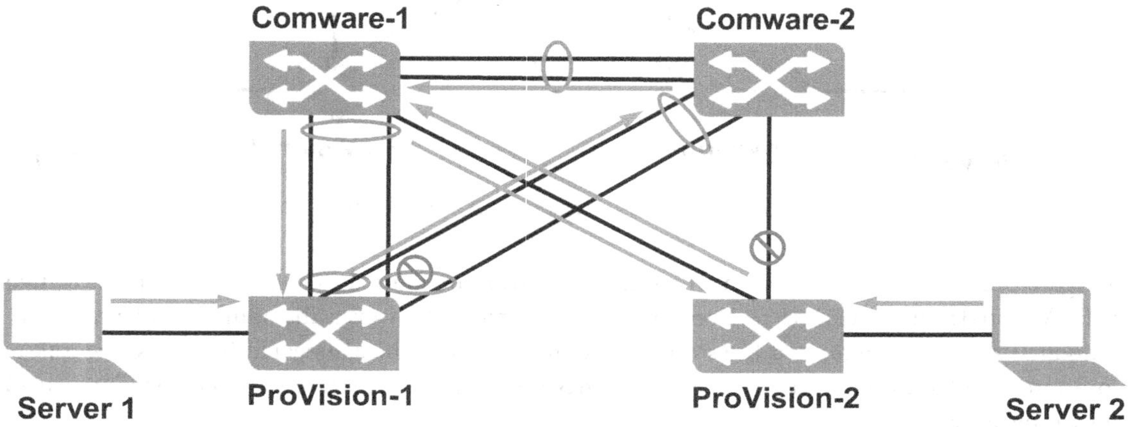


Figure 8-8: Traffic flow across the misconfigured aggregated links (2)

If Comware-1 selects the interface that correctly connects to ProVision-1, ProVision-1 will start the process again, leading to a broadcast storm.

As more conversations like this occur, connectivity is disrupted as far as the VLAN extends.

## Link Aggregation Control Protocol (LACP)

IEEE 802.3ad, Link Aggregation Control Protocol (LACP), provides a solution to the issues that can occur when a manual aggregated link is misconfigured. This protocol requires switches to exchange messages, called LACP data units (DUs), to establish the aggregated link (as illustrated in Figure 8-9). These messages include a system ID, which uniquely identifies the switch. Each LACP peer checks the messages that arrive on each link in the aggregation. An aggregated link establishes correctly only if the incoming system ID and operational key matches on all the links. The system ID identifies the switch and the operational key identifies the link aggregation group. In this way, each peer ensures that all links are connected to the same other peer and aggregated.

- Uses messages to establish the link aggregation
- Ensures that all links are connected to the same peer (system ID) and link aggregation (operational key)
- Ensures compatibility for other settings
- Manages adding and removing links

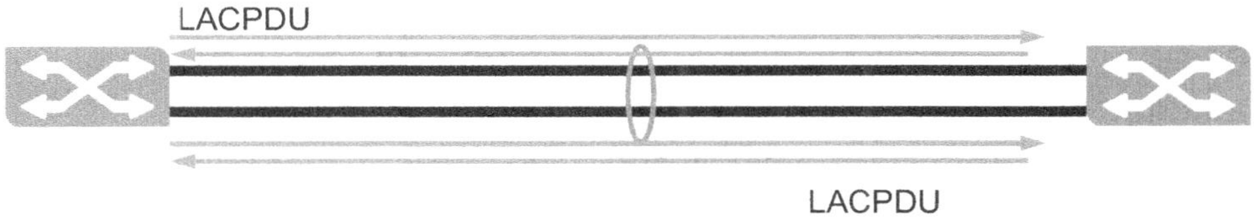


Figure 8-9: LACP

Consider what would have happened if you had used LACP for the misconfigured aggregated links in the previous example. For aggregated link 1 on ProVision-1, LACP would have detected that the system ID for the peer on the two physical interfaces was different. One of the physical interfaces would not have been allowed to become active.

Similarly, if Comware-1 were using LACP on its aggregated links, LACP would have detected a different operational key from the peer on the two links. Again, LACP would have prevented both links from becoming active.

LACP allows the peers to exchange other information about the aggregated link as well. Using LACP, ProVision switches check for compatibility with:

- Media type and speed
- Duplex mode—LACP requires full-duplex operation

ProVision switches then allow only compatible links to join the aggregated link. Comware LACP aggregated links continue to check for compatibility with the settings in Table 8-3 on page 265.

As you learned before, a reference port determines which port's settings are used for the aggregation; other ports must match or be excluded. With LACP, the switch with the reference port is determined by system priority (smaller value has higher priority; the priority is derived from a priority value and MAC address). That switch then selects the reference port by port priority (smaller value has higher priority; the priority is derived from a priority value and port ID). Comware switches also make sure to select a reference port that has class two settings that match the BAGG interface and preferentially that operate in full-duplex mode.

LACPDUs allow the devices to manage the logical aggregated link. LACP manages adding or removing physical links. It also handles failovers, ensuring that all frames are delivered in order even when traffic needs to move to a new link.

## LACP operational modes

LACP has two operational modes, as illustrated in Figure 8-10.

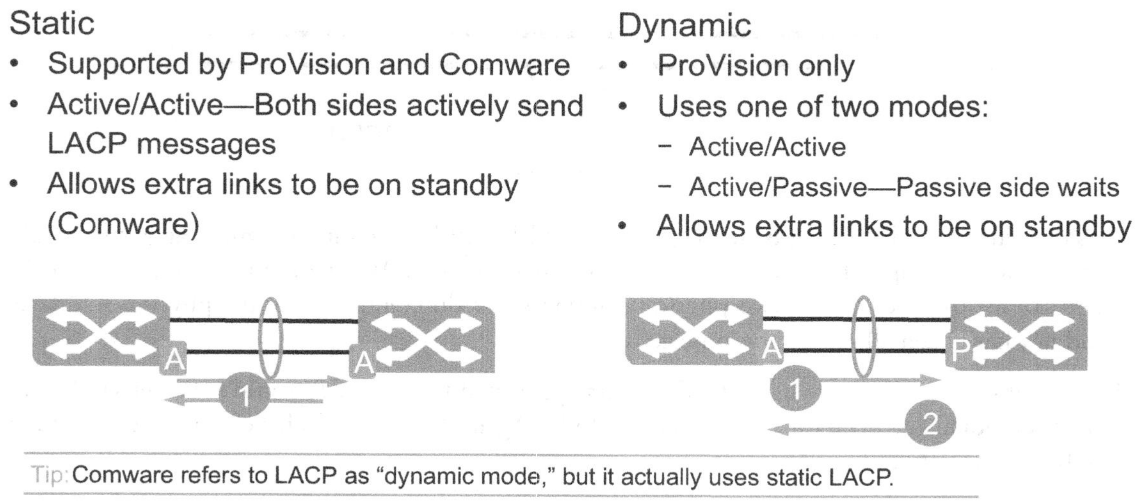


Figure 8-10: LACP operational modes

## Static mode

You set up a static mode LACP aggregated link in much the same way as you do a manual aggregated link. However, when you create the aggregated link interface (whether a Comware BAGG or ProVision trunk), you enable LACP on that interface.

The physical interfaces that are assigned to that aggregated link then operate in active LACP state. That is, both sides of the link send LACPDUs to each other and set up the aggregated link as you learned on the previous page.

Both ProVision and Comware switches support this mode. However, Comware refers to static mode LACP as "dynamic mode" for the aggregated link. This can be confusing, but remember that Comware's dynamic mode is actually static LACP.

---

**Important**

The term "dynamic LACP" as used on Comware switches actually refers to the static LACP operational mode.

---

With static LACP, Comware switches can have standby (unselected) links beyond the maximum allowed in the aggregated link. ProVision switches do not support this capability with static LACP.

## Dynamic mode

LACP actually defines two states, and dynamic LACP allows a port to operate in one or the other:

- **Active**—Transmits LACPDUs to advertise that it can create an aggregated link
- **Passive**—Listens for LACPDUs and responds with LACPDUs only after it receives one from an active port

Both ProVision and Comware switches support passive mode. ProVision switches support dynamic mode, but Comware switches do not.

To set up a dynamic aggregated link, you enable LACP active or passive on the switch interfaces that you want to function as part of an aggregated link. But you do not actually create a link aggregation (trunk) interface. You do the same to the switch on the other side of the link. As long as one or both sides of each link are in active mode, the aggregated link automatically establishes. (If both sides are in a passive state, then neither will initiate the exchange process.) LACP automatically determines which links are between the same two devices and places those links in the same aggregated link.

When you use dynamic LACP, ProVision switches can have standby links beyond the maximum allowed in the aggregation. For example, if the switch supports only eight links, you might be able to add 12. Links over the maximum are standby links, which cannot be used to forward traffic. However, if a link fails, LACP automatically adds one of the standby links to the aggregated link.

A dynamic LACP aggregated link does not have a logical interface in the switch configuration. Therefore, you cannot assign nondefault STP, VLAN, and other such settings to it. (You can use a GVRP, a protocol for dynamically learning VLANs, to add VLANs to the dynamic aggregated link. However, this protocol is not covered in this study guide.) For this reason, static LACP is typically recommended.

## Example configuration: Set up an LACP aggregated link

This section outlines the steps for adding an LACP aggregated link between Comware-1 and ProVision-2 on the example network shown in Figure 8-11. This example network uses the settings established in "Example configuration: Set up a manual aggregated link" on page 256.

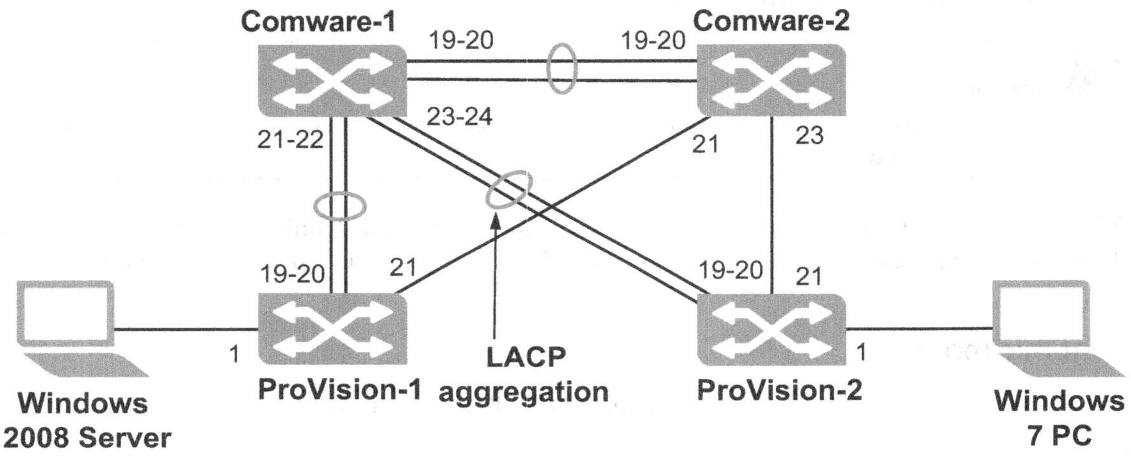


Figure 8-11: Configure an LACP aggregated link

### Configure an LACP aggregated link

1. Access the Comware-1 CLI and move to the system view.
2. Create BAGG 3. Enable LACP on this interface. (The command uses the `dynamic` option, but the mode is actually static LACP; Comware switches do not support dynamic LACP.)

```
[Comware-1] interface Bridge-Aggregation 3

[Comware-1-Bridge-Aggregation3] link-aggregation mode dynamic

[Comware-1-Bridge-Aggregation3] quit
```

**Note**
You cannot change the link-aggregation mode after you assign interfaces to the BAGG.

3. Assign the ports that connect to ProVision-2 to this bridge interface. Remember to enable the new port.

```
[Comware-1] interface g1/0/23
[Comware-1-GigabitEthernet1/0/23] port link-aggregation group 3
[Comware-1-GigabitEthernet1/0/23] interface g1/0/24
[Comware-1-GigabitEthernet1/0/24] port link-aggregation group 3
[Comware-1-GigabitEthernet1/0/24] undo shutdown
[Comware-1-GigabitEthernet1/0/24] quit
```

4. Configure the VLAN settings on the bridge interface.

```
[Comware-1] interface Bridge-Aggregation 3
[Comware-1-Bridge-Aggregation3] port link-type trunk
[Comware-1-Bridge-Aggregation3] port trunk permit vlan all
[Comware-1-Bridge-Aggregation3] quit
```

5. Access the ProVision-2 CLI. Move to the global configuration context.
6. Create a static LACP aggregated link on the two ports that connect to Comware-1.

```
ProVision-2(config)# trunk 19-20 trk1 lacp
```

7. Enable interface 20.

```
ProVision-2(config)# interface 20 enable
```

8. Tag VLAN 11 and 12 on the aggregated link (trk1).

```
ProVision-2(config)# vlan 11 tagged trk1
ProVision-2(config)# vlan 12 tagged trk1
```

## View the aggregated link

You will now examine the aggregated link.

1. View the trunk on ProVision-2.

```
ProVision-2(config)# show trunks
```

```
Load Balancing Method: L3-based (Default), L2-based if non-IP traffic

 Port    | Name          Type         Group  Type
 ------ + ------------- --------- + ------ --------
 19      |               100/1000T | Trk1   LACP
 20      |               100/1000T | Trk1   LACP
```

2. LACP provides more information about the aggregated link. View LACP information on ProVision-2.

```
ProVision-2(config)# show lacp
                          LACP

     LACP    Trunk   Port             LACP    Admin  Oper
Port Enabled Group   Status  Partner Status  Key    Key
---- ------- ------- ------- ------- ------- ------ -----
19   Active  Trk1    Up      Yes     Success 0      562
20   Active  Trk1    Up      Yes     Success 0      562
```

3. You can also view information specific to the local side of the aggregated link. For example, you can see this ProVision-2's system ID and that the links are aggregated.

```
ProVision-2# show lacp local
LACP Local Information.
System ID: d4c9ef-84fe80

           LACP                 Tx     Rx Timer
 Port Trunk Mode     Aggregated  Timer  Expired
 ---- ------ -------- ----------- ------ --------
 19   Trk1   Active   Yes         Slow   No
 20   Trk1   Active   Yes         Slow   No
```

4. And you can view information about the peer. On ProVision-2 note that both peer interfaces have the same system ID and operational key.

```
ProVision-2# show lacp peer
LACP Peer Information.
System ID: d4c9ef-84fe80
```

```
Local Local                           Port      Oper  LACP    Tx
Port  Trunk  System ID     Port  Priority  Key   Mode    Timer
----- ------ ------------- ----- --------- ----- ------- -----
19    Trk1   d07e28-cec94f 23    32768     3     Active  Slow
20    Trk1   d07e28-cec94f 24    32768     3     Active  Slow
```

5. View log messages on ProVision-2.

```
ProVision-2(config)# show log -r
```

The `-r` option shows the most recent logs first, so the events occurred chronologically from the bottom of the list.

You can see that the interfaces were blocked by LACP, added to the link aggregation interface (Trk1), blocked by STP, and then brought online as a part of the aggregated link.

```
 Keys:   W=Warning   I=Information
         M=Major     D=Debug E=Error
----  Reverse event Log listing: Events Since Boot  ----
I 02/13/15 05:32:33 00001 vlan: VLAN11 virtual LAN enabled
I 02/13/15 05:32:28 00076 ports: port 20 in Trk1 is now on-line
I 02/13/15 05:32:28 00435 ports: port 20 is Blocked by STP
I 02/13/15 05:32:28 00435 ports: port 20 is Blocked by LACP
I 02/13/15 05:32:23 00076 ports: port 19 in Trk1 is now on-line
I 02/13/15 05:32:23 00002 vlan: VLAN11 virtual LAN disabled
I 02/13/15 05:32:23 00435 ports: port 19 is Blocked by STP
I 02/13/15 05:32:23 00078 ports: trunk Trk1 is now active
I 02/13/15 05:32:23 00435 ports: port 19 is Blocked by LACP
I 02/13/15 05:32:20 00077 ports: port 19 is now off-line
I 02/13/15 05:32:03 00076 ports: port 19 is now on-line
I 02/13/15 05:32:03 00435 ports: port 19 is Blocked by STP
I 02/13/15 05:31:46 00076 ports: port 19 is now on-line
I 02/13/15 05:31:43 00435 ports: port 19 is Blocked by STP
```

6. Access the Comware-1 CLI.

7. Verify the status of the aggregated link on Comware-1.

```
[Comware-1] display link-aggregation verbose Bridge-Aggregation 3
Loadsharing Type: Shar -- Loadsharing, NonS -- Non-Loadsharing
Port Status: S -- Selected, U -- Unselected
Flags:  A -- LACP_Activity, B -- LACP_Timeout, C -- Aggregation,
        D -- Synchronization, E -- Collecting, F -- Distributing,
        G -- Defaulted, H -- Expired
Aggregation Interface: Bridge-Aggregation3
Aggregation Mode: Dynamic
Loadsharing Type: Shar
System ID: 0x8000, d07e-28ce-c94f
Local:
Port          Status  Priority Oper-Key  Flag
--------------------------------------------------------------------
GE1/0/23      S       32768     3          {ACDEF}
GE1/0/24      S       32768     3          {ACDEF}
Remote:
Actor     Partner Priority Oper-Key SystemID                Flag
----------------------------------------------------------------------------
GE1/0/23 21       0          562       0xfe80, d4c9-ef84-fe80 {ACDEF}
GE1/0/24 22       0          562       0xfe80, d4c9-ef84-fe80 {ACDEF}
```

8. Note in the output that this BAGG is dynamic.

   The system ID is the same as the peer ID that you observed on ProVision-2.

   You should see S, selected, status for both interfaces. You should also see that LACP has successfully established both interfaces as part of the link. Look for flags ACDEF, all of which should be displayed. These flags indicate that the link has LACP activity (A), has been successfully added to the aggregation (C), has synchronized with the peer (D), and is sending and receiving traffic (E and F). When you see these flags on both links, you know that the aggregated link is load balancing on both links.

You can also see information from the other side of the link. The peer system ID is the same as the local system ID that you observed on ProVision-2.

9. Also view a summary of the aggregated links on Comware-1.

```
<Comware-1> display link-aggregation summary

Aggregation Interface Type:

BAGG -- Bridge-Aggregation, RAGG -- Route-Aggregation

Aggregation Mode: S -- Static, D -- Dynamic

Loadsharing Type: Shar -- Loadsharing, NonS -- Non-Loadsharing

Actor System ID: 0x8000, d07e-28ce-c94f

AGG       AGG  Partner ID              Select Unselect Share

Interface Mode                         Ports  Ports    Type

------------------------------------------------------------

BAGG1     S    none                     2      0       Shar

BAGG2     S    none                     2      0       Shar

BAGG3     D    0xfe80, d4c9-ef84-fe80 2        0       Shar
```

## Observe load sharing

This section shows load sharing across the links.

1. Verify connectivity in VLAN 1, from ProVision-2, ping Comware-1 (10.1.1.1), Comware-2 (10.1.1.2), and ProVision-1 (10.1.1.3).

```
ProVision-2(config)# ping 10.1.1.1

10.1.1.1 is alive, time = 2 ms

ProVision-2(config)# ping 10.1.1.2

10.1.1.2 is alive, time = 5 ms

ProVision-2(config)# ping 10.1.1.3

is alive, time = 2 ms
```

2. View the ARP table and the MAC forwarding table for VLAN 1 on ProVision-2.

   Notice that ProVision-2 has learned Comware-1's and Comware-2's MAC addresses on Trk1. (VLAN 1 is a part of instance 0, in which Comware-1 is the root, so traffic is using the Trk1 interface.)

```
ProVision-2(config)# show arp

 IP ARP table

  IP Address       MAC Address       Type    Port
  ---------------  ----------------- ------- ----
  10.1.1.1         d07e28-cec955     dynamic Trk1
  10.1.1.2         d07e28-ced324     dynamic Trk1
  10.1.1.3         d4c9ef-851680     dynamic Trk1
ProVision-2(config)# show mac-address vlan 1

 Status and Counters - Address Table - VLAN 1

  MAC Address   Port
  ------------- -------
  d07e28-cec955 Trk1
  d07e28-cec98a Trk1
  d07e28-cec98b Trk1
  d07e28-ced324 Trk1
  d4c9ef-851680 Trk1
```

Now you will observe how traffic flows over the physical interfaces within the aggregated link.

3. On ProVision-2, clear the interface statistics and start up the interface menu display.

```
ProVision-2(config)# clear statistics global
ProVision-2(config)# show interface display
```

4. On ProVision-2, use the <DOWN> arrow key until you can see the Trk 1interface is displayed.

You should already see traffic traversing both links in a trunk from things like STP BPDUs and miscellaneous traffic from the Windows 7 and 2008 computers (like NetBIOS and IP6 management traffic if enabled). Keep this console window open.

```
                            Status and Counters - Port Counters
                                                                Flow  Bca*
 Port  Total Bytes  Total Frames Errors Rx  Drops Tx  Ctrl Lim*
----- ----------- ------------  ---------  --------  ---- ---*
16              0            0          0         0  off    0
17              0            0          0         0  off    0
```

```
18              0           0           0           0  off   0
19-Trk1      6723          41           0           0  off   0
20-Trk1      1750           8           0           0  off   0
21           5437          34           0           0  off   0
22              0           0           0           0  off   0
23              0           0           0           0  off   0
24              0           0           0           0  off   0
25              0           0           0           0  off   0
26              0           0           0
 Actions->    Back      Show details      Reset      Help
```

5. Access the Comware-1 CLI. Return to user view.
6. Send 500 pings to ProVision-2.

```
<Comware-1> ping -c 500 10.1.1.4
```

7. On ProVision-2, look at the statistics for the interfaces in Trk1. You should see statistics for both ports. The statistics for one of the ports will probably increase more rapidly depending on how your switches choose links for the conversation.

```
                            Status and Counters - Port Counters

                                                                Flow  Bca*
   Port   Total Bytes  Total Frames   Errors Rx Drops Tx Ctrl  Lim*
  -----   -----------  -------------   --------- -------- ---- ---*
  16                0              0          0         0  off   0
  17                0              0          0         0  off   0
  18                0              0          0         0  off   0
  19-Trk1      20,841            143          0         0  off   0
  20-Trk1        8659             51          0         0  off   0
  21           14,431             87          0         0  off   0
  22                0              0          0         0  off   0
  23                0              0          0         0  off   0
  24                0              0          0         0  off   0
```

```
25                  0               0           0         0   off    0
26                  0               0           0         0   off    0
 Actions->    Back      Show details      Reset      Help
```

8. Now start another conversation. Access a terminal session with Comware-2 and return to user view.

9. Begin pinging ProVision-2.

```
<Comware-2> ping -c 500 10.1.1.4
```

10. On ProVision-2, look at the statistics for the interfaces in Trk1. With another conversation, you are more likely to see traffic on both links.

```
Status and Counters - Port Counters

                                                               Flow Bca*
Port    Total Bytes Total Frames  Errors Rx  Drops Tx  Ctrl Lim*
------  ---------- -------------  ---------  --------  ---- ---*
16                0             0          0         0  off   0
17                0             0          0         0  off   0
18                0             0          0         0  off   0
19-Trk1      67,542           437          0         0  off   0
20-Trk1      34,626           312          0         0  off   0
21           36,011           135          0         0  off   0
22                0             0          0         0  off   0
23                0             0          0         0  off   0
24                0             0          0         0  off   0
25                0             0          0         0  off   0
26                0             0          0         0  off   0
Actions->    Back      Show details      Reset      Help
```

11. Press **Enter** on the `Back` option to back out of the display. Then press **Ctrl+c.**

12. On Comware-1 and on Comware-2, press **Ctrl+c** to end the pings.

Save the current configuration on each of the four switches. For ProVision, use the `write memory` command. For Comware, use the `save [force]` command.

## Learning check 3

Answer the following questions to determine how well you understand how LACP is used to establish aggregated links on ProVision and Comware switches.

1. What is the difference between setting up an LACP aggregated link and a manual aggregated link on:
   - Comware switches
   - ProVision switches
2. What ProVision command show whether an interface has successfully been added to an aggregated link and can pass traffic?
3. What Comware commands shows whether an interface has successfully been added to an aggregated link and can pass traffic? Which command shows the LACP flags?
4. Which LACP flags indicate that the link has been successfully added to the aggregated link and can pass traffic? If you see these flags on multiple links in an aggregation, you know that the aggregation is load balancing.

LACP lets switches communicate information about the link and negotiate whether a link becomes part of the aggregation. In the example configuration, the two interfaces on each switch were compatible, so both links successfully joined the aggregation.

In the following output, LACP fails on the slower link. You can see in the Comware **display link-aggregation** output that the second link is not synched. Due to this failure, both sides of the link will only use one physical interface in the aggregated link. Connectivity is not disrupted (although the admin should correct the problem and add a valid link to the aggregation).

```
ProVision-2(config)# show lacp

                              LACP

     LACP      Trunk  Port            LACP      Admin  Oper

Port Enabled   Group  Status Partner  Status    Key    Key

---- -------   ------ ------- ------- -------   -----  ------

19   Active    Trk1   Up       Yes     Success   0      562

20   Active    Trk1   Blocked  Yes     Failure   0      562

<Comware-1> display link-aggregation verbose bridge-aggregation 3

Loadsharing Type: Shar -- Loadsharing, NonS -- Non-Loadsharing

Port Status: S -- Selected, U -- Unselected
```

```
Flags:  A -- LACP_Activity, B -- LACP_Timeout, C -- Aggregation,
        D -- Synchronization, E -- Collecting, F -- Distributing,
        G -- Defaulted, H -- Expired
Aggregation Interface: Bridge-Aggregation3
Aggregation Mode: Dynamic
Loadsharing Type: Shar
System ID: 0x8000, d07e-28ce-c94f
Local:
Port        Status  Priority Oper-Key  Flag
------------------------------------------------------------------
GE1/0/23    S       32768    3         {ACDEF}
GE1/0/24    U       32768    5         {AC}
Remote:
Actor     Partner Priority Oper-Key  SystemID               Flag
------------------------------------------------------------------
GE1/0/23 19       0        562       0xfe80, d4c9-ef84-fe80 {ACDEF}
GE1/0/24 20       0        562       0xfe80, d4c9-ef84-fe80 {ACEF}
```

5. You checked ProVision-2's MAC forwarding table. What forwarding interface did this switch learn for other switches?
6. When you pinged ProVision-2 from other switches, did you see traffic on both physical interfaces? Did you see exactly the same amount on both interfaces?

## Learning check 3 answers

1. What is the difference between setting up an LACP aggregated link and a manual aggregated link on:
   - Comware switches
   - ProVision switches

   The process is very similar on both types of switches. On Comware switches, you just need to use the `link-aggregation mode dynamic` command in the BAGG interface. On the ProVision switches, you use the `lacp` option with the `trunk` command.

2. What ProVision command shows whether an interface has been successfully added to an aggregated link and can pass traffic?

   The `show lacp [local | peer]` command shows this information. You should verify that the LACP status is Success. If you use the `local` or `peer` option, you can see more information about the status on the local interfaces or the peer interfaces.

3. What Comware commands show whether an interface has successfully been added to an aggregated link and can pass traffic? Which command shows the LACP flags?

   The Comware commands are:

   - `display link-aggregation summary`
   - `display link-aggregation verbose bridge-interface <ID>`

     The `verbose` command shows the flags.

4. Which LACP flags indicate that the link has been successfully added to the aggregated link and can pass traffic?

   The flags are ACEDF.

5. You checked ProVision-2's MAC forwarding table. What forwarding interface did this switch learn for other switches?

   The MAC address forwarding table learns the link aggregation interface as the forwarding interface for traffic, not individual physical interfaces.

6. When you pinged ProVision-2 from other switches, did you see traffic on both physical interfaces? Did you see exactly the same amount on both interfaces?

   You should have seen traffic on both interfaces, but the amount might not have been equal. Aggregated links load-share traffic over all active physical interfaces in the aggregated link, but, as you will learn in a moment, the sharing tends to become more even when more devices are communicating (as in a real network).

## Load-sharing traffic over an aggregated link

You will now explore how the link aggregation decides which link to use for specific traffic. Refer to Figure 8-12 as you review the information that follows.

The aggregated link selects a link for each *conversation*, or one-way communication, between a source and a destination. For example, when a workstation sends an Ethernet frame to a server, a conversation begins. The switch hashes the source and destination address and, based on the hash, assigns the conversation to a link within the aggregation. All subsequent frames from the same workstation to the same server are part of that conversation and assigned to the same physical link, avoiding out-of-order packet delivery. The aggregated link will only select a new link for the conversation if the current one fails. If you are using LACP, LACP ensures that all packets are delivered in order as the link fails over.

The link aggregation uses a hash to select a link for each *conversation:*

- Conversation defined by the load-sharing mode
- Link for a conversation stays constant
- Traffic tends to be balanced more evenly as conversations increase

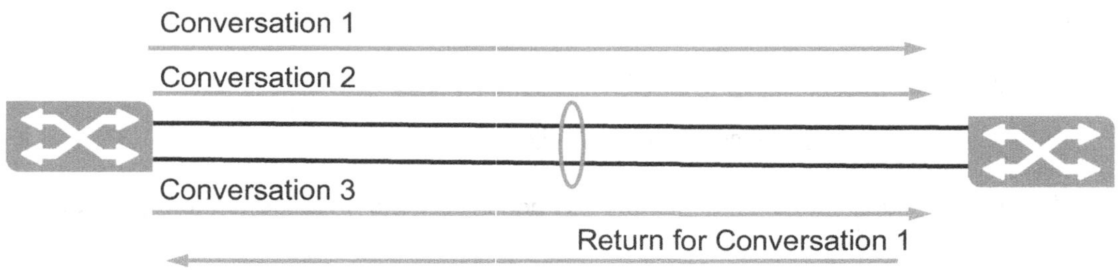


Figure 8-12: Load-sharing traffic over an aggregated link

Which addresses are used to define the source and destination for a conversation depends on the aggregated link's load-sharing mode. The default mode uses source MAC address and destination MAC address, but HP switches support other options, as indicated in Table 8-4.

Table 8-4: Options for load-sharing mode

| Option | ProVision | Comware |
|---|---|---|
| Source MAC address only | | X |
| Destination MAC address only | | X |
| Source and destination MAC address | X—Used for all non-IP traffic | X—Default |
| Source IP address only | | X |
| Destination IP address only | | X |
| Source and destination IP address (or IPv6 address)<br>*Called Layer 3 on ProVision | X—Default for IP (or IPv6) traffic | X |
| Source IP address and source port number | | X |
| Destination IP address and destination port number | | X |
| Source and destination IP (or IPv6) address and source and destination port number<br>*Called Layer 4 on ProVision | X | |

### Load-sharing multiple conversations

The hash assigns conversations to links arbitrarily, not based on round-robin. In other words, a conversation between client 1 and server 1 begins and is assigned to link 2 in the aggregation. When another conversation between client 2 and server 2 begins, it *could* be is assigned to the same link. However, when an aggregated link carries many conversations, the conversations tend to spread out over the links more or less evenly.

Even when all physical links carry precisely the same *number* of conversations, though, the links might experience significant differences in traffic flow. This is because different conversations will have different bandwidth requirements, different levels of burstiness, and so on. In a test environment with a relatively small number of source/destination pairs, you might see dramatic differences in traffic distribution.

In a real-world network, which has larger numbers of conversations each carrying variable bandwidth, the links within an aggregation will tend to carry roughly even amounts of traffic. Although distribution will never be perfectly equal, it will be good enough for practical purposes—while still allowing the switch to quickly assign traffic to a link at line rate.

### Considering the other sides of the conversation

Each switch is responsible for assigning a conversation to a link locally. The return traffic is another conversation, and the connected switch might assign it to a different link even when that switch is using the same load-sharing mode. However, this behavior does not typically cause issues. Each switch considers the aggregated link as a whole as the logical interface, learns MAC addresses on the link aggregation interface, and accepts traffic on any active link.

## Other options for aggregated links

In this chapter you have focused on aggregated links between switches, but other use cases exist, as you can see in Figure 8-13.

For example, you can establish an aggregated link between a switch and a server with multiple NICs. You can use a manual aggregated link or an LACP aggregated link, but to choose the correct option, you need to understand the server settings and capabilities. When the server is a virtual host for virtual machines (VMs), the virtual switch configuration determines the mode for the aggregated link—some VMware and Citrix XenServer versions do not support LACP—and even whether redundant links should be part of an aggregated link at all.

Similarly, you can configure aggregated links between a switch and a storage device that uses Ethernet.

You should attend other HP training for more details on these topics or refer to the appropriate documentation. When you search for information, keep in mind that the feature might be called bonding or NIC teaming rather than link aggregation.

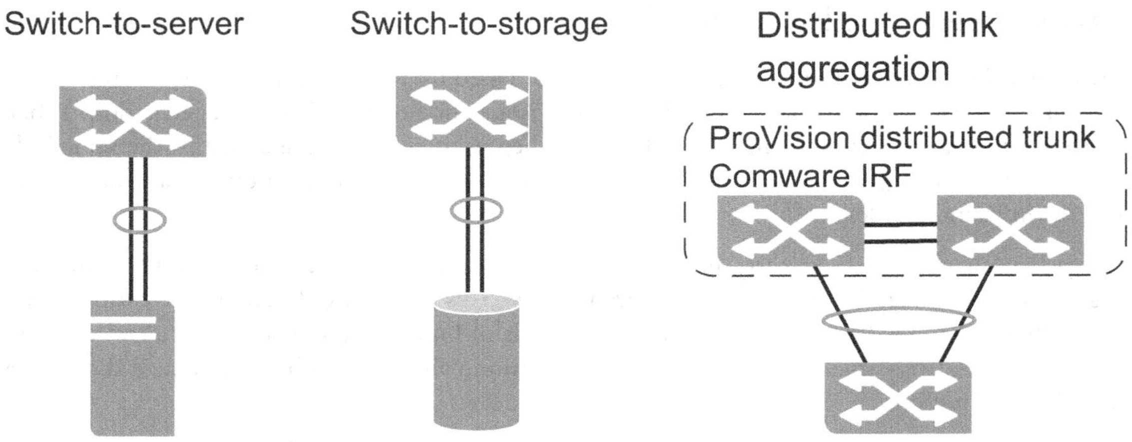


Figure 8-13: Other options for aggregated links

Finally, note that you can break the rule about how aggregated links must connect the same switches in special circumstances. You can create distributed trunks between a downstream device and two ProVision switches. This feature is covered in the *Deploying HP FlexNetwork Core Technologies* course. Later in this study guide, you will learn how an aggregated link can connect a downstream device and multiple switches in the same Comware Intelligent Resilient Framework (IRF) virtual device.

## Summary

This chapter has introduced you to aggregated links, which combine multiple physical interfaces into a single logical interface that load-shares traffic across it.

You learned how to create manual, or static, aggregated links on ProVision and Comware switches. You then learned about the advantages of LACP for managing the aggregated link, facilitating failover, and ensuring that links are compatible and connected to the same peer. You then practiced setting up an LACP aggregated link.

## Learning check 4

Review what you learned in this chapter by answering the questions below.

1. What is the ProVision command for adding interfaces to an aggregated static link?
   a. `trunk <int-id-list> trk<ID>`
   b. `interface <int-id-list> trk<ID>`
   c. `interface bridge-aggregation <number>`
   d. `port trunk`

2. You want to configure a static LACP aggregated link on a ProVision switch. The trunk will use ID 1. The aggregation should include interfaces A1 and A2. What is the command?

3. You want to configure a static LACP aggregated link on a Comware switch. The aggregation should include interfaces g1/0/1 and g1/0/2. What are the commands?

4. What is the default load-sharing mode for Comware switches for link aggregation? What is the default mode for ProVision switches?

## Learning check 4 answers

1. What is the ProVision command for adding interfaces to an aggregated static link?

   a. `trunk <int-id-list> trk<ID>`

   b. `interface <int-id-list> trk<ID>`

   c. `interface bridge-aggregation <number>`

   d. `port trunk`

2. You want to configure a static LACP aggregated link on a ProVision switch. The trunk will use ID 1. The aggregation should include interfaces A1 and A2. What is the command?

   `trunk a1,a2 trk1 lacp`

3. You want to configure a static LACP aggregated link on a Comware switch. The aggregation should include interfaces g1/0/1 and g1/0/2. What are the commands?

   `interface bridge-aggregation 1`

   `interface g1/0/1`

   `port link-aggregation group 1`

   `interface g1/0/2`

   `port link-aggregation group 1`

4. What is the default load-sharing mode on Comware switches for aggregated link? What is the default mode on ProVision switches?

   The default mode for Comware is source MAC address and destination MAC address. The default mode for ProVision is source IP address and destination IP address for IP traffic and source MAC address and destination MAC address for non-IP traffic.

# 9 IP Routing

## EXAM OBJECTIVES

✓ Configure static routes on HP switches

✓ Interpret IP routing tables on HP switches

✓ Configure a basic Open Shortest Path First (OSPF) solution on HP switches

## INTRODUCTION

Chapter 5 "VLANs" explained how routing enables communications between devices in different subnets (VLANs). Many HP ProVision switches support routing, even though it is disabled by default, and the vast majority of HP Comware switches support routing. This chapter teaches you more about IP routing.

### Assumed knowledge

Before reading this chapter, you should have a basic understanding of:

- Layer 2 and Layer 3 in the Open Systems Interconnection (OSI)
- IP addressing
- VLANs

### IP routes

In the network shown in Figure 9-1, the Comware-1 switch is using its routing table to route traffic for the network. You will now look at what the Comware-1 switch is doing in more detail.

Switches acting at Layer 2 make switching decisions based on destination MAC addresses. When a Layer 3 routing switch, such as Comware-1, receives IP traffic that is destined to its own MAC address, though, it looks at the destination IP address inside the Ethernet frame. If the destination IP address does not belong to it, the routing switch knows that it needs to route the traffic.

The routing switch uses the destination IP address to decide how to forward the traffic. It looks in its IP routing table and finds the route with the most specific match to the packet's destination IP address—in other words, the route with a matching destination network that has the longest subnet mask.

In the example network shown in Figure 9-1, the server needs to send a packet to the client. The server directs the frame to Comware-1's MAC address because Comware-1 is its default gateway. Comware-1 receives the frame and decapsulates it. Comware-1 then matches the destination IP address, 10.1.12.12, to IP route 10.1.12.0/24 in its routing table.

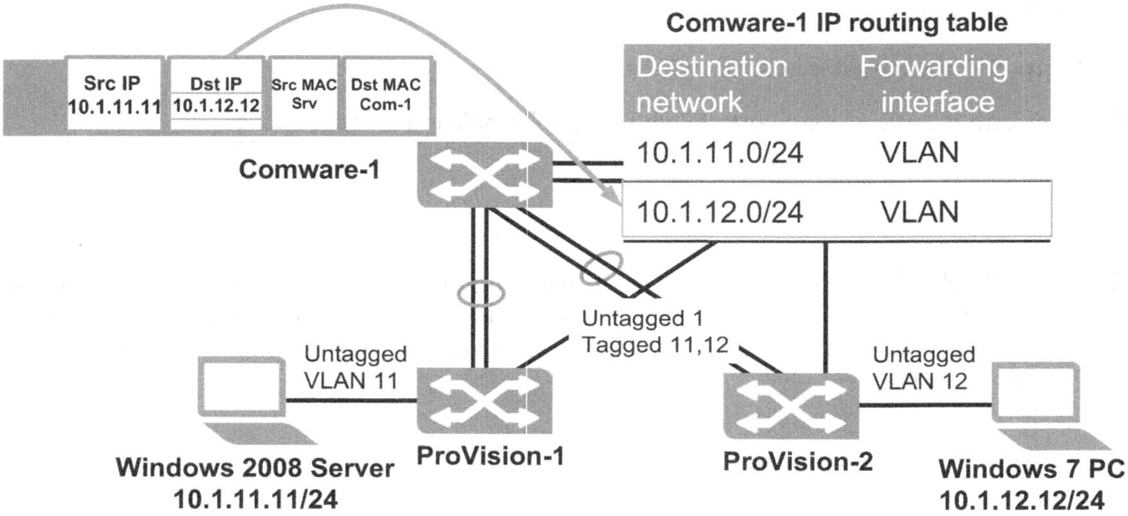


Figure 9-1: IP routes

HP ExpertOne If you want to view a video about IP routing, access the HP Aurasma application on your smartphone and use the application to focus on Figure 9-1. This action will launch the video.

## Direct IP routes

The Comware-1 routing table, shown in Figure 9-2, includes direct routes, which are sometimes called connected routes. Direct routes are for local networks on which the routing device has an IP address itself. Direct routes are associated with a Layer 3 forwarding interface such as a VLAN. Such direct routes are automatically added to the routing table when you configure the IP address on the device.

On ProVision switches, routing is disabled by default. If you want the switch to use its direct routes (or any other routes) to route traffic for other devices, you must enable IP routing.

- Routes to local networks
- No configuration necessary

**IP routing table**

| Comware-1 IP addresses | Destination network | Forwarding interface |
|---|---|---|
| 10.1.1.1/24 | 10.1.1.0/24 | VLAN 1 |
| 10.1.11.1/24 | 10.1.11.0/24 | VLAN11 |
| 10.1.12.1/24 | 10.1.12.0/24 | VLAN12 |

Note: Remember to enable IP routing on ProVision switches.

Figure 9-2: Direct IP routes

## Indirect IP routes—default

An indirect route, on the other hand, is a route to a remote network, which does not exist on the routing device. You must configure this route on the device manually, or the device must learn it dynamically with a routing protocol. As Figure 9-3 shows, an indirect route includes a destination network address and the next hop—the next device that will forward the packet to its final destination. This next hop is sometimes called the gateway.

The next hop IP address should be on a network connected to the routing switch. The routing switch automatically determines the forwarding interface based on which interface connects to the next hop, as shown in Figure 9-3.

- Routes to non-local networks
- Must be:
  - Configured manually (static)
  - Learned dynamically with a protocol

Default route:

- Special type of indirect route
- Applies to all traffic without a specific route

| Destination network | Next hop | Forwarding interface |
|---|---|---|
| 0.0.0.0/0 | 10.100.1.1 | VLAN 1000 |
| 10.1.1.0/24 | | VLAN 1 |
| 10.1.11.0/24 | | VLAN 11 |
| 10.1.12.0/24 | | VLAN 12 |
| 10.1.100.0/30 | | VLAN 1000 |

**Indirect route, default** → 0.0.0.0/0

Figure 9-3: Indirect IP routes—default

You have already created one special type of indirect route: a default route. As you recall (and as shown in Figure 9-3), a default route is a route to 0.0.0.0/0.The default route applies to all IP traffic for which the device does not have a more specific route. For example, Comware-1 might have a default route to an external router. It uses its direct routes for all the local traffic because these routes are more specific. It only uses the default route when it must route traffic for which it does not know another route.

## Topology that requires indirect routes

In a small, simple network topology, direct routes with a few default routes work well. For example, in a simple network topology, all the VLANs might use the same default router, which routes traffic between its connected networks.

The network topology shown in Figure 9-4, however, is somewhat more complex. There is a Layer 3 hop between each switch, and each switch acts as a routing switch. This network topology requires indirect routes. For example, ProVision-1 in Figure 9-4 routes traffic for VLAN 11, subnet 10.1.11.0/24. Subnet 10.1.12.0/24 is not local on this switch, so ProVision-1 requires an indirect (remote) route to it.

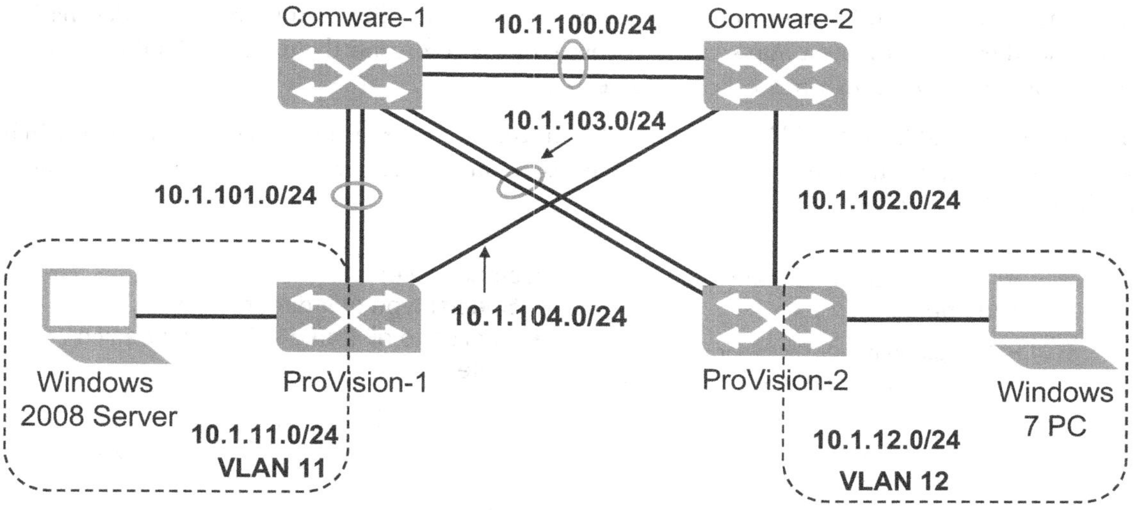


Figure 9-4: Topology that requires indirect routes

HP ExpertOne If you want to view a video about more advanced IP routing concepts, access the HP Aurasma application on your smartphone and use the application to focus on Figure 9-4. This action will launch the video.

In the network shown in Figure 9-4, each switch-to-switch link is associated with a different IP subnet. You have two options for making this association:

- Dedicated VLAN interface—You can create a VLAN that is specific to the switch-to-switch link—for example, VLAN 102 for the link between ProVision-2 and Comware-2. On both sides of the link, you assign *only* this VLAN to the switch port. You do *not* assign any other ports to this VLAN. You then assign an IP address to this VLAN, making it a Layer 3 interface. You can use this option on Comware and ProVision switches, as you can see in Figure 9-5. This is the only option for ProVision switches.

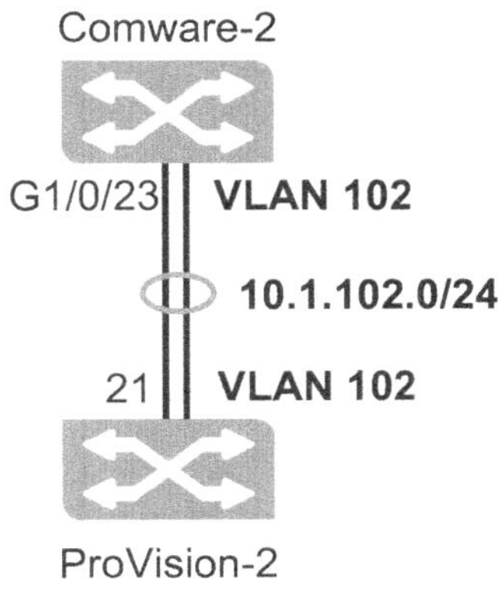


Figure 9-5: Using dedicated VLANs

- Route mode interface—On a Comware switch, you can change the physical interface to operate in route mode. The physical interface itself becomes the Layer 3 interface, and you assign the IP address directly to it. For example, on the network shown in Figure 9-5, you could configure Comware-2's g1/023 port in route mode and assign IP address 10.1.102.2/24 to it. A route mode interface acts like an interface on a router. It does not switch traffic; instead, it routes traffic to other Layer 3 interfaces. A route mode does not transmit or process any Rapid Spanning Tree Protocol (RSTP) or Multiple Spanning Tree Protocol (MSTP), Bridge Protocol Data Units (BPDUs), and it terminates any spanning tree. It has a unique association with the subnet configured on it. Just as you cannot configure an IP address in the same subnet on more than one VLAN interface on a switch, you cannot configure an IP address in the same subnet on more than one route mode interface. You can use the route mode interface on the Comware side of a link and a VLAN on the ProVision side, as illustrated in Figure 9-6. Just make sure that the ProVision switch assigns the VLAN as untagged because the Comware route mode port does not accept VLAN tags.

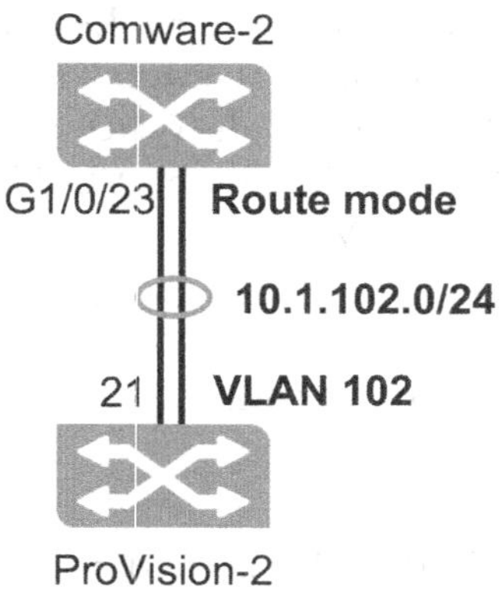


Figure 9-6: Using a Comware route mode interface

No matter how you set up the topology, you might choose to use a topology like this for several reasons:

- You want to eliminate RSTP/MSTP on the switch-to-switch links. In this new topology, VLANs assigned to edge ports are terminated at each switch and not carried on to the switch-to-switch links. Therefore, Layer 2 loops cannot be introduced. You have seen how MSTP provides some load-sharing of traffic. But using routing to determine traffic's path instead can result in more optimal paths.

**Note**
This topology eliminates the need for spanning tree on the switch-to-switch links. But you would still run spanning tree on the switches to prevent accidental loops downstream.

- You want to configure the default router role on different switches. In the network shown in Figure 9-4, ProVision-1 is a default router for devices connected to it, and ProVision-2 is a default router for devices connected to it. The burden is distributed, and the negative effect of one switch failing is minimized.

## Example configuration: Set up a network topology to explore IP routing

This section sets up a network topology that will be used later in this chapter to illustrate IP routing. It also explores additional concepts related to MSTP.

### Configure the VLAN and IP topology on Comware-1

The example network is using the topology shown in Figure 9-7, which was established in Chapter 8. See "Example configuration: Set up an LACP aggregated link" on page 272.

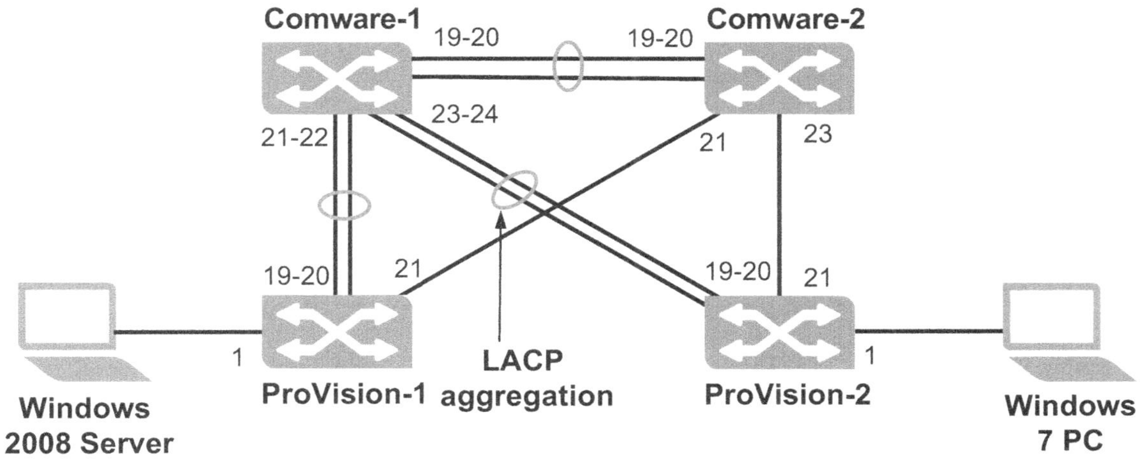


Figure 9-7: Configure an LACP aggregated link

In the new VLAN and IP topology, shown in Table 9-1, Comware-1 assigns a different VLAN and subnet for each connection to another switch. Table 9-1 indicates the VLAN and IP address for each connection. The instructions below outline how to make these changes to the example network.

Table 9-1: Comware-1 VLANs and IP addresses

| Connection to | VLAN | IP address |
|---|---|---|
| Comware-2 | 100 | 10.1.100.1/24 |
| ProVision-1 | 101 | 10.1.101.1/24 |
| ProVision-2 | 103 | 10.1.103.1/24 |

1. Access the Comware-1 CLI and move to system view.
2. Remove VLANs 11 and 12.

```
[Comware-1] undo vlan 11
[Comware-1] undo vlan 12
```

3. Create three VLANs: 100, 101, and 103.

```
[Comware-1] vlan 100
[Comware-1-vlan100] vlan 101
[Comware-1-vlan101] vlan 103
```

4. Change each BAGG interface to an access port and assign the interface to the VLAN indicated in Table 9-1.

```
[Comware-1-vlan103] interface Bridge-Aggregation 1
[Comware-1-Bridge-Aggregation1] port link-type access
[Comware-1-Bridge-Aggregation1] port access vlan 100
[Comware-1-Bridge-Aggregation1] interface Bridge-Aggregation 2
[Comware-1-Bridge-Aggregation2] port link-type access
[Comware-1-Bridge-Aggregation2] port access vlan 101
[Comware-1-Bridge-Aggregation2] interface Bridge-Aggregation 3
[Comware-1-Bridge-Aggregation3] port link-type access
[Comware-1-Bridge-Aggregation3] port access vlan 103
[Comware-1-Bridge-Aggregation3] quit
```

5. Verify the VLAN configuration.

```
[Comware-1] display vlan 100
 VLAN ID: 100
 VLAN Type: static
 Route Interface: not configured
 Description: VLAN 0100
 Name: VLAN 0100
 Tagged   Ports: none
Untagged Ports:
    Bridge-Aggregation1
    GigabitEthernet1/0/19    GigabitEthernet1/0/20
[Comware-1] display vlan 101
 VLAN ID: 101
 VLAN Type: static
 Route Interface: not configured
 Description: VLAN 0101
```

```
 Name: VLAN 0101
 Tagged   Ports: none
Untagged Ports:
    Bridge-Aggregation2
    GigabitEthernet1/0/21    GigabitEthernet1/0/22
[Comware-1] display vlan 103
 VLAN ID: 103
 VLAN Type: static
 Route Interface: not configured
 Description: VLAN 0103
 Name: VLAN 0103
 Tagged   Ports: none
Untagged Ports:
    Bridge-Aggregation3
    GigabitEthernet1/0/23    GigabitEthernet1/0/24
```

6. Create a Layer 3 interface for each VLAN and assign it the IP address indicated in Table 9-1.

```
[Comware-1] interface vlan-interface 100
[Comware-1-Vlan-interface100] ip address 10.1.100.1 24
[Comware-1-Vlan-interface100] quit

[Comware-1] interface vlan-interface 101
[Comware-1-Vlan-interface101] ip address 10.1.101.1 24
[Comware-1-Vlan-interface101] quit

[Comware-1] interface vlan-interface 103
[Comware-1-Vlan-interface103] ip address 10.1.103.1 24
[Comware-1-Vlan-interface103] quit
```

7. Verify the configuration. Note that Comware-1 still has the VLAN 1 IP address, but that VLAN is down. Verify that the three new IP interfaces are up.

```
[Comware-1] display interface brief
```

```
The brief information of interface(s) under route mode:
Interface  Link  Protocol-link  Protocol type  Main IP
NULL0      UP    UP(spoofing)   NULL           --
Vlan1      DOWN  DOWN           ETHERNET       10.1.1.1
Vlan100    UP    UP             ETHERNET       10.1.100.1
Vlan101    UP    UP             ETHERNET       10.1.101.1
Vlan103    UP    UP             ETHERNET       10.1.103.1
<-output omitted->
```

## Configure the VLAN and IP topology on Comware-2

Comware-2 also assigns a different VLAN and subnet for each connection to another switch. You will begin by making the connections just to Comware-1 and ProVision-1. Table 9-2 indicates the VLAN and IP address for each connection.

Table 9-2: Comware-2 VLANs and IP addresses

| Connection to | VLAN | IP address |
|---|---|---|
| Comware-1 | 100 | 10.1.100.2/24 |
| ProVision-1 | 104 | 10.1.104.2/24 |

8. Access a terminal session with Comware-2 and move to system view.
9. Remove VLANs 11 and 12.

```
[Comware-2] undo vlan 11
[Comware-2] undo vlan 12
```

10. Create two VLANs: 100 and 104.

```
[Comware-2] vlan 100
[Comware-2-vlan100] vlan 104
[Comware-2-vlan104] quit
```

11. Change the connections to Comware-1 and ProVision-1 to access ports. Assign the VLANs indicated in Table 9-2.

```
[Comware-2] interface Bridge-Aggregation 1
[Comware-2-Bridge-Aggregation1] port link-type access
```

```
[Comware-2-Bridge-Aggregation1] port access vlan 100
[Comware-2-Bridge-Aggregation1] quit
[Comware-2] interface g1/0/21
[Comware-2-GigabitEthernet1/0/21] port link-type access
[Comware-2-GigabitEthernet1/0/21] port access vlan 104
[Comware-2-GigabitEthernet1/0/21] quit
```

12. Verify your VLAN configuration.

```
[Comware-2] display vlan 100
 VLAN ID: 100
 VLAN Type: static
 Route Interface: not configured
 Description: VLAN 0100
 Name: VLAN 0100
 Tagged   Ports: none
 Untagged Ports:
    Bridge-Aggregation1
    GigabitEthernet1/0/19     GigabitEthernet1/0/20
[Comware-2] display vlan 104
 VLAN ID: 104
 VLAN Type: static
 Route Interface: not configured
 Description: VLAN 0104
 Name: VLAN 0104
 Tagged   Ports: none
 Untagged Ports:
    GigabitEthernet1/0/21
```

13. Create a Layer 3 interface for each VLAN and assign it the IP address indicated in Table 9-2.

```
[Comware-2] interface vlan-interface 100
[Comware-2-Vlan-interface100] ip address 10.1.100.2 24
[Comware-2-Vlan-interface100] quit
[Comware-2] interface vlan-interface 104
[Comware-2-Vlan-interface104] ip address 10.1.104.2 24
[Comware-2-Vlan-interface104] quit
```

14. Verify the configuration.

```
[Comware-2] display interface brief
The brief information of interface(s) under route mode:
Interface   Link   Protocol-link   Protocol type Main IP
NULL0       UP     UP(spoofing)    NULL           --
Vlan1       DOWN   DOWN             ETHERNET       10.1.1.2
Vlan100     UP     UP               ETHERNET       10.1.100.2
Vlan104     UP     UP               ETHERNET       10.1.104.2
<-output omitted->
```

15. You have set up VLAN 100, subnet 10.1.100.0/24, on both sides of the link. You should be able to ping Comware-1 from Comware-2.

```
[Comware-2] ping 10.1.100.1
```

## Configure the VLAN and IP topology on ProVision-1

ProVision-1 also assigns a different VLAN and subnet for each connection to another switch. Table 9-3 indicates the VLAN and IP address for each connection. You will now make these changes.

Table 9-3: ProVision-1 VLANs and IP addresses

| Connection to | VLAN | IP address |
|---|---|---|
| Comware-1 | 101 | 10.1.101.3/24 |
| Comware-2 | 104 | 10.1.104.3/24 |
| Servers | 11 | 10.1.11.1/24 |

16. Access a terminal session with ProVision-1. Move to the global configuration context.
17. Remove VLAN 12.

```
ProVision-1(config)# no vlan 12
```

18. Remove VLAN 11 from the connections to Comware-1 and Comware-2.

```
ProVision-1(config)# no vlan 11 tagged trk1,21
```

19. Create the two new VLANs. Assign each to the proper connection as an untagged VLAN.

```
ProVision-1(config)# vlan 101
ProVision-1(vlan-101)# untagged trk1
ProVision-1(vlan-101)# exit
ProVision-1(config)# vlan 104
ProVision-1(vlan-104)# untagged 21
ProVision-1(vlan-104)# exit
```

20. Verify your VLAN configuration.

```
ProVision-1(config)# show vlan 101
 Status and Counters - VLAN Information - VLAN 101
  VLAN ID : 101
  Name : VLAN101
  Status : Port-based
  Voice : No
  Jumbo : No
  Port Information Mode     Unknown VLAN Status
  ---------------- -------- ------------ ----------
  Trk1             Untagged Learn        Up
ProVision-1(config)# show vlan 104
 Status and Counters - VLAN Information - VLAN 104
  VLAN ID : 104
  Name : VLAN104
  Status : Port-based
```

```
  Voice : No
  Jumbo : No
  Port Information Mode     Unknown VLAN Status
  ---------------- -------- ------------ ----------
  21               Untagged Learn        Up
ProVision-1(config)# show vlan 11
 Status and Counters - VLAN Information - VLAN 11
  VLAN ID : 11
  Name : VLAN11
  Status : Port-based
  Voice : No
  Jumbo : No
  Port Information Mode     Unknown VLAN Status
  ---------------- -------- ------------ ----------
  1                Untagged Learn        Up
```

21. Assign each VLAN the IP addresses indicated in Table 9-3.

```
ProVision-1(config)# vlan 101 ip address 10.1.101.3/24
ProVision-1(config)# vlan 104 ip address 10.1.104.3/24
ProVision-1(config)# vlan 11 ip address 10.1.11.1/24
```

22. Verify the configuration.

```
ProVision-1(config)# show ip
 Internet (IP) Service
  IP Routing : Disabled
  Default Gateway :
  Default TTL     : 64
  Arp Age         : 20
  Domain Suffix   :
  DNS server      :
```

```
             |                                        Proxy ARP
VLAN         | IP Config IP Address Subnet Mask   Std  Local
------------ + --------- ---------- ------------- ---- -----
DEFAULT_VLAN | Manual    10.1.1.3   255.255.255.0 No   No
VLAN11       | Manual    10.1.11.1  255.255.255.0 No   No
VLAN101      | Manual    10.1.101.3 255.255.255.0 No   No
VLAN104      | Manual    10.1.104.3 255.255.255.0 No   No
```

23. You have set up VLAN 101 (subnet 10.1.100.0/24) and VLAN 104 (subnet 10.1.104.0/24) on both sides of the link. Ping Comware-1 and Comware-2 from ProVision-1.

```
ProVision-1(config)# ping 10.1.101.1
10.1.101.1 is alive, time = 3 ms
ProVision-1(config)# ping 10.1.104.2
The destination address is unreachable
```

24. View the IST roles for the connections to Comware-1 and Comware-2. (The new VLANs are part of the IST.)

```
ProVision-1(config)# show spanning-tree trk1,21 instance ist
<-output omitted->
                                                       Designated
Port Type      Cost  Priority Role      State      Bridge
---- --------- ----- -------- --------- ---------- -------------
21   100/1000T 20000 128      Alternate Blocking   d07e28-cec94f
Trk1           20000  64      Root      Forwarding d07e28-cec94f
```

## Fix the issue

As you see from the output above, MSTP blocks the link to Comware-2. Every link in the new topology has its own VLAN, so loops will not form within VLANs. But MSTP builds the topology based on link regardless of the VLAN. You need to disable spanning tree on the links between ProVision-1, Comware-1, and Comware-2 to allow routing over all links. However, you want to keep spanning tree enabled to prevent against accidental loops. You will disable spanning tree specifically on the switch-to-switch ports.

25. On ProVision-1, implement BPDU filters to disable spanning-tree on trk1 and on the interface that connects to Comware-2.

```
ProVision-1(config)# spanning-tree trk1,21 bpdu-filter
```

**Note**

You will see a warning because the BPDU filter could cause loops to occur. You already checked your topology to ensure that you will not introduce a loop.

You can similarly filter BPDUs on Comware switches by disabling STP on a specific port or BAGG interface.

26. Access the Comware-1 CLI.

27. Disable spanning tree on BAGG interface 1. Again, you see warnings.

```
[Comware-1] interface Bridge-Aggregation 1
[Comware-1-Bridge-Aggregation1] stp disable
```

For best practices you would continue to filter BPDUs on all interfaces that are dedicated to one VLAN.

28. Return to the terminal session with ProVision-1. Attempt to ping Comware-2 again. The ping should now be successful.

```
ProVision-1(config)# ping 10.1.104.2
10.1.104.2 is alive, time = 3 ms
```

You will now look at another way to resolve this issue: using route-mode ports on Comware switches. Begin by removing the BPDU filters on ProVision-1 and Comware-1 and verifying that ProVision-1 can no longer reach Comware-2 on its direct link.

29. Remove the BPDU filter on ProVision-1.

```
ProVision-1(config)# no spanning-tree trk1,21 bpdu-filter
```

30. Access the Comware-1 CLI. Remove the BPDU filter on Comware-1.

```
[Comware-1-Bridge-Aggregation1] stp enable
[Comware-1-Bridge-Aggregation1] quit
```

31. Verify that ProVision-1 can no longer ping Comware-2.

```
ProVision-1(config)# ping 10.1.104.2
The destination address is unreachable
```

You will now configure the route-mode port on Comware-2. First you must delete VLAN 104. You will assign the IP address directly to the physical port instead of to the VLAN interface.

32. Access the terminal session with Comware-2 and make sure that you are in the system view.

33. Remove VLAN 104.

```
[Comware-2] undo vlan 104
```

34. Move to the port that connects to ProVision-1. Set the link mode to routed.

```
[Comware-2] interface g1/0/21

[Comware-2-GigabitEthernet1/0/21] port link-mode route
```

35. Assign the 10.1.104.2/24 IP address to the port.

```
[Comware-2-GigabitEthernet1/0/21] ip address 10.1.104.2 24

[Comware-2-GigabitEthernet1/0/21] quit
```

Comware-2 now terminates the spanning tree on its side of the link. Therefore, ProVision-1 will transition the link to designated (it treats the link as an edge port) and begin forwarding on it.

36. Verify that ProVision-1 can now ping Comware-2.

```
ProVision-1(config)# ping 10.1.104.2

10.1.104.2 is alive, time = 3 ms
```

Think about what you have learned about RSTP/MSTP and route mode ports on Comware switches. In the real world, you would probably want to configure all of the Comware links in this topology in route mode. You can even configure aggregated links in routed mode. You would simply make the physical ports operate in route mode, create a route-aggregation interface (RAGG) instead of a bridge-aggregation interface (BAGG), and assign the physical ports to the RAGG with the `port link-aggregation group` command. (For the HP 5800 Switch Series, support for RAGG interfaces began in Comware 5 software F1305P01.)

## Finish configuring the IP topology on Comware-2

You will now add Comware-2's connection to ProVision-2, as shown in Table 9-4.

Table 9-4: Comware-2 IP addresses

| Connection to | VLAN or physical interface | IP address |
|---|---|---|
| Comware-1 | VLAN 100 | 10.1.100.2/24 |
| ProVision-1 | G1/0/21 | 10.1.104.2/24 |
| ProVision-2 | G1/0/23 | 10.1.102.2/24 |

37. Access the terminal session with Comware-2 and make sure that you are in system view.
38. Move to the physical port that connects to ProVision-2, make the port a routed port, and assign it the proper IP address.

```
[Comware-2] interface g1/0/23
[Comware-2-GigabitEthernet1/0/23] port link-mode route
[Comware-2-GigabitEthernet1/0/23] ip address 10.1.102.2 24
[Comware-2-GigabitEthernet1/0/23] quit
```

## Configure the VLAN and IP topology on ProVision-2

ProVision-2 also assigns a different VLAN and subnet to each connection to another switch. Table 9-5 indicates the VLAN and IP address for each connection. You will now make these changes.

Table 9-5: ProVision-2 VLANs and IP addresses

| Connection to | VLAN | IP address |
|---|---|---|
| Comware-1 | 103 | 10.1.103.4/24 |
| Comware-2 | 102 | 10.1.102.4/24 |
| Clients | 12 | 10.1.12.1/24 |

39. Access a terminal session with ProVision-2. Move to the global configuration mode.
40. Remove VLAN 11.

```
ProVision-2(config)# no vlan 11
```

41. Remove VLAN 12 from the connections to Comware-1 and Comware-2.

```
ProVision-2(config)# no vlan 12 tagged trk1,21
```

42. Create the two new VLANs. Assign each to the proper connection as an untagged VLAN.

```
ProVision-2(config)# vlan 102
ProVision-2(vlan-102)# untagged 21
ProVision-2(vlan-102)# exit
ProVision-2(config)# vlan 103
ProVision-2(vlan-103)# untagged trk1
ProVision-2(vlan-102)# exit
```

43. Verify your VLAN configuration.

```
ProVision-2(config)# show vlan 102

 Status and Counters - VLAN Information - VLAN 102

  VLAN ID : 102
  Name : VLAN102
  Status : Port-based
  Voice : No
  Jumbo : No

  Port Information Mode     Unknown VLAN Status
  ---------------- -------- ------------ ----------
  21               Untagged Learn        Up

ProVision-2(config)# show vlan 103

 Status and Counters - VLAN Information - VLAN 103

  VLAN ID : 103
  Name : VLAN103
  Status : Port-based
  Voice : No
  Jumbo : No

  Port Information Mode     Unknown VLAN Status
  ---------------- -------- ------------ ----------
  Trk1             Untagged Learn        Up
ProVision-2(config)# show vlan 12

 Status and Counters - VLAN Information - VLAN 12

  VLAN ID : 12
  Name : VLAN12
  Status : Port-based
  Voice : No
  Jumbo : No
```

```
  Port Information Mode     Unknown VLAN Status
  ---------------- -------- ------------ ----------
  1                Untagged Learn        Up
```

44. Assign each VLAN the IP addresses indicated in Table L9.1-5.

```
ProVision-2(config)# vlan 102 ip address 10.1.102.4/24
ProVision-2(config)# vlan 103 ip address 10.1.103.4/24
ProVision-2(config)# vlan 12 ip address 10.1.12.1/24
```

45. Verify the configuration.

```
ProVision-2(config)# show ip
 Internet (IP) Service
  IP Routing : Disabled
<-output omitted->
                |                                                  Proxy ARP
  VLAN          | IP Config  IP Address  Subnet Mask     Std  Local
  ------------ + ---------- ----------- -------------- ----------
  DEFAULT_VLAN | Manual      10.1.1.4    255.255.255.0  No    No
  VLAN12       | Manual      10.1.12.1   255.255.255.0  No    No
  VLAN102      | Manual      10.1.102.4  255.255.255.0  No    No
  VLAN103      | Manual      10.1.103.4  255.255.255.0  No    No
```

46. You have set up VLAN 103 (subnet 10.1.103.0/24) and VLAN 102 (subnet 10.1.102.0/24) on both sides of the link. You should be able to ping Comware-1 and Comware-2 from ProVision-2.

```
ProVision-2(config)# ping 10.1.103.1
10.1.103.1 is alive, time = 3 ms
ProVision-2(config)# ping 10.1.102.2
10.1.102.2 is alive, time = 3 ms
```

## Configure a static IP address on the client

When you changed the logical topology, you disrupted connectivity between the Windows client, its default gateway, and the DHCP server in the example network. This section outlines how to assign the client a static IP address.

47. Access the Windows 7 PC desktop.
48. Select **TCP/IP version 4** and click **Properties**.
49. Change the client's IP address to 10.1.12.12/24. Set the default gateway address to 10.1.12.1.
50. After you have set the address, remember to click **OK** in the Properties window and then **Close**.
51. Open a command prompt. Validate that you have set the IP address correctly.

```
ipconfig
```

---

**Note**

If the client is not using the new IP address, make sure that you clicked **Close** in the Properties window.

---

52. Ping the default gateway and verify success.

```
ping 10.1.12.1
```

## Learning check 1

Answer the following questions to assess what you learned about STP and IP routing in the example configuration.

1. How were you able to disable spanning tree on switch-to-switch links (implement BPDU filters) without introducing broadcast storms? Why did you need to disable spanning tree on these links in your initial setup?

2. Later you configured Comware-2's physical interfaces in route mode. Why were the BPDU filters no longer necessary then?

3. Which IP addresses can ProVision-1 ping? Which IP addresses can it not ping, and why not?

## Answers to learning check 1

1. How were you able to disable spanning tree on switch-to-switch links (implement BPDU filters) without introducing broadcast storms? Why did you need to disable spanning tree on these links in your initial setup?

   In the new topology, VLANs are isolated on one switch only. Because the broadcast domain does not extend over the redundant switch-to-switch links, broadcast storms cannot occur.

You needed to disable spanning tree on these links (filter BPDUs) so that the redundant switch-to-switch links are available for routing. Otherwise, MSTP puts some of the interfaces in a blocking state.

2. Later you configured Comware-2's physical interfaces in route mode. Why were the BPDU filters no longer necessary then?

   A route-mode interface is a Layer 3 interface. It terminates the spanning tree domain, so it does not send BPDUs and does not process them. In the topology, terminating the spanning tree on the Comware-2 connections is enough to eliminate loops. In other topologies you might still need to implement BPDU filters on ProVision switches or set Comware-1 interfaces to route mode.

3. Which IP addresses can ProVision-1 ping? Which IP addresses can it not ping, and why not?

   ProVision-1 can ping IP addresses on the networks that are directly connected to it:

   - Comware-1 at 10.1.101.1
   - Comware-2 at 10.1.104.2
   - The server at 10.1.11.11

   ProVision-1 cannot ping any other IP addresses, including other IP addresses on Comware-1 and Comware-2.

## Indirect IP routes

Figure 9-8 shows a simplified network with just ProVision-1, ProVision-2, and Comware-1. This simplified network does not include any redundant connections through Comware-2. You will now learn how to set up routes in this topology.

| | Destination network | Next hop |
|---|---|---|
| ProVision-1 | 10.1.12.0/24 | 10.1.101.1 |
| Comware-1 | 10.1.12.0/24 | 10.1.103.4 |

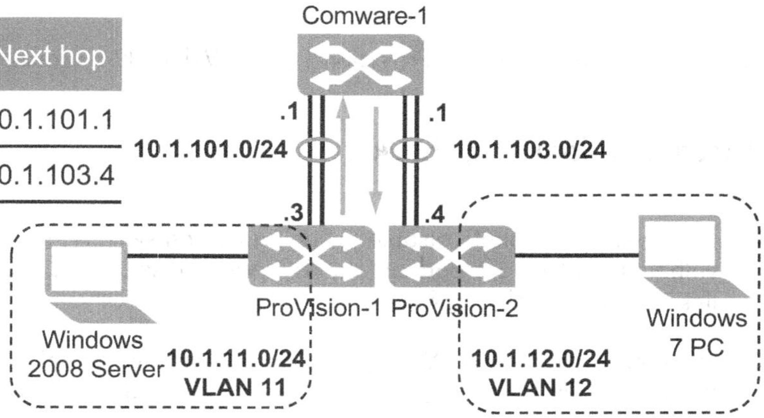


Figure 9-8: Indirect IP routes

You need to create specific indirect routes (as opposed to just direct and default routes). For example, you might create a route to 10.1.12.0/24 on ProVision-1 so that this switch can route server traffic to clients.

In addition to the destination network address, the route specifies the next hop. It specifies *only* the next hop; as far as the local switch's routing table is concerned, the rest of the path does not exist.

The next hop for ProVision-1's route to 10.1.12.0/24 is Comware-1. Comware-1 has several IP addresses. For the next hop address, you should specify the IP address on the subnet that connects the two switches: 10.1.101.1. If you specified a different IP address on Comware-1, ProVision-1 would not know how to reach that address. A route with an unreachable next hop is not added to the active IP routing table, and the switch cannot use it.

To finish the path to 10.1.12.0/24, you must also create a route to 10.1.12.0/24 on Comware-1. Again you specify the next hop in the path. This next hop is ProVision-2 and, specifically, IP address 10.1.103.1 on ProVision-2.

Now ProVision-1 can route traffic to 10.1.12.0/24 successfully.

Of course, you would need to create a return route for traffic to enable full connectivity.

For example, you want to set up connectivity between servers in subnet 10.1.11.0/24 and clients in subnet 10.1.12.0/24. You have already configured a route to 10.1.12.0/24 on ProVision-1 and Comware-1. As Figure 9-9 shows, you would need to configure these routes for the return traffic:

- ProVision-2: A route to 10.1.11.0/24 through next hop 10.1.103.1.
- Comware-1: A route to 10.1.11.0/24 through next hop 10.1.104.1.

| | Destination network | Next hop |
|---|---|---|
| ProVision-2 | 10.1.11.0/24 | 10.1.103.1 |
| Comware-1 | 10.1.11.0/24 | 10.1.101.3 |

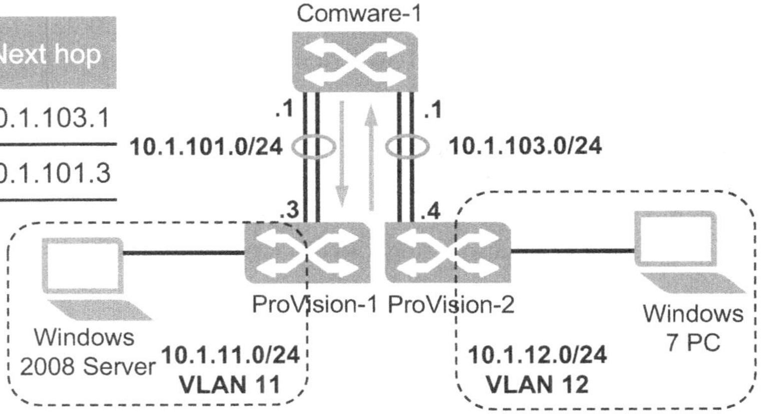


Figure 9-9: Indirect routes for return traffic to 10.1.11.0/24

## Static IP routing

Switches can learn indirect routes statically or dynamically through a routing protocol. For now, you are focusing on static routes, which are configured manually by the administrator. When you use static routing, you must configure the routes manually on each router or routing switch in the path. You must take care to specify the next hops correctly, and you must take care to create all the necessary paths.

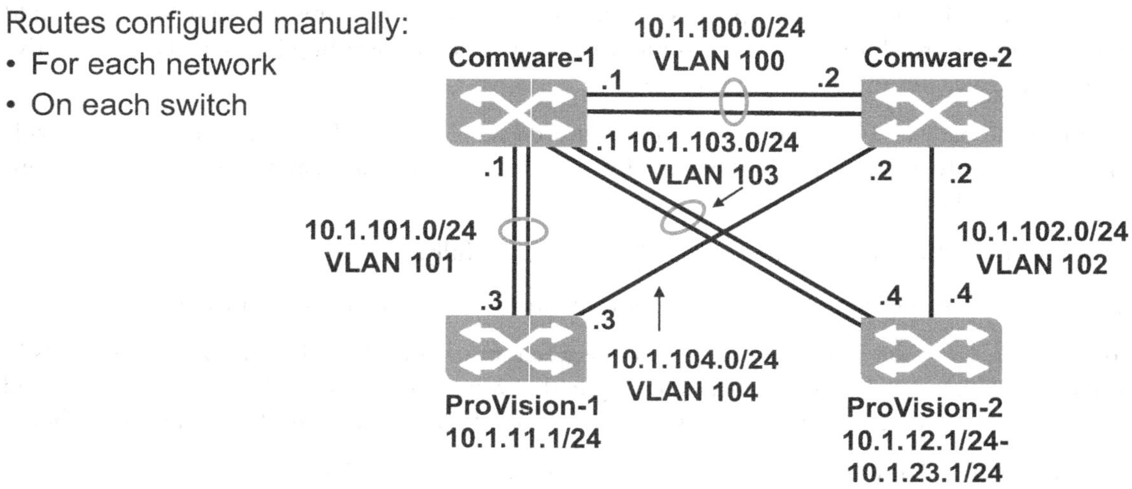


Figure 9-10: Static IP routing

For example, the routes that you examined previously for the example network shown in Figure 9-10 establish connectivity between 10.1.11.0/24 and 10.1.12.0/24. However, ProVision-1 in the example network shown in Figure 9-10 still cannot route traffic to the 10.1.102.0/24 subnet on ProVision-2. You might not care about routing traffic to a subnet devoted to a switch-to-switch link.

But what if ProVision-2 supported several other subnets with devices in them? For example, it might support 10.1.13.0/24, 10.1.14.0/24, and so on, up to 10.1.23.1/24. You would need to create a route for each on each routing device in the pathway. (Or, if subnets are contiguous, you could create a route to a larger destination subnet that includes the other subnets within it.)

As you can see, the process can be laborious.

As a final note, the network has redundancy, but it is not currently taking advantage of the redundant links.

## Example configuration: Create static IP routes

This section builds on the example network established in "Example configuration: Set up a network topology to explore IP routing" on page 294. In this section, you will learn how to set up static routes on ProVision-1, ProVision-2, and Comware-1 to enable connectivity between the server in 10.1.11.0/24 and the PC in 10.1.12.0/24, as shown in Figure 9-11.

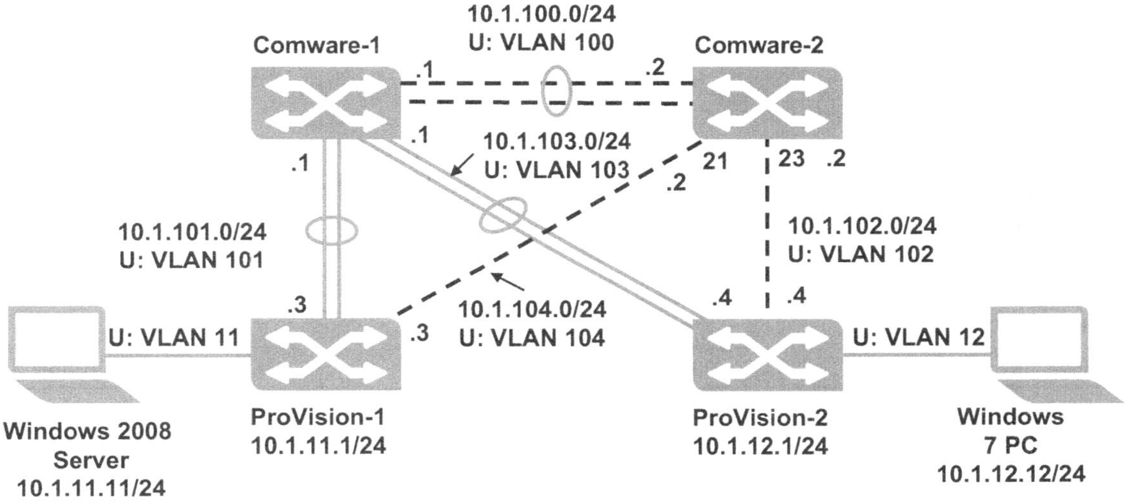


Figure 9-11: Example configuration: Create static IP routes

You will use default routes on the ProVision switches, which are using Comware-1 as their single next hop for all other destinations. You will create routes to specific destination networks on Comware-1, which needs to route some traffic to ProVision-1 and some traffic to ProVision-2.

## Configure the static routes

Figure 9-12 shows only the links for the routes, and Table 9-6 lists the routes required to establish the path between the server and Windows PC.

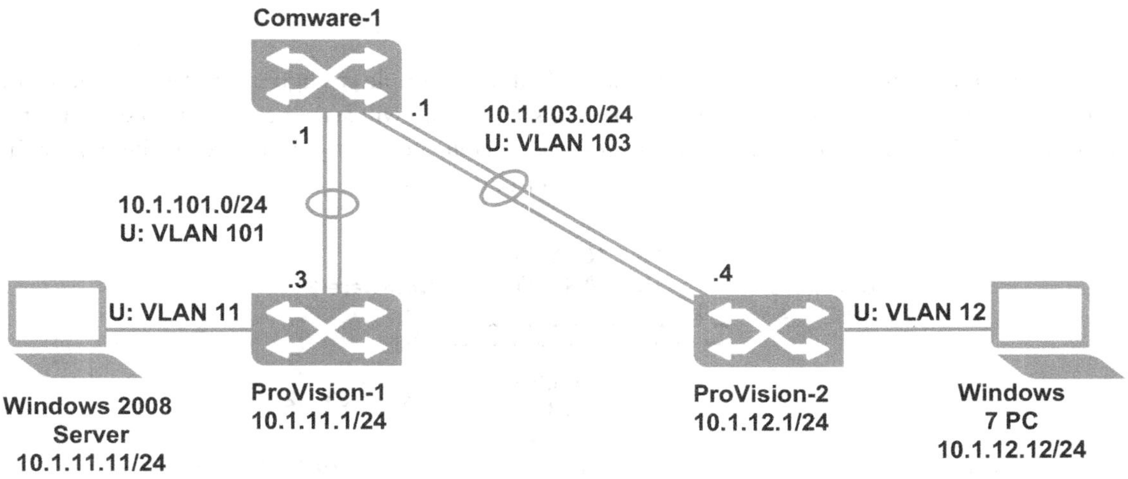


Figure 9-12: Network topology showing only the links for the route

Table 9-6: Static routes

| Switch | Destination network | Next hop |
|---|---|---|
| ProVision-1 | 10.1.12.0/24 or 0.0.0.0/0 | 10.1.101.1 |
| Comware-1 | 10.1.12.0/24 | 10.1.103.4 |
| | 10.1.11.0/24 | 10.1.101.3 |
| ProVision-2 | 10.1.11.0/24 or 0.0.0.0/0 | 10.1.103.1 |

1. Access the ProVision-1 CLI and move to the global configuration mode context.
2. Enable IP routing.

```
ProVision-1(config)# ip routing
```

3. Create a default route through Comware-1 at 10.1.101.1.

```
ProVision-1(config)# ip route 0.0.0.0/0 10.1.101.1
```

4. View the IP routing table.

```
ProVision-1(config)# show ip route

                        IP Route Entries

 Destination     Gateway    VLAN  Type       Sub-Type Metric  Dist.
 -------------- -------- ---- --------- --------- ------- -----
 0.0.0.0/0      10.1.101.1 101  static              1       1
 10.1.11.0/24   VLAN11       11  connected           1       0
 10.1.101.0/24  VLAN101     101  connected           1       0
```

```
10.1.104.0/24 VLAN104     104 connected          1       0
127.0.0.0/8   reject          static             0       0
127.0.0.1/32  lo0             connected          1       0
```

## Windows server

In this section, you will use trace routes to learn more about the topology. A trace route lists each routing hop between the device that executes the trace route and the destination, as shown in Figure 9-13. (Intervening devices at Layer 2 do not show up in the trace route.)

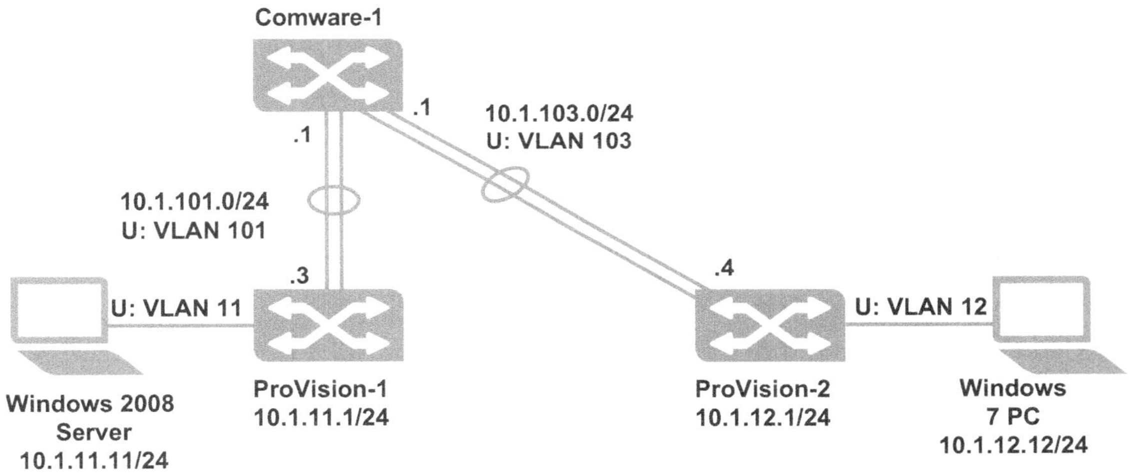


Figure 9-13: Network topology showing only the links for the route

5. Access the Windows 2008 server desktop and open a command prompt.
6. Execute a trace route to the Windows 7 PC (10.1.12.12).

```
tracert -d 10.1.12.12

Tracing route to 10.1.12.12 over a maximum of 30 hops

  1    <1 ms    <1 ms     1 ms  10.1.11.1
  2     *        *        *     Request timed out.
  3     *        *        *     Request timed out.
```

7. Notice that the trace fails after one hop. Press **<Ctrl+C>** to end it.
8. Why did the routing fail?

   ProVision-1 routes the ping to Comware-1, but Comware-1 does not yet know how to reach the destination.

9. Access the Comware-1 CLI and move to system view.
10. Examine the routing table to validate that Comware-1 has no route to 10.1.12.0/24.

```
[Comware-1] display ip routing-table

Routing Tables: Public

      Destinations : 8        Routes : 8

Destination/Mask  Proto   Pre  Cost NextHop      Interface

10.1.100.0/24     Direct  0    0    10.1.100.1   Vlan100

10.1.100.1/32     Direct  0    0    127.0.0.1    InLoop0

10.1.101.0/24     Direct  0    0    10.1.101.1   Vlan101

10.1.101.1/32     Direct  0    0    127.0.0.1    InLoop0

10.1.103.0/24     Direct  0    0    10.1.103.1   Vlan103

10.1.103.1/32     Direct  0    0    127.0.0.1    InLoop0

127.0.0.0/8       Direct  0    0    127.0.0.1    InLoop0

127.0.0.1/32      Direct  0    0    127.0.0.1    InLoop0
```

11. Create the static route to 10.1.12.0/24 through ProVision-2 at 10.1.103.4. (Refer to Figure 9-14 as needed.)

```
[Comware-1] ip route-static 10.1.12.0 24 10.1.103.4
```

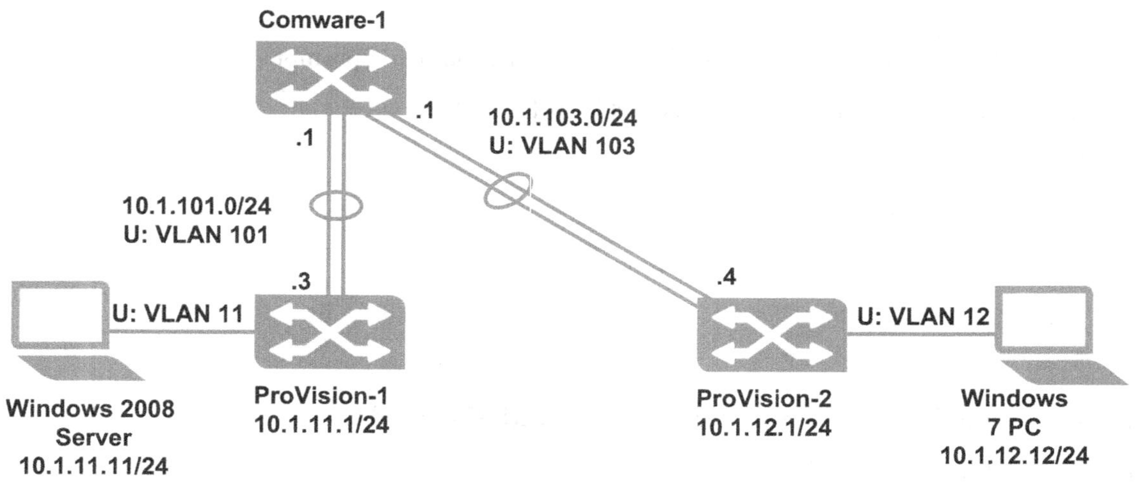


Figure 9-14: Network topology showing only the links for the route

12. Also create the static route for return traffic to 10.1.11.0/24. The next hop is ProVision-1 at 10.1.101.3.

```
[Comware-1] ip route-static 10.1.11.0 24 10.1.101.3
```

13. View the new routes in the IP routing table.

```
[Comware-1] display ip routing-table

Routing Tables: Public

        Destinations : 10       Routes : 10

Destination/Mask Proto   Pre  Cost NextHop      Interface

10.1.11.0/24     Static  60   0    10.1.101.3   Vlan101

10.1.12.0/24     Static  60   0    10.1.103.4   Vlan103

10.1.100.0/24    Direct  0    0    10.1.100.1   Vlan100

10.1.100.1/32    Direct  0    0    127.0.0.1    InLoop0

10.1.101.0/24    Direct  0    0    10.1.101.1   Vlan101

10.1.101.1/32    Direct  0    0    127.0.0.1    InLoop0

10.1.103.0/24    Direct  0    0    10.1.103.1   Vlan103

10.1.103.1/32    Direct  0    0    127.0.0.1    InLoop0

127.0.0.0/8      Direct  0    0    127.0.0.1    InLoop0

127.0.0.1/32      Direct 0    0    127.0.0.1    InLoop0
```

14. Access the Windows 2008 server and do the trace route again.

```
tracert -d 10.1.12.12

Tracing route to 10.1.12.12 over a maximum of 30 hops

  1    <1 ms    <1 ms    <1 ms  10.1.11.1

  2     1 ms     1 ms     1 ms  10.1.101.1

  3     *        *        *     Request timed out.

  4     *        *        *     Request timed out.
```

**Note**

If you do not see the 10.1.101.1 hop, make sure that Comware-1 has TTLs and ICMP unreachables enabled. Enter these commands on Comware-1 in system view: `ip ttl-expires enable` and `ip unreachables enable`.

15. Now after two hops, the trace route begins to time out. Press **<Ctrl+c>** to end it.

    Why does the trace route fail? The trace route ICMP reaches ProVision-2, but ProVision-2 does not know how to route return traffic.

16. Access the ProVision-2 CLI and move to the global configuration mode context.

17. Enable IP routing.

```
ProVision-2(config)# ip routing
```

18. Create a default route through Comware-1 at 10.1.103.1.

```
ProVision-2(config)# ip route 0.0.0.0/0 10.1.103.1
```

19. View the IP routing table

```
ProVision-2(config)# show ip route

                 IP Route Entries

Destination     Gateway      VLAN Type      Sub-Type  Metric Dist.
------------- ----------- ---- -------- --------- ------ -----
0.0.0.0/0      10.1.103.1  103  static              1      1
10.1.12.0/24   VLAN12      12   connected           1      0
10.1.102.0/24  VLAN102     102  connected           1      0
10.1.103.0/24  VLAN103     103  connected           1      0
127.0.0.0/8    reject           static              0      0
127.0.0.1/32   lo0              connected           1      0
```

20. Access the Windows 2008 server and start the trace route again. It should be successful.

```
tracert -d 10.1.12.12

Tracing route to 10.1.12.12 over a maximum of 30 hops

  1    <1 ms    <1 ms    <1 ms  10.1.11.1
  2     1 ms     1 ms     1 ms  10.1.101.1
  3    <1 ms    <1 ms    <1 ms  10.1.103.4
  4     4 ms    <1 ms    <1 ms  10.1.12.12

Trace complete.
```

**Note**

If you do not see the 10.1.103.4 hop, check the route on ProVision-2. If you do not see the 10.1.12.12 hop, check that the Windows client has the correct default gateway (use `ipconfig`).

Save the current configuration on each of the switches. For ProVision, use the `write memory` command. For Comware, use the `save [force]` command.

## Learning check 2

1. To enable a ping between the server and the client, on which switches did you need to configure static routes?

2. What would happen if ProVision-1 loses its connection to Comware-1? Will the server and PC still be able to reach each other?

### Learning check 2 answers

1. To enable a ping between the server and the client, on which switches did you need to configure static routes?

   You needed to configure a static route on ProVision-1, Comware-1, and ProVision-2. Every routing hop in the pathway requires a route, so both ProVision-1 and Comware-1 required a route to 10.1.12.0/24 to enable the server traffic to reach the client. Because the client return traffic also required routing, you had to set up routes to 10.1.11.0/24 on ProVision-2 and Comware-1.

2. What would happen if ProVision-1 loses its connection to Comware-1? Will the server and PC still be able to reach each other?

   No. When ProVision-1 loses its connection to Comware-1, its 10.1.101.0/24 subnet becomes unavailable. ProVision-1 can no longer reach the next hop in its static route, so the route is no longer available. Even though ProVision-1 still has a physical pathway to ProVision-2 through Comware-2, it cannot use this pathway because it has no route.

## Manage redundant static routes

ProVision-1 IP routing table

| Destination network | Next hop | Adm Dist | Adm Dist |
|---|---|---|---|
| 0.0.0.0/0 | 10.1.101.1 | 1 | 1 |
| 0.0.0.0/0 | 10.1.104.2 | 1 | 10 |

Load balance

Use second route as backup only

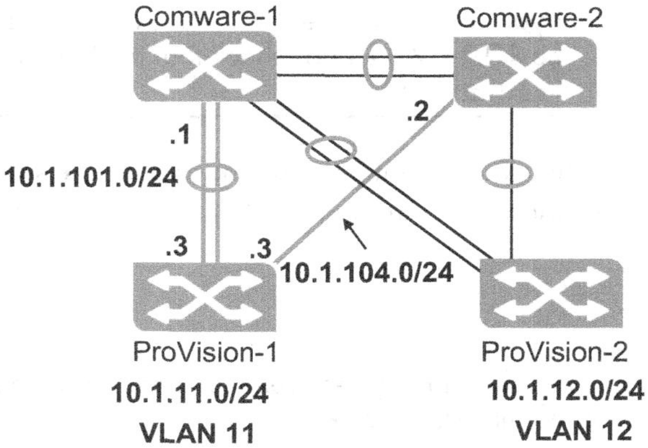


Figure 9-15: Manage redundant static routes

You can add to the resiliency of your network by setting up redundant routes. The redundant routes can also introduce load-sharing. For example, you can create a second default route on ProVision-1 through Comware-2 at IP address 10.1.104.2. (Of course, Comware-2 needs to have the correct routes so that it can route the traffic successfully.)

As you can see in Figure 9-15, ProVision-1 has two routes to the same destination network. It needs to decide whether it will add the first route, the second, or both to its routing table. Assuming that both routes have a valid next hop, the switch makes this decision based on these route properties:

- **Administrative distance**—Routing devices use administrative distance to distinguish between routes that are learned in different ways such as through different routing protocols. You will examine administrative distance in more detail when you learn about dynamic routing protocols. Note, though, that you can set a static route's administrative distance when you create the route. The default administrative distance for a static route is 1 on ProVision and 60 on Comware. Because you can set different administrative distances for different static routes, you can use administrative distance to distinguish between static routes. The route with the lowest administrative distance is preferred. Comware uses the terms *preference* or *precedence* for administrative distance.

- **Metric**—Routing devices use metrics to distinguish between routes that are learned in the same way such as statically or with the same routing protocol. Generally, these routes have the same administrative distance, so metric breaks the tie and permits the routing device to choose a route for the routing table. The route with the lower metric is preferred. The default metric for a static route is 1 on ProVision and 0 on Comware. You set the metric for a ProVision static route when you create the route. Comware static routes always use 0. Metric is sometimes referred to as cost, and you will see "Cost" in the Comware routing table.

Consider some examples. You could add a second default route on ProVision-1 that uses the default metric and administrative distance just like the first default route. Because the default routes tie for administrative distance and metric, ProVision-1 adds both routes to its routing table and uses both routes. The switch uses a hashing mechanism to decide which traffic to assign to each route. Thus, it load-shares the traffic over the links to Comware-1 and Comware-2.

**Note**

The switch must support Equal Cost Multi-Pathing (ECMP) routes in order to add both routes. By default, both ProVision and Comware switches support this feature. Refer to your switch documentation for details on the number of ECMP routes permitted. If the number of ECMP routes exceeds the limit, the switch uses a factor such as next hop address to break the tie.

In the example network topology, ProVision-1 has an aggregated link to Comware-1 (2 Gbps), but just a single Gbps link to Comware-2. You might decide, therefore, that you do not want to load share the traffic but instead prefer the route through Comware-1. In that case, you set the administrative distance on the route through Comware-2 higher than the distance on the route through Comware-1.

ProVision switches let you adjust both the metric and the administrative distance. (Comware only lets you adjust the administrative distance.) As you learned, the ProVision switch first compares the static routes' administrative distance and chooses the route with the lower distance. But if the routes have the same distance, it chooses the route with the lower metric. So you could set a higher metric on the route through Comware-2 rather than a higher administrative distance.

In either case, ProVision-1 now adds only the route through Comware-1 to its routing table. The other route still exists, but it is not active. If, however, the link to Comware-1 fails and that route is removed from the IP routing table, ProVision-1 starts using the route through Comware-2.

It is also important for you to understand that planning redundancy involves examining the entire path. For example, the redundant route on ProVision-1 lets the server continue to reach the client if ProVision-1's link to Comware-1 fails. But, as you can see in Figure 9-16, Comware-1 also needs a redundant route for 10.1.11.0/24 so that it can continue to return traffic from the client.

Remember redundancy for return traffic

Comware-1 IP routing table

| Destination network | Next hop | Dist (Pref) |
|---|---|---|
| 10.1.11.0/24 | 10.1.101.3 | 60 |
| 10.1.11.0/24 | 10.1.100.2 | 70 |

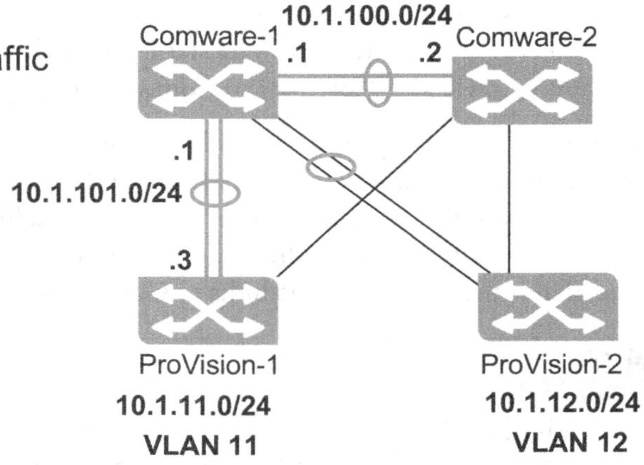


Figure 9-16: Additional redundant routes

## Example configuration: Create redundant static IP routes

This section builds on the network topology established in "Example configuration: Create static IP routes" on page 313. The example network is shown in Figure 9-17. In this section you will learn how to set up routes to enable redundancy for the ProVision-1 links.

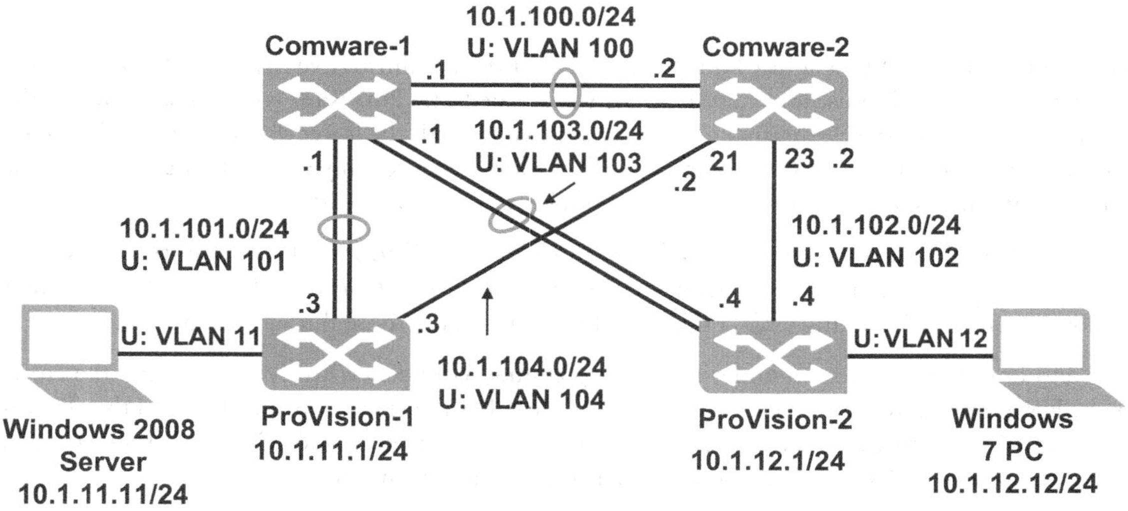


Figure 9-17: Create static routes

## Set up static routes on Comware-2

You will configure ProVision-1 to route traffic through Comware-2, so you must set up routing on Comware-2.

1. Access the Comware-2 CLI and move to the system view.
2. Create a static route to 10.1.12.0/24 through ProVision-2 at 10.1.102.4.

```
[Comware-2] ip route-static 10.1.12.0 24 10.1.102.4
```

3. Create a static route to 10.1.11.0/24 through ProVision-1 at 10.1.104.3.

```
[Comware-2] ip route-static 10.1.11.0 24 10.1.104.3
```

4. Check the routes.

```
[Comware-2] display ip routing

Routing Tables: Public

        Destinations : 10        Routes : 10

Destination/Mask   Proto  Pre  Cost NextHop      Interface

10.1.11.0/24       Static 60   0    10.1.104.3   GE1/0/21

10.1.12.0/24       Static 60   0    10.1.102.4   GE1/0/23

10.1.100.0/24      Direct 0    0    10.1.100.2   Vlan100

10.1.100.2/32      Direct 0    0    127.0.0.1    InLoop0

10.1.102.0/24      Direct 0    0    10.1.102.2   GE1/0/23

10.1.102.2/32      Direct 0    0    127.0.0.1    InLoop0

10.1.104.0/24      Direct 0    0    10.1.104.2   GE1/0/21

10.1.104.2/32      Direct 0    0    127.0.0.1    InLoop0

127.0.0.0/8        Direct 0    0    127.0.0.1    InLoop0

127.0.0.1/32       Direct 0    0    127.0.0.1    InLoop0
```

5. Trace routes use pings with successively higher TTLs to trace each hop in the route. They also require routers to send ICMP unreachable messages when they receive a ping with an expired TTL. By default, Comware devices ignore ICMP messages with TTLs and do not respond with ICMP unreachable messages. To allow your trace routes to succeed, you must enable these functions on the Comware-2.

```
[Comware-2] ip ttl-expires enable

[Comware-2] ip unreachables enable
```

## Create a second default route on ProVision-1

You will now create a default route on ProVision-1 through Comware-2. You will first assign the route to the default metric and administrative distance and observe the behavior. You will then assign the route a higher metric and observe the behavior.

1. Access the ProVision-1 CLI and move to the global configuration mode context.
2. Create a default route through Comware-2 at 10.1.104.2.

```
ProVision-1(config)# ip route 0.0.0.0/0 10.1.104.2
```

3. View the IP routing table and verify that you see both routes. As you see, they have the same metric and administrative distance.

```
ProVision-1(config)# show ip route

                        IP Route Entries

Destination     Gateway     VLAN Type      Sub-Type  Metric Dist.
-------------- ----------- ---- --------- --------  ------ ----
0.0.0.0/0       10.1.101.1  101            static    1      1
0.0.0.0/0       10.1.104.2  104            static    1      1
10.1.11.0/24    VLAN11       11            connected 1      0
10.1.101.0/24   VLAN101     101            connected 1      0
10.1.104.0/24   VLAN104     104            connected 1      0
127.0.0.0/8     reject                     static    0      0
127.0.0.1/32    lo0                       connected  1      0
```

4. Access the Windows 2008 server desktop and open a command prompt.
5. Execute a trace route to the Windows 7 PC (10.1.12.12).

```
tracert -d 10.1.12.12

Tracing route to 10.1.12.12 over a maximum of 30 hops

  1    <1 ms    <1 ms    1 ms  10.1.11.1
  2    <1 ms    <1 ms    1 ms  10.1.104.2
  3    <1 ms    <1 ms    1 ms  10.1.103.4
  4    <1 ms    <1 ms    1 ms  10.1.12.12
```

6. Execute a trace route to 10.1.12.1.

```
tracert -d 10.1.12.1

Tracing route to 10.1.12.12 over a maximum of 30 hops

  1    <1 ms    <1 ms    1 ms  10.1.11.1

  2    <1 ms    <1 ms    1 ms  10.1.101.1

  3    <1 ms    <1 ms    1 ms  10.1.12.1
```

   As you see, ProVision-1 is using both routes, assigning some traffic to one and other traffic to the other. ProVision-1 has a higher bandwidth connection to Comware-1, so now you will make ProVision-1 prefer the route through Comware-1.

7. Return to the ProVision-1 CLI.
8. Delete the current route through Comware-2. Then add the route with a metric of 10.

```
ProVision-1(config)# no ip route 0.0.0.0/0 10.1.104.2

ProVision-1(config)# ip route 0.0.0.0/0 10.1.104.2 metric 10
```

9. View the IP routing table and see that only the route through Comware-1 is present. The other route does not display in the active routing table because it has a higher metric and is not preferred.

```
ProVision-1(config)# show ip route

                      IP Route Entries

Destination    Gateway    VLAN Type      Sub-Type  Metric  Dist.
-------------- ---------- ---- --------- --------- ------- ----
0.0.0.0/0      10.1.101.1  101           static    1       1
10.1.11.0/24   VLAN11       11           connected 1       0
10.1.101.0/24  VLAN101     101           connected 1       0
10.1.104.0/24  VLAN104     104           connected 1       0
127.0.0.0/8    reject                    static    0       0
127.0.0.1/32   lo0                       connected 1       0
```

10. You will now observe failover to the higher metric route. Disable the links to Comware-1.

```
ProVision-1(config)# interface 19,20 disable
```

11. View the IP routing table and see that it now contains the route through Comware-2.

```
ProVision-1(config)# show ip route

                        IP Route Entries

Destination     Gateway      VLAN Type    Sub-Type  Metric  Dist.
--------------- ----------- ---- ------- --------- ------- -----
0.0.0.0/0       10.1.104.2  104          static    10       1
10.1.11.0/24    VLAN11       11          connected  1       0
10.1.104.0/24   VLAN104     104          connected  1       0
127.0.0.0/8     reject                   static     0       0
127.0.0.1/32    lo0                      connected  1       0
```

12. Access the Windows 2008 server desktop and open a command prompt.
13. Execute a trace route to the Windows 7 PC (10.1.12.12).

```
tracert -d 10.1.12.12

Tracing route to 10.1.12.12 over a maximum of 30 hops

  1    <1 ms    <1 ms     1 ms  10.1.11.1
  2    <1 ms    <1 ms     1 ms  10.1.104.2
  3    *   *        *     Request timed out
```

14. As you see the trace route fails. Press **<Ctrl+c>** to end it.

    You already saw that ProVision-1 started using the redundant route. Why is connectivity still disrupted?

## Create a redundant route on Comware-1

As shown in Figure 9-18, connectivity is disrupted because ProVision-2 routes return traffic to Comware-1, and Comware-1 cannot route traffic back to 10.1.11.0/24. It needs a redundant route through Comware-2, as well.

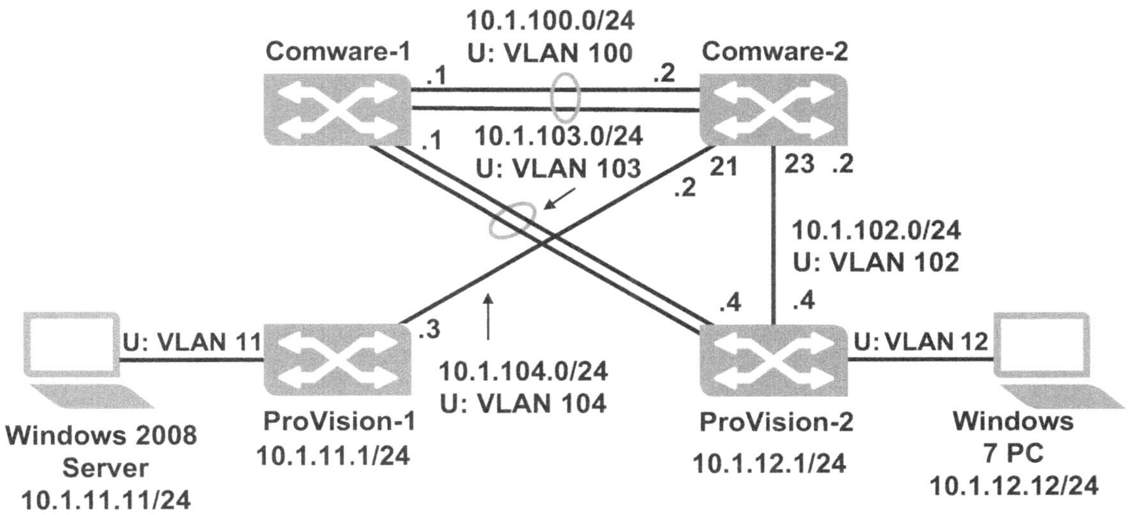


Figure 9-18: Network topology with one link down

1. Access the Comware-1 CLI. Move to system view.
2. Examine the routing table to validate that Comware-1 has no active route to 10.1.11.0/24.

```
[Comware-1] display ip routing

Routing Tables: Public

        Destinations : 7        Routes : 7

Destination/Mask   Proto  Pre  Cost  NextHop      Interface

10.1.12.0/24       Static 60   0     10.1.103.4   Vlan103

10.1.100.0/24      Direct 0    0     10.1.100.1   Vlan100

10.1.100.1/32      Direct 0    0     127.0.0.1    InLoop0

10.1.103.0/24      Direct 0    0     10.1.103.1   Vlan103

10.1.103.1/32      Direct 0    0     127.0.0.1    InLoop0

127.0.0.0/8        Direct 0    0     127.0.0.1    InLoop0

127.0.0.1/32       Direct 0    0     127.0.0.1    InLoop0
```

3. Create the redundant static route to 10.1.11.0/24 through Comware-2 at 10.1.100.2. Set the administrative distance, preference, to 70, which is higher than the default. (Comware does not allow you to set a metric for a static route.)

```
[Comware-1] ip route-static 10.1.11.0 24 10.1.100.2 preference 70
```

4. View the new route in the IP routing table.

```
[Comware-1] display ip routing

Routing Tables: Public

        Destinations : 8        Routes : 8

Destination/Mask Proto  Pre  Cost NextHop      Interface

10.1.11.0/24     Static 70   0    10.1.100.2   Vlan100

10.1.12.0/24     Static 60   0    10.1.103.4   Vlan103

10.1.100.0/24    Direct 0    0    10.1.100.1   Vlan100

10.1.100.1/32    Direct 0    0    127.0.0.1    InLoop0

10.1.103.0/24    Direct 0    0    10.1.103.1   Vlan103

10.1.103.1/32    Direct 0    0    127.0.0.1    InLoop0

127.0.0.0/8      Direct 0    0    127.0.0.1    InLoop0

127.0.0.1/32     Direct 0    0    127.0.0.1    InLoop0
```

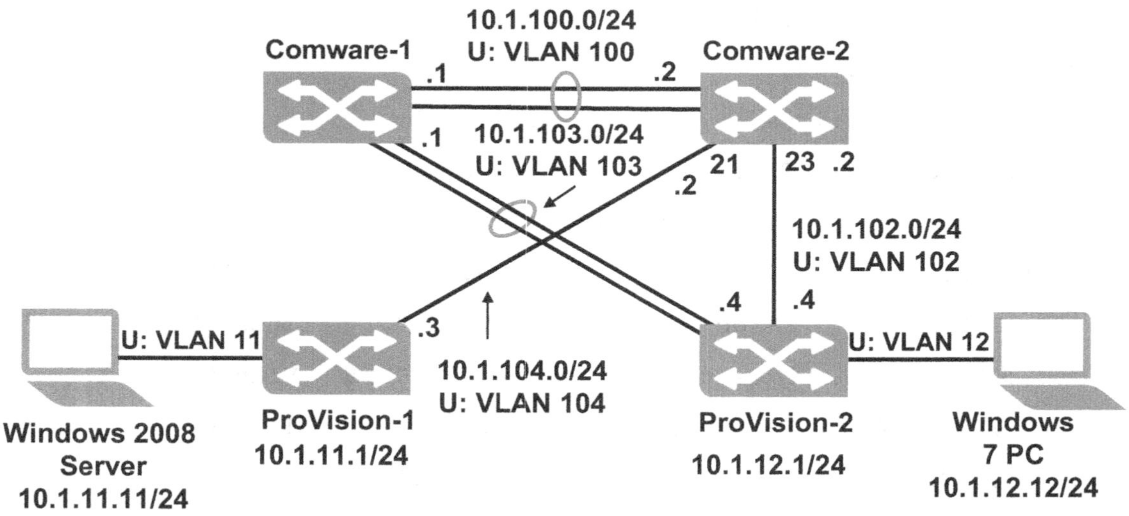


Figure 9-19: Network topology with one link down

5. Access the Windows 2008 server and start the trace route again. It is now successful. (Refer to Figure 9-19 as needed.)

```
tracert -d 10.1.12.12
```

```
Tracing route to 10.1.12.12 over a maximum of 30 hops

  1    <1 ms    <1 ms     1 ms  10.1.11.1

  2    <1 ms    <1 ms     1 ms  10.1.104.2

  3    <1 ms    <1 ms     1 ms  10.1.103.4

  4    <1 ms    <1 ms     1 ms  10.1.12.12
```

6. Return to the terminal session with ProVision-1. Restore the failed link.

```
ProVision-1(config)# interface 19,20 enable
```

7. Observe that Comware-1 is using its preferred route through ProVision-1 again.

```
[Comware-1] display ip routing

Routing Tables: Public

        Destinations : 10       Routes : 10

Destination/Mask   Proto  Pre  Cost  NextHop       Interface

10.1.11.0/24       Static 60   0     10.1.101.3    Vlan101

10.1.12.0/24       Static 60   0     10.1.103.4    Vlan103

10.1.100.0/24      Direct 0    0     10.1.100.1    Vlan100

10.1.100.1/32      Direct 0    0     127.0.0.1     InLoop0

10.1.101.0/24      Direct 0    0     10.1.101.1    Vlan101

10.1.101.1/32      Direct 0    0     127.0.0.1     InLoop0

10.1.103.0/24      Direct 0    0     10.1.103.1    Vlan103

10.1.103.1/32      Direct 0    0     127.0.0.1     InLoop0

127.0.0.0/8        Direct 0    0     127.0.0.1     InLoop0

127.0.0.1/32       Direct 0    0     127.0.0.1     InLoop0
```

Save the current configuration on each of the switches. For ProVision, use the `write memory` command. For Comware, use the `save [force]` command.

## Learning check 3

1. In the example configuration, routes were set up to allow ProVision-1 to use its link to Comware-1 during normal operation and its link to Comware-2 as a backup. Now assume that you want to do the same on ProVision-2. What steps must you complete?

2. The network illustrated in Figure 9-20 and Figure 9-21 is using the routes that are established in the example configuration. The link between ProVision-1 and Comware-1 fails. Use Figure 9-20 to trace the traffic flow for traffic from the server to the PC. Use Figure 9-21 to trace the flow from the PC back to the server.

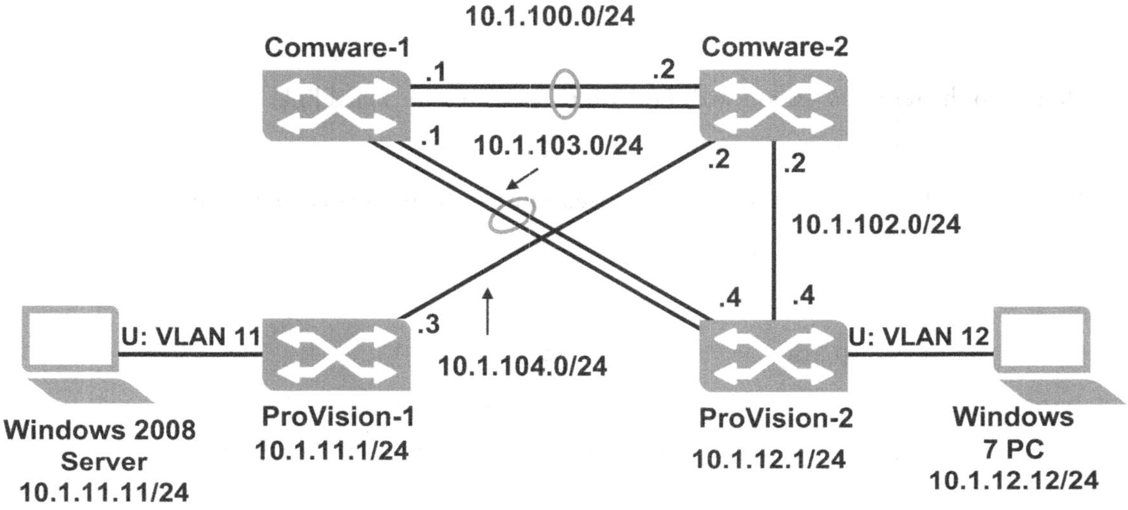


Figure 9-20: Traffic flow from the server to the PC

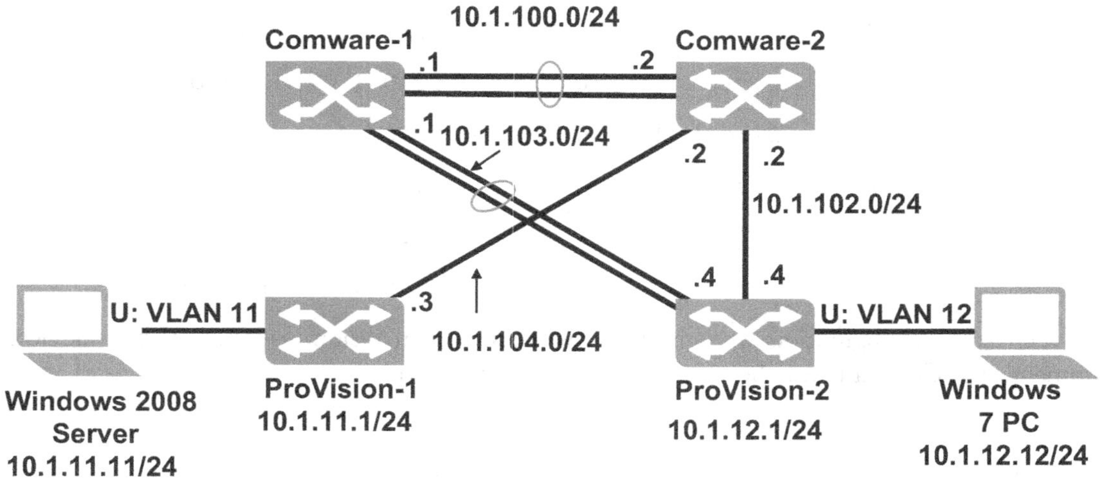


Figure 9-21: Traffic flow from the PC to the server

3. Now consider a network in which the bandwidth for all switch-to-switch connections is the same. You set up the redundant routes on ProVision-1 and ProVision-2 to have the same metric and administrative distance. What is the traffic flow from the PC to the server during normal operation? Use dashed lines on Figure 9-22 to indicate load-sharing.

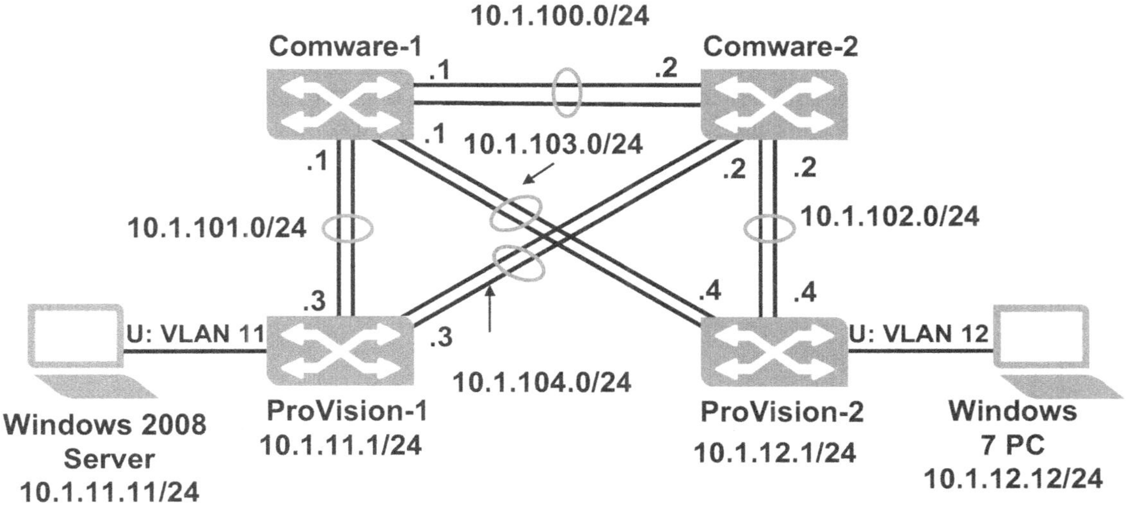


Figure 9-22: Traffic flow from the PC to the server

4. Consider the scenario presented in question 3. Suppose the link between ProVision-1 and Comware-1 fails, as you can see in Figure 9-23. What is the traffic flow from the PC to the server now? Is this flow ideal?

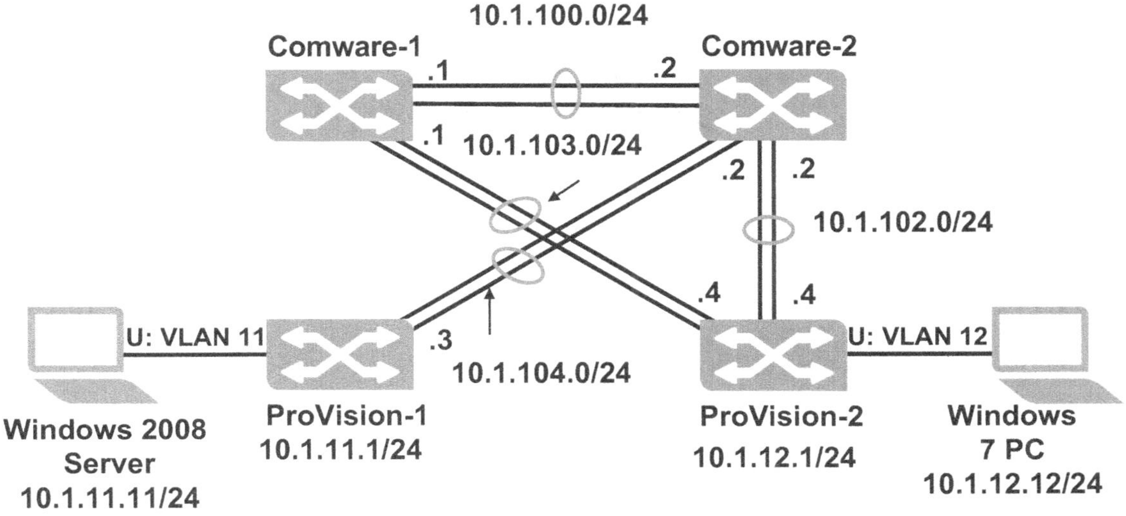


Figure 9-23: Traffic flow from the PC to the server

## Learning check 3 answers

1. In the example configuration, routes were set up to allow ProVision-1 to use its link to Comware-1 during normal operation and its link to Comware-2 as a backup. Now assume that you want to do the same on ProVision-2. What steps must you complete?

   You need to create a second default route through Comware-2 on ProVision-1. You might set the administrative distance higher than the route through Comware-1 to favor the Comware-1 route. You must make sure that Comware-2 has the proper routes. You must create a route on Comware-1 to 10.1.11.0/24 through Comware-2, again setting the administrative distance (preference) higher than the existing route.

   You might want to provide failover in case the Comware-1 link on one ProVision switch fails and the Comware-2 link on the other ProVision switch fails. To achieve this redundancy, you must set up redundant routes on Comware-2 to 10.1.11.0/24 and 10.1.12.0/24 through Comware-1.

2. The network illustrated in Figure 9-20 and Figure 9-21 is using the routes that established in the example configuration. The link between ProVision-1 and Comware-1 fails. Use Figure 9-20 to trace the traffic flow for traffic from the server to the PC. Use Figure 9-21 to trace the flow from the PC back to the server.

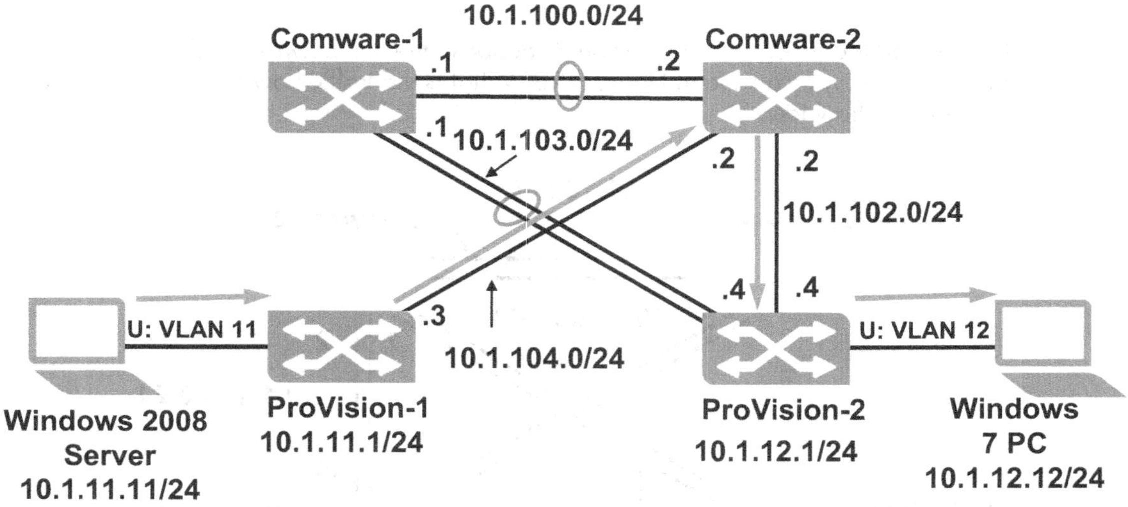


Figure 9-20 (answer): Traffic flow from the server to the PC

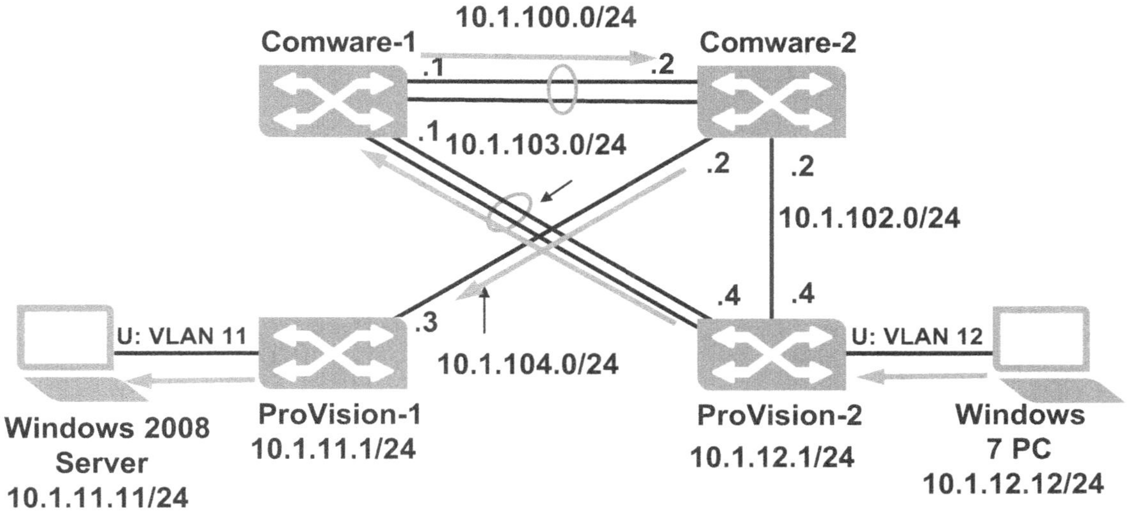


Figure 9-21 (answer): Traffic flow from the PC to the server

3. Now consider a network in which the bandwidth for all switch-to-switch connections is the same. You set up the redundant routes on ProVision-1 and ProVision-2 to have the same metric and administrative distance. What is the traffic flow from the PC to the server during normal operation? Use dashed lines on Figure 9-22 to indicate load-sharing.

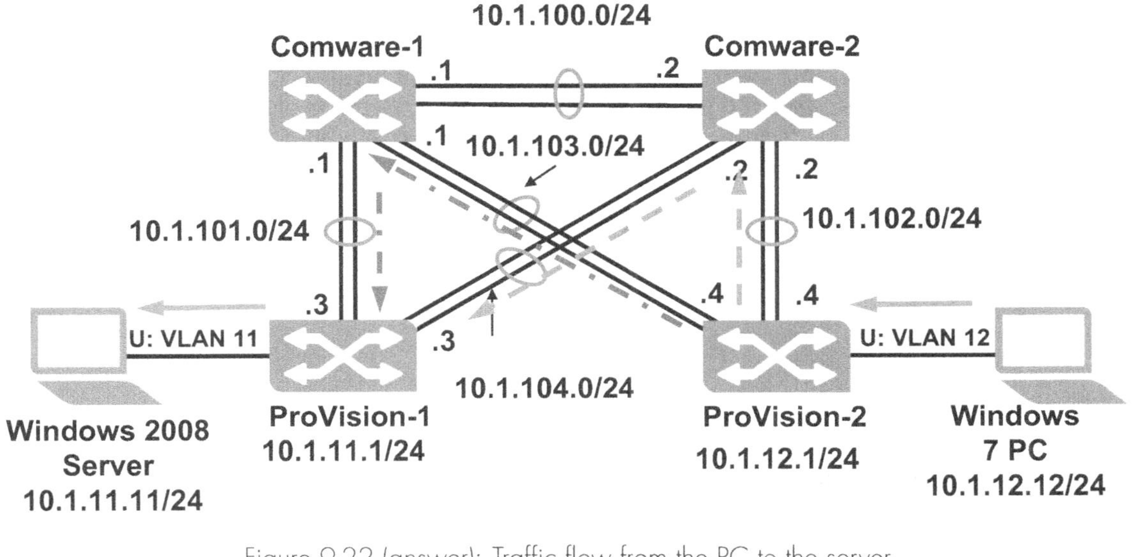


Figure 9-22 (answer): Traffic flow from the PC to the server

4. Consider the scenario presented in question 3. Suppose the link between ProVision-1 and Comware-1 fails, as you can see in Figure 9-23. What is the traffic flow from the PC to the server now? Is this flow ideal?

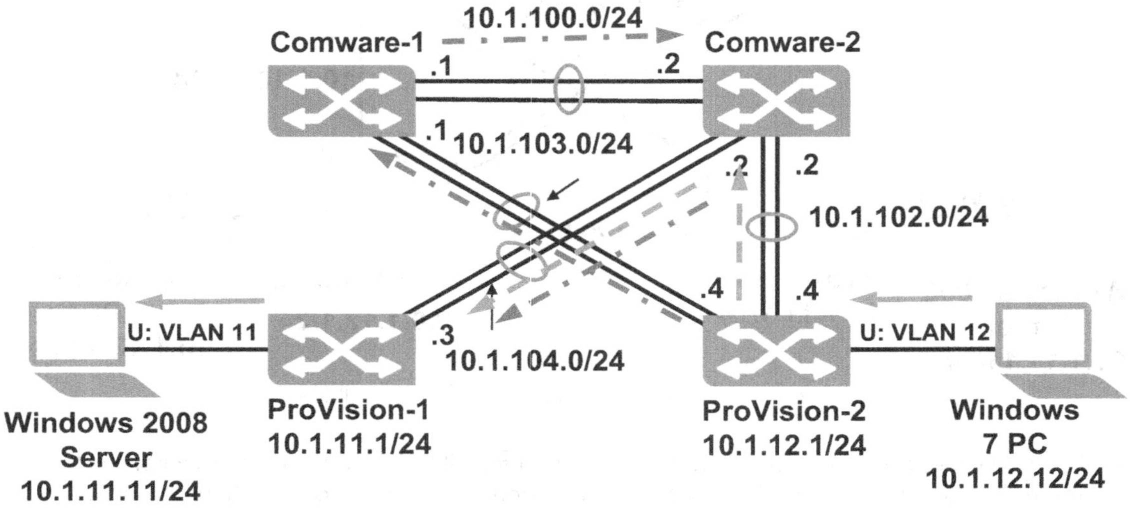


Figure 9-23 (answer): Traffic flow from the PC to the server

The flow in this failover situation is not ideal. It would be better if ProVision-2 used only the route to Comware-2 because Comware-1 has to send the traffic back to Comware-2 anyway. Sending all traffic through Comware-2 would also eliminate the asymmetric routing in which the PC-to-server traffic takes a different path from the server-to-PC traffic. (One reason that asymmetric routing is undesirable is that it can sometimes, although not often, lead to the flooding of unknown unicasts. In any case, it is better to use symmetric routing.)

However, static routing does not let ProVision-1 change its routes in response to topology changes elsewhere.

## Dynamic routing protocols

Static routes have a role to play in routing design, but they tend not to work well in a redundant network with multiple paths. As you just learned, you might find it difficult to plan all of the routes that are necessary for all failover situations. Also the static routes on one device cannot adjust based on changes in the topology in other areas of the network, interfering with the solution's resiliency. You could also accidentally introduce routing loops, which can disrupt connectivity.

For these reasons, you should typically use a dynamic routing protocol for more complex network topologies. In addition, as networks grow, configuring static routes on each routing device becomes infeasible (as well as complex and error prone).

Routing protocols eliminate the need to create routes manually. Instead routers communicate with each other to discover available routes and the best paths to destinations automatically.

Depending on the protocol, routers exchange different types of information and learn routes differently:

- **Distance vector**—Routers advertise routes that they know, and the advertisements propagate across all routers participating in the protocol. Routers select routes with the hop count. Routing Information Protocol (RIP) is an example of a distance vector protocol.
- **Link state**—Routers advertise information about their connected Layer 3 interfaces, also called links, and the cost (or metric) associated with that interface. These link state advertisements propagate across participating routers. Routers use the advertisements to construct a topology of the network, consisting of all the known links, the routers connected to the links, and the associated costs. Using the topology, routers calculate their own lowest cost route to each destination subnet. OSPF is a link state protocol.

**Important**

Link state protocols such as OSPF tend to choose better routing paths and to converge faster than distance vector protocols such as RIP.

You can also classify dynamic routing protocols based on how they are used:

- **Interior Gateway Protocol (IGP)**—Facilitates the exchange of routing information among routers under the same organizational control, or in other words within the same autonomous system (AS). A common IGP is OSPF.
- **Exterior Gateway Protocol (EGP)**—Facilitates exchange of routing information among routers in different ASes. Border Gateway Protocol v4 (BGP4), a common example, is used by ISPs to route traffic on the Internet backbone.

This study guide focuses on routing within a private network, so you will look at OSPF, an IGP.

## Basic OSPF setup

As mentioned earlier, OSPF is a link state protocol. OSPF routers advertise their links (connected subnets) to each other using link state advertisements (LSAs), build a topology based on the advertisements, and calculate best routes based on this topology.

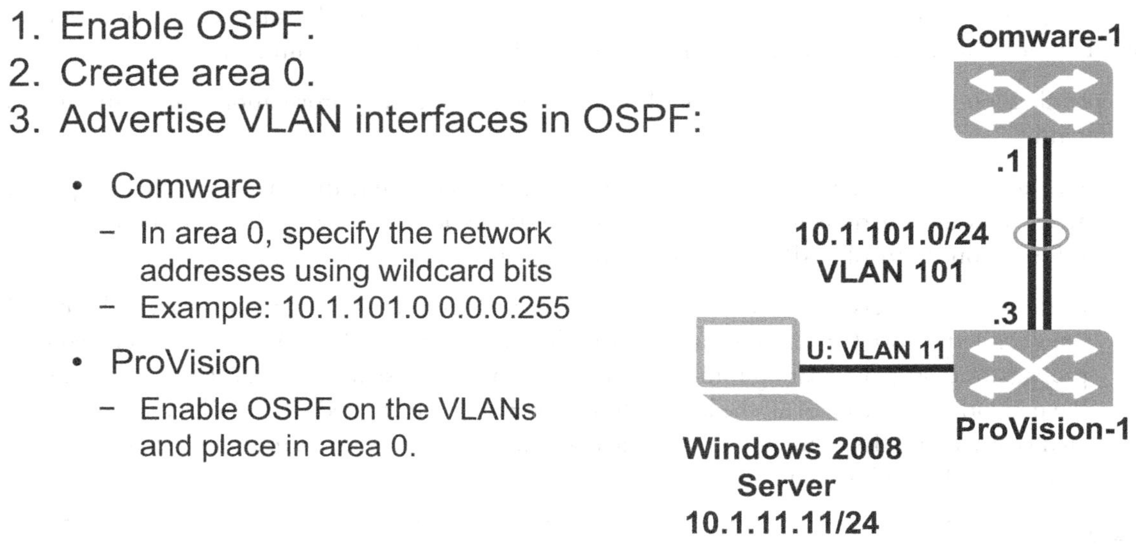


Figure 9-24: Basic OSPF setup

If you want to view a video about OSPF, access the HP Aurasma application on your smartphone and use the application to focus on Figure 9-24. This action will launch the video.

For this study guide, you will just look at the steps required to set up a basic, functional OPSF solution, as shown in Figure 9-24. You should attend *Deploying HP FlexNetwork Core Technologies* if you want to learn more about topics such as these:

- How OSPF routers discover their neighbors
- How OSPF routers advertise topology information, build topologies, and choose routes
- How to configure router IDs and use loopback interfaces to enhance stability
- How to configure metrics (costs) to help routing switches choose the best paths
- How to import external routes into OSPF
- How to set up multi area solutions

To set up a basic OSPF solution, follow these steps:

1. Enable OSPF.
2. Create area 0.

   Area 0 is a special area called the backbone. You can learn about the purpose of the backbone area in multi area topologies in *Deploying HP FlexNetwork Core Technologies*. For a single area topology like yours, area 0 is a choice (in case you want to move to a multi area topology later).

3. Enable OSPF on the VLAN interfaces. (If your Comware switch has Layer 3 physical interfaces, you would enable OSPF on them as well.) Enabling OSPF on an interface has two effects:
   - The routing switch sends OSPF messages on the interface and can form neighbor relationships with other routers on the same interface. Neighbors that fully exchange their topology information are called *fully adjacent*.
   - The routing switch advertises this interface in OSPF. For example, if the interface is associated with 10.1.11.0/24, the routing switch advertises that it has a link to 10.1.11.0/24.

VLANs that need to be advertised in OSPF include:

- Those that support endpoints (like VLAN 11 on ProVision-1)
- Those that connect routing switches (like VLAN 101 on ProVision-1 and Comware-1)

You enable OSPF as follows:

- **Comware**—Specify the interfaces' network addresses in OSPF area 0. When you specify the network address, you use wildcard bits, which are the reverse of a subnet mask. For example, for VLAN 101 on Comware-1, which has IP address 10.1.101.1 255.255.255.0, you would specify network 10.1.101.0 0.0.0.255.
- **ProVision**—Access the VLAN and enter a command that enables OSPF and places the VLAN in area 0.

An optional step, but one that you should typically perform is setting a router ID. The router ID does not have to be a valid IP address on the switch, but you often should use an IP address on a loopback interface.

After you complete these steps, your routing switches will automatically create a topology of your network and learn routes to every subnet within it. If a link fails, the routing switches inform each other of the change and learn new routes.

Table 9-7 outlines settings that must match on OSPF neighbors before they can achieve adjacency.

Table 9-7: Settings that must match for OSPF neighbors to achieve adjacency

| Setting |
|---|
| Network address and mask |
| Area ID |
| Area type (OSPF defines different types of areas covered in other training) |
| Authentication method and key if authentication is used |
| Network type |
| Hello timer |

Table 9-8 provides information about LSA types used in a single area OSPF configuration.

Table 9-8: LSA types used in single area configurations

| Type | Name | Information advertised | Advertised by | Scope |
|---|---|---|---|---|
| 1 | Router | The OSPF links on a routing device (transit and stub networks) | Each OSPF routing device | Area |
| 2 | Network | The routers on broadcast or NBMA networks | Designated Router (DR)* for the network | Area |

**Note**

The DR for a subnet running OSPF (OSPF link) is elected based on priority (higher priority value wins) and, in the case of a tie, higher router ID.

When a link has only two routers, as in the design in your topology, the DR and BDR do not usually matter. If the link has multiple routers, the DR establishes a full adjacency with all routers on the link, as does the BDR. The other routers on the link see each other as neighbors but do not establish full adjacencies nor exchange LSAs with each other (only with the DR and BDR). Instead the DR handles forwarding LSAs and updates.

## Choosing the preferred route with static and dynamic routing

You were introduced to the concepts of metric and administrative distance earlier. You will now return to these concepts so that you can understand what happens if your routing switch has a static route and dynamic route to the same destination. Refer to Figure 9-25 as you review the information that follows.

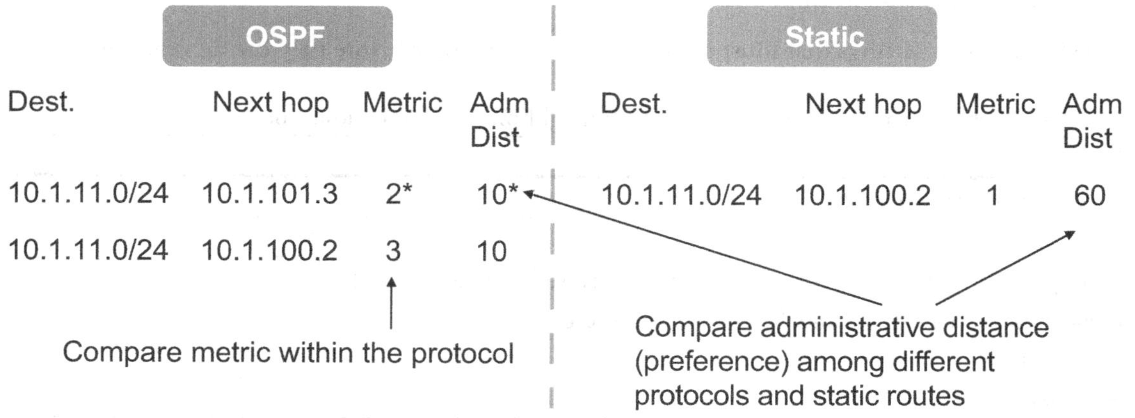


Figure 9-25: Choosing the preferred route with static and dynamic routing

- **Metric**—Routing devices use metrics to distinguish between routes that are learned in the same way. Dynamic routing protocols each have their own way of defining a route's metric. An OSPF routing switch might automatically determine the metric for a link based on the link speed, or you can set a metric (OSPF) on a VLAN interface. The switch advertises that metric when it advertises the link. When a switch calculates a route to a destination network, it adds up the metric on all the links on the way to that network. If OSPF learns two routes to the same destination, it chooses the route with the lowest metric to propose to the IP routing table (or multiple routes with the same lowest metric). Remember: metrics are locally significant to static routes or to a particular routing protocol. In other words, the switch will compare the metric of two OSPF routes, but it will not compare the metric of an OSPF route to the metric of a static route.
- **Administrative distance**—It is the administrative distance that allows the switch to choose between a static route and an OSPF (or other IGP) route to the same destination. You can set a static route's administrative distance when you create the route. Each routing protocol is associated with an administrative distance, which applies to all routes learned with that protocol.

Remember: the switch first compares routes within a routing protocol based on metric. *Then* it examines the administrative distance across the static routes and the routes proposed by each routing protocol. The route with the *lowest administrative distance is always chosen* even if it has a higher metric than a route with a higher administrative distance. This is because metrics are not compared between routing protocols.

Table 9-9 shows the default administrative distances used for static routes and OSPF routes on ProVision and Comware switches. As you see, Comware switches prefer internal OSPF routes (the type that you are using) to static routes, and ProVision switches prefer static routes to all types of OSPF routes.

Table 9-9: Default administrative distances (preferences)

| Routing method | Comware | ProVision |
|---|---|---|
| Static | 60 | 1 |
| OSPF internal | 10 | 110 |
| OSPF external | 150 | 110 |

## Example configuration: Setting up a basic OSPF configuration

This section builds on the example network established in "Example configuration: Create redundant static IP routes" on page 322. This section outlines the steps for setting up OSPF on the network shown in Figure 9-26.

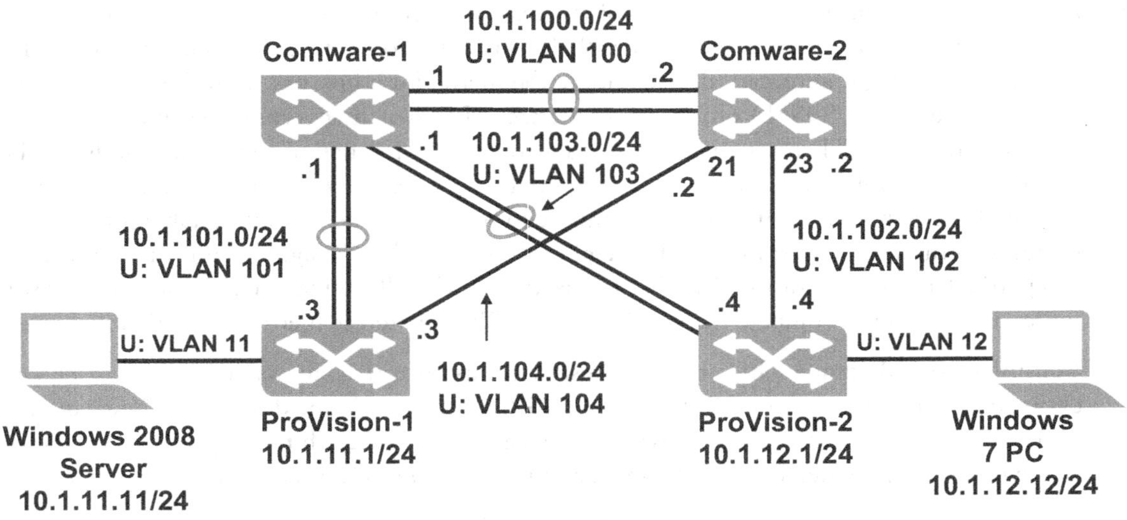


Figure 9-26: Set up a basic OSPF configuration

## Configure OSPF on ProVision-1

1. Access the ProVision-1 CLI. Move to global configuration mode.
2. Set the router ID to 10.0.0.3. In this case, the router ID simply identifies the ProVision-1 to other OSPF routers. It is not an IP address. (Other HP training gives guidelines for configuring the router ID on a loopback interface so that you can ping it.)

```
ProVision-1(config)# ip router-id 10.0.0.3
```

3. Move to the OSPF configuration mode.

```
ProVision-1(config)# router ospf
```

4. Create area 0.

```
ProVision-1(ospf)# area 0
```

5. Enable OSPF.

```
ProVision-1(ospf)# enable
```

6. Access each of the switch's active VLANs. Enable OSPF and place the VLAN in area 0.

```
ProVision-1(ospf)# vlan 11
ProVision-1(vlan-11)# ip ospf area 0
ProVision-1(vlan-11)# vlan 101
ProVision-1(vlan-101)# ip ospf area 0
```

```
ProVision-1(vlan-101)# vlan 104
ProVision-1(vlan-104)# ip ospf area 0
ProVision-1(vlan-104)# exit
```

7. If you want, you can make VLAN 11 a passive interface, which prevents the switch making neighbors on the interface.

```
ProVision-1(config)# vlan 11 ip ospf passive
```

8. Verify that OSPF is enabled on the VLAN interfaces. Note that area 0 is also called the backbone area.

```
ProVision-1(config)# show ip ospf interface

OSPF Interface Status

IP Address  Status   Area ID   State Auth-type Cost  Pri Passive
---------- ------- --------- ----- --------- ----- --- ------
10.1.11.1  enabled  backbone WAIT  none      1     1   no
10.1.101.3 enabled  backbone WAIT  none      1     1   no
10.1.104.3 enabled  backbone WAIT  none      1     1   no
```

## Configure OSPF on Comware-1

1. Access the Comware-1 CLI and move to system view.
2. Enable an OSPF process. At the same time, set the router ID. (Comware switches support multiple processes with different IDs. You just need one process, and you will use ID 1.)

```
[Comware-1] ospf 1 router-id 10.0.0.1
```

3. Create area 0.

```
[Comware-1-ospf-1] area 0
```

4. Assign all subnets connected to the switch to this area. Remember that you use wildcard bits rather than a subnet mask.

```
[Comware-1-ospf-1-area-0.0.0.0] network 10.1.100.0 0.0.0.255
[Comware-1-ospf-1-area-0.0.0.0] network 10.1.101.0 0.0.0.255
[Comware-1-ospf-1-area-0.0.0.0] network 10.1.103.0 0.0.0.255
```

5. Comware-1 and ProVision-1 are both running OSPF on 10.1.101.0/24. Verify that they are communicating with each other as neighbors (or peers). Also that they have achieved the Full state, which indicates that they have shared topology information.

```
[Comware-1-ospf-1-area-0.0.0.0] display ospf peer

                    OSPF Process 1 with Router ID 10.0.0.1

                          Neighbor Brief Information

 Area: 0.0.0.0

 Router ID    Address      Pri Dead-Time  Interface    State

 10.0.0.3     10.1.101.3  1   32         Vlan101      Full/DR
```

## Configure OSPF on Comware-2

1. Access the Comware-2 CLI and move to system view.
2. Enable an OSPF process and set the router ID.

```
[Comware-2] ospf 1 router-id 10.0.0.2
```

3. Create area 0.

```
[Comware-2-ospf-1] area 0
```

4. Assign all subnets connected to the switch to this area.

```
[Comware-2-ospf-1-area-0.0.0.0] network 10.1.100.0 0.0.0.255
[Comware-2-ospf-1-area-0.0.0.0] network 10.1.102.0 0.0.0.255
[Comware-2-ospf-1-area-0.0.0.0] network 10.1.104.0 0.0.0.255
```

5. Comware-2 should now be OSPF neighbors with Comware-1 and ProVision-1, which are already running OSPF on the same networks. Verify and check for the Full state.

   If you do not see both peers, wait about 30 seconds or a minute and try again.

```
[Comware-2-ospf-1-area-0.0.0.0] display ospf peer

                    OSPF Process 1 with Router ID 10.0.0.2

                          Neighbor Brief Information

 Area: 0.0.0.0

 Router ID   Address      Pri Dead-Time Interface    State

 10.0.0.1    10.1.100.1  1   37         Vlan100      Full/DR

 10.0.0.3    10.1.104.3  1   38         GE1/0/21     Full/DR
```

## Configure OSPF on ProVision-2

1. Access the ProVision-2 CLI and move to global configuration mode.
2. Set the router ID.

    ```
    ProVision-2(config)# ip router-id 10.0.0.4
    ```

3. Move to the OSPF configuration mode.

    ```
    ProVision-2(config)# router ospf
    ```

4. Create area 0.

    ```
    ProVision-2(ospf)# area 0
    ```

5. Enable OSPF.

    ```
    ProVision-2(ospf)# enable
    ```

6. Access each of the switch's active VLANs. Enable OSPF and place the VLAN in area 0. Note that if you do not specify an area, the switch automatically uses area 0.

    ```
    ProVision-2(ospf)# vlan 12
    ProVision-2(vlan-12)# ip ospf
    ProVision-2(vlan-12)# vlan 102
    ProVision-2(vlan-102)# ip ospf
    ProVision-2(vlan-102)# vlan 103
    ProVision-2(vlan-103)# ip ospf
    ProVision-2(vlan-103)# exit
    ```

7. If you want, you can make VLAN 12 a passive interface.

    ```
    ProVision-2(config)# vlan 12 ip ospf passive
    ```

8. Verify that ProVision-2 is OSPF neighbors with the two connected switches, Comware-1 and Comware-2. Also verify that the state is Full for each.

    ```
    ProVision-2(config)# show ip ospf neighbor

    OSPF Neighbor Information

                                                      Rxmt          Helper
    Router ID  Pri IP Address   NbIfState State  QLen  Events Status
    ---------- --- ----------- --------- ------ ----- ------ ------
    10.0.0.2   1   10.1.102.2  DR        FULL   0     6      None
    10.0.0.1   1   10.1.103.1  DR        FULL   0     6      None
    ```

## Verify the solution

You can now verify that the switches have learned the routes.

1. View routes on ProVision-1. You should see that the switch has learned a route to every subnet in your topology. You should also see that the switch has multiple routes to some destinations such as 10.1.12.0/24. Also notice the "ospf" route type and distance.

```
ProVision-1(config)# show ip route

                IP Route Entries

Destination    Gateway      VLAN Type      Sub-Type  Metric Dist.
-------------- ------------ ---- --------- --------- ------- ----
0.0.0.0/0      10.1.101.1   101  static              1       1
10.1.11.0/24   VLAN11       11   connected           1       0
10.1.12.0/24   10.1.101.1   101  ospf      IntraArea 3       110
10.1.12.0/24   10.1.104.2   104  ospf      IntraArea 3       110
10.1.100.0/24  10.1.101.1   101  ospf      IntraArea 2       110
10.1.100.0/24  10.1.104.2   104  ospf      IntraArea 2       110
10.1.101.0/24  VLAN101      101  connected           1       0
10.1.102.0/24  10.1.104.2   104  ospf      IntraArea 2       110
10.1.103.0/24  10.1.101.1   101  ospf      IntraArea 2       110
10.1.104.0/24  VLAN104      104  connected           1       0
127.0.0.0/8    reject            static              0       0
127.0.0.1/32   lo0               connected           1       0
```

2. ProVision-1 still has its static default route. When will ProVision-1 use the OSPF routes and when will it use the static default route?

   ProVision-1 will use the OSPF routes for routing traffic to all the subnets in your network because these routes are more specific. ProVision-1 will only use the static default route when it needs to route traffic to a destination for which it does not have a specific route.

3. View IP routes on ProVision-2 and see that it has learned routes to all subnets. You can filter the display for just OSPF routes.

```
ProVision-2(config)# show ip route ospf

                        IP Route Entries
```

```
Destination    Gateway     VLAN Type  Sub-Type   Metric  Dist.
-------------- ----------- ---- ----- ---------- ------- -----
10.1.11.0/24   10.1.102.2  102  ospf  IntraArea  3       110
10.1.11.0/24   10.1.103.1  103  ospf  IntraArea  3       110
10.1.100.0/24  10.1.102.2  102  ospf  IntraArea  2       110
10.1.100.0/24  10.1.103.1  103  ospf  IntraArea  2       110
10.1.101.0/24  10.1.103.1  103  ospf  IntraArea  2       110
10.1.104.0/24  10.1.102.2  102  ospf  IntraArea  2       110
```

4. View IP routes on Comware-1 and see that it has learned routes to all networks. Notice that Comware-1 has learned just one route to 10.1.12.0/24. Notice that OSPF is listed as the protocol, and the preference (administrative distance) is 10 for Comware.

```
[Comware-1] display ip routing
Routing Tables: Public
        Destinations : 12       Routes : 14
Destination/Mask Proto  Pre  Cost   NextHop      Interface
10.1.11.0/24     OSPF   10   2      10.1.101.3   Vlan101
10.1.12.0/24     OSPF   10   2      10.1.103.4   Vlan103
10.1.100.0/24    Direct 0    0      10.1.100.1   Vlan100
10.1.100.1/32    Direct 0    0      127.0.0.1    InLoop0
10.1.101.0/24    Direct 0    0      10.1.101.1   Vlan101
10.1.101.1/32    Direct 0    0      127.0.0.1    InLoop0
10.1.102.0/24    OSPF   10   2      10.1.100.2   Vlan100
                 OSPF   10   2      10.1.103.4   Vlan103
10.1.103.0/24    Direct 0    0      10.1.103.1   Vlan103
10.1.103.1/32    Direct 0    0      127.0.0.1    InLoop0
10.1.104.0/24    OSPF   10   2      10.1.101.3   Vlan101
                 OSPF   10   2      10.1.100.2   Vlan100
127.0.0.0/8      Direct 0    0      127.0.0.1    InLoop0
127.0.0.1/32     Direct 0    0      127.0.0.1    InLoop0
```

5. You can also view just OSPF routes on Comware switches. View OSPF routes on Comware-2. Note that you can see which switch, identified by its router ID, advertised each route. For example, ProVision-2 (10.0.0.4) advertises 10.1.12.0/24, which is connected to it.

```
<Comware-2>display ospf routing

         OSPF Process 1 with Router ID 10.0.0.2

             Routing Tables

 Routing for Network

Destination     Cost  Type     NextHop        AdvRouter     Area

10.1.102.0/24   1     Transit  10.1.102.2     10.0.0.2      0.0.0.0

10.1.103.0/24   2     Transit  10.1.102.4     10.0.0.1      0.0.0.0

10.1.103.0/24   2     Transit  10.1.100.1     10.0.0.1      0.0.0.0

10.1.104.0/24   1     Transit  10.1.104.2     10.0.0.3      0.0.0.0

10.1.11.0/24    2     Stub     10.1.104.3     10.0.0.3      0.0.0.0

10.1.12.0/24    2     Stub     10.1.102.4     10.0.0.4      0.0.0.0

10.1.100.0/24   1     Transit  10.1.100.2     10.0.0.1      0.0.0.0

10.1.101.0/24   2     Transit  10.1.100.1     10.0.0.3      0.0.0.0

10.1.101.0/24   2     Transit  10.1.104.3     10.0.0.3      0.0.0.0

Total Nets: 9

Intra Area: 9  Inter Area: 0  ASE: 0  NSSA: 0
```

6. Access the Windows 2008 server desktop and open a command prompt.

7. Execute a trace route to the Windows 7 PC (10.1.12.12).

```
tracert -d 10.1.12.12

Tracing route to 10.1.12.12 over a maximum of 30 hops

  1    <1 ms    <1 ms     1 ms  10.1.11.1

  2    <1 ms    <1 ms     1 ms  10.1.104.2

  3    <1 ms    <1 ms     1 ms  10.1.102.4

<1 ms    <1 ms     1 ms  10.1.12.12
```

8. Execute a trace route to 10.1.12.1.

```
tracert -d 10.1.12.1

Tracing route to 10.1.12.1 over a maximum of 30 hops

  1    <1 ms    <1 ms     1 ms  10.1.11.1

  2    <1 ms    <1 ms     1 ms  10.1.101.1

  3    <1 ms    <1 ms     1 ms  10.1.12.1
```

Save the current configuration on each of the switches. For ProVision, use the `write memory` command. For Comware, use the `save [force]` command.

## Summary

This chapter taught you how to create a variety of static routes. You also learned to use metrics and administrative distance to affect which routes the switch adds to its routing table.

You considered the complexity of using static routes within a complicated, redundant topology, and you learned about the advantages of a dynamic routing protocol for such environments. You then learned how to set up a basic OSPF solution. You also learned how to interpret an IP routing table and trace routing paths.

## Learning check 4

Answer the following question to assess how well you understand the concepts presented in this chapter.

1. Which switches in the example network are now neighbors with each other? Why do some switches become neighbors and others do not?
2. Which neighbor state should you look for to ensure that the neighbors have connected successfully and exchanged their topologies?
3. You need to enable OSPF on VLAN interfaces on both ProVision and Comware switches. What are the differences in the configuration?
4. ProVision-1 does not have any neighbors on 10.1.11.0/24 (VLAN 11). Similarly, ProVision-2 does not have any neighbors on 10.1.12.0/24 (VLAN 12).
   - Why did you still enable OSPF on these VLAN interfaces?
   - Are there any drawbacks to enabling OSPF on these VLAN interfaces?
5. You noticed that ProVision-1 is load-sharing traffic to 10.1.12.0/24 over routes through Comware-1 and Comware-2. Comware-1, on the other hand, is using one route for this network.
   - Why has ProVision-1 added two routes to 10.1.12.0/24 to its routing table?
   - Why does Comware-1 have only one route to 10.1.12.0/24 in its table?
   - How could you make ProVision-1 prefer its 10.1.12.0/24 route through Comware-1?
6. You did not delete the static routes on Comware-1. Why is Comware-1 using OSPF routes to forward traffic to 10.1.12.0/24 and other destinations in your local topology?
7. You did not delete the static routes on ProVision-1. Why is ProVision-1 using OSPF routes to forward traffic to 10.1.12.0/24 and other destinations in your local topology?

## Learning check 4 answers

1. Which switches in the example network are now neighbors with each other? Why do some switches become neighbors and others do not?

   ProVision-1 is neighbors with Comware-1 and Comware-2.

   ProVision-2 is neighbors with Comware-1 and Comware-2.

   Comware-1 is neighbors with Comware-1, ProVision-1, and ProVision-2.

   Comware-2 is neighbors with Comware-2, ProVision-1, and ProVision-2.

   OSPF routers become neighbors with the other routers on the same subnet. In this topology, the neighbors have direct Ethernet links because you dedicate a subnet to each of those links. But in other topologies, OSPF routers could become neighbors with non directly connected routers as long as they are on the same subnet.

2. What neighbor state should you look for to ensure that the neighbors have connected successfully and exchanged their topologies?

   Full

3. You need to enable OSPF on VLAN interfaces on both ProVision and Comware switches. What are the differences in the configuration?

   On ProVision switches, you enable OSPF from the VLAN context. You set the area in the same command. On Comware switches, you specify the network address for a VLAN interface (or other Layer 3 interface) in the OSPF area view. You configured each VLAN network separately in the example configuration, but you could configure a range of networks (like 10.1.0.0 0.0.255.255).

4. ProVision-1 does not have any neighbors on 10.1.11.0/24 (VLAN 11). Similarly, ProVision-2 does not have any neighbors on 10.1.12.0/24 (VLAN 12).

   - Why did you still enable OSPF on these VLAN interfaces?
   - Are there any drawbacks to enabling OSPF on these VLAN interfaces?

   You wanted the switches to advertise the subnets in OSPF so that other routing switches could learn routes to them.

   If you enable OSPF on the interface, it is possible that you could leak routing information to endpoints or establish an OSPF neighbor relationship with an unauthorized device. That unauthorized device could then disrupt your routing topology.

You can set the VLAN interface as a passive interface to prevent adjacencies from forming. Alternatively, you could leave the interface out of the OSPF configuration and have the routing switch redistribute connected (direct) routes. The switch would then advertise the subnet like an external route. You can attend the *Deploying HP FlexNetwork Core Technologies* course to learn more about these options.

5. You noticed that ProVision-1 is load-sharing traffic to 10.1.12.0/24 over routes through Comware-1 and Comware-2. Comware-1, on the other hand, is using one route for this network.

- Why has ProVision-1 added two routes to 10.1.12.0/24 to its routing table?
- Why does Comware-1 have only one route to 10.1.12.0/24 in its table?
- How could you make ProVision-1 prefer its 10.1.12.0/24 route through Comware-1?

As you learned, OSPF chooses between routes based on metric. Both ProVision routes to 10.1.12.0/24 have the same metric, so both are added to the routing table. ProVision-1 actually has a higher bandwidth connection to Comware-1, but OSPF is running on the Layer 3 interface (the VLAN), and the metric for every VLAN is the same by default. The paths through Comware-1 and Comware-2 have the same number of Layer 3 hops (each hop is a VLAN interface), so they have the same metric. OSPF uses path cost, not hops, to choose routes, but because every interface has the same cost in this network, hop count dictates the cost.

The situation is similar on Comware-1. Comware-1, as well as all other routing switches, is using the same cost on every VLAN interface. However, the route to 10.1.11.0/24 through ProVision-1 has fewer hops than the route through Comware-2 or ProVision-2, so its metric is lower. Comware-1 prefers that route.

If you want to make ProVision-1 prefer its 10.1.12.0/24 route through Comware-1, you should set the metric (OSPF cost) for the VLAN that connects to Comware-1 lower. Set the OSPF cost for the VLAN that connects to Comware-2 higher.

6. You did not delete the static routes on Comware-1. Why is Comware-1 using OSPF routes to forward traffic to 10.1.12.0/24 and other destinations in your local topology?

   By default, Comware switches assign administrative distance 10 to internal OSPF routes. They assign administrative distance 60 to static routes. Therefore, Comware-1 is using the OSPF routes rather than the static routes.

7. You did not delete the static routes on ProVision-1. Why is ProVision-1 using OSPF routes to forward traffic to 10.1.12.0/24 and other destinations in your local topology?

   ProVision switches prefer static routes (administrative distance 1) to OSPF routes (administrative distance 110). However, ProVision-1 does not have any specific static routes, only default routes. Default routes only apply if the routing table does not have a specific route to the destination.

# 10 HP Intelligent Resilient Framework (IRF)

## EXAM OBJECTIVES

✓ Describe how HP Intelligent Resilient Framework (IRF) works and the advantages that it provides

✓ Configure and verify a simple IRF virtual device

## INTRODUCTION

This chapter introduces HP IRF, an HP technology that allows multiple HPE Comware switches or routers in the same series to act as a single virtual device. IRF helps to provide fast failover, scalability, manageability, and high availability.

**Note**

IRF is available on Comware switches. Some ProVision switches support a similar technology, called backplane stacking, which is covered in the *Deploying HP FlexNetwork Core Technologies* course.

### Assumed knowledge

Before reading this chapter, you should have a basic understanding of:

- Link aggregation
- VLANs

### IRF use case

You might be familiar with management stacking, which allows you to manage multiple switches from a single management interface. IRF gives this benefit but goes much further. It lets two or more Comware devices in the same series function as a single virtual switch.

This chapter includes an example configuration in which the Comware-1 and Comware-2 switches shown in Figure 10-1 are combined into a single IRF virtual device.

The switches in an IRF virtual device can be managed as a single entity. You can also treat the switches as a single entity from the point of designing Layer 2 and Layer 3 protocols. IRF, therefore, introduces many benefits in management simplicity, resiliency, operational efficiency, and design simplicity at any layer of the network. You will explore these benefits in more detail throughout this chapter.

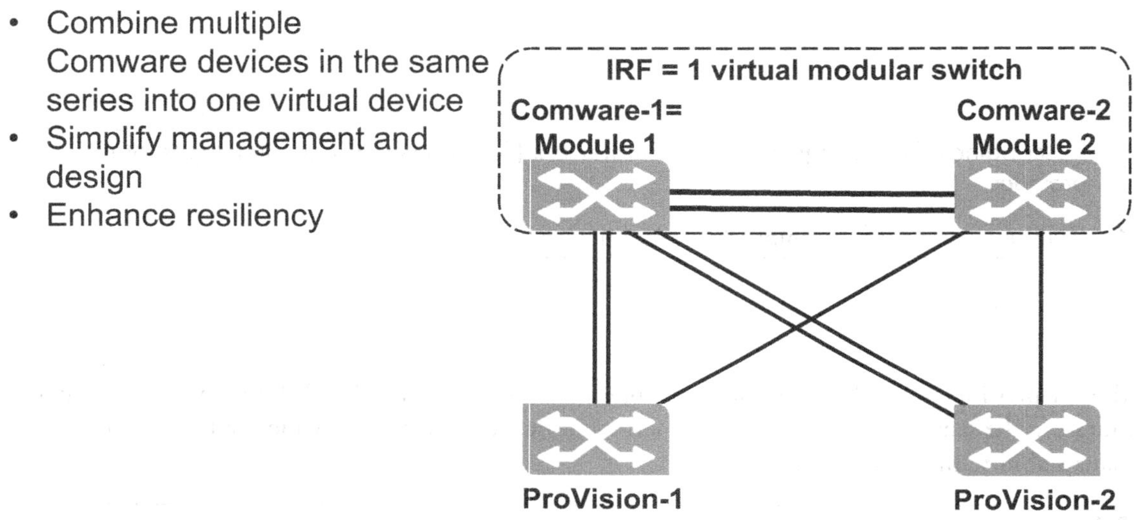


Figure 10-1: IRF use case

In effect, an IRF virtual device is a single modular switch at the core/aggregation layer of the network; that "switch" is distributed physically in two chassis.

Figure 10-2 illustrates this analogy. A modular switch has one active management and control plane in a management module and multiple active forwarding planes in its interface modules. These interface modules also proxy portions of the control plane. An IRF virtual device has one active management and control plane in a master switch and multiple active forwarding planes in members that are like interface modules. The members also proxy portions of the control plane, and their management and control planes are on standby.

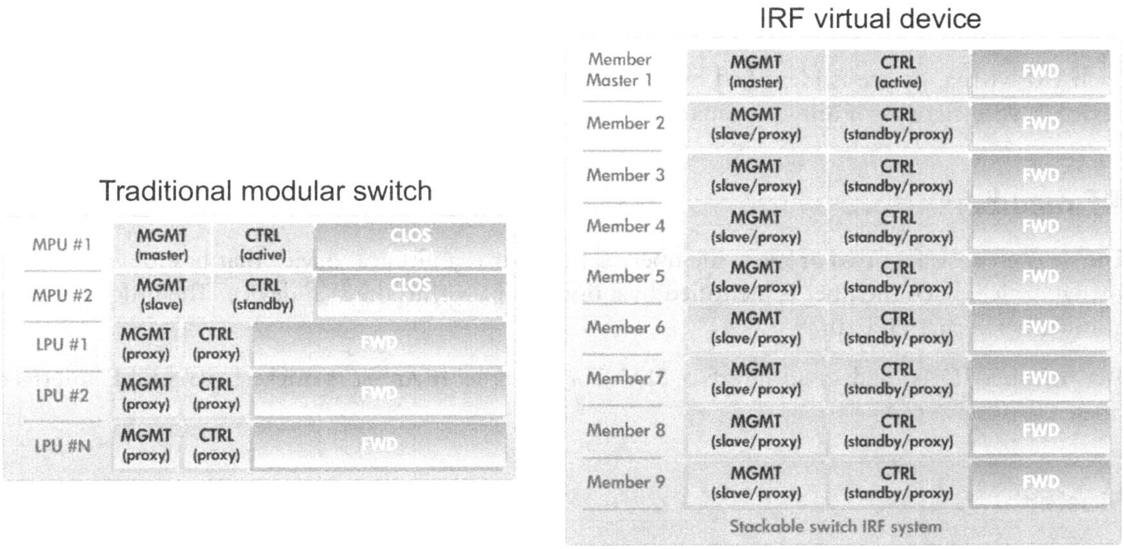


Figure 10-2: IRF analogy

Figure 10-2 shows an IRF virtual device with fixed-port, or stackable, switches, but the IRF can consist of modular switches. In that case, one modular switch's management module is active and others are on standby. All modular switches contribute their interface modules to the IRF device.

## IRF virtual device

Figure 10-3 shows the components of an IRF virtual device in more detail.

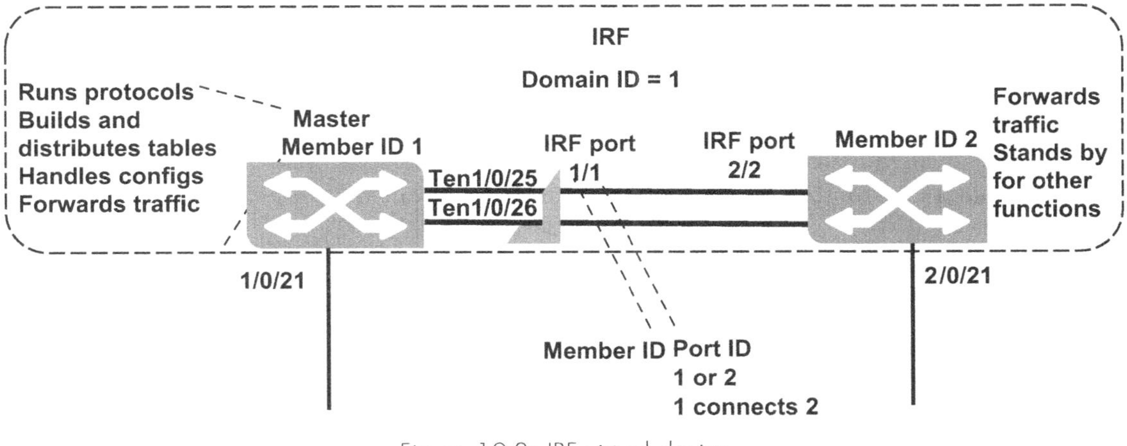


Figure 10-3: IRF virtual device

## IRF domain ID

Each IRF virtual device is identified by a domain ID. It is best practice to assign every IRF virtual device in your network a unique domain ID.

## IRF members

The device consists of two or more members; a member is a physical device that has joined the IRF virtual device. Each member is identified by a unique ID within the IRF device. The unique ID is a number between 1 and 10.

The member ID is used in all interface IDs. For example, member 2 might have a GbE interface labeled 2/0/21.

## Master

One member is the master. In Figure 10-3, the member with ID 1 is master. (The master does not automatically have this ID; you will learn how a master is elected a bit later.)

The master acts as the brain of the virtual device. It provides the active control plane, which means that it handles protocol functions. Only the master builds MAC addressing tables, ARP tables, routing tables, and so on. The master also runs all IP routing protocols.

The master provides the active management plane, which interfaces with the CLI, Web browser interface, and Simple Network Management Protocol (SNMP) solutions to manage the IRF virtual switch configuration.

In addition to providing the device intelligence, the master has active physical interfaces, and it receives and forwards traffic.

### Slaves

All other members are slaves. You can think of the slaves as interface modules in a modular switch. They do not have their own active control and management functions. Instead they receive the configuration for their interfaces from the master. They also receive relevant portions of the MAC addressing tables and routing tables from the master. They can then receive and forward traffic correctly.

Unlike simple interface modules, though, each slave retains the ability to take over the management and control functions if the master fails. Therefore, even though slaves do not have active management and control functions, they do receive copies of the current configuration and store those.

## IRF links

IRF members connect to each other over IRF links. You can think of these links a bit like the backplane paths in a modular switch. These links carry:

- IRF protocol traffic for managing the IRF device
- Data traffic that arrives on one member's interface that needs to be forwarded on a different member's interface

IRF links consist of a logical IRF port on one member connected to a logical port on another member. Each member supports up to two logical IRF ports with IDs 1 and 2. The full ID for the IRF port is member ID/port ID. For example, as shown in Figure 10-4, member 1 has IRF ports 1/1 and 1/2.

Port 1 on one member *must* always connect to port 2 on another member. Member 1 and member 2 can connect on ports 1/1 and 2/2 or on ports 1/2 and 2/1. But they can never connect on ports 1/1 and 2/1 or 1/2 and 2/2.

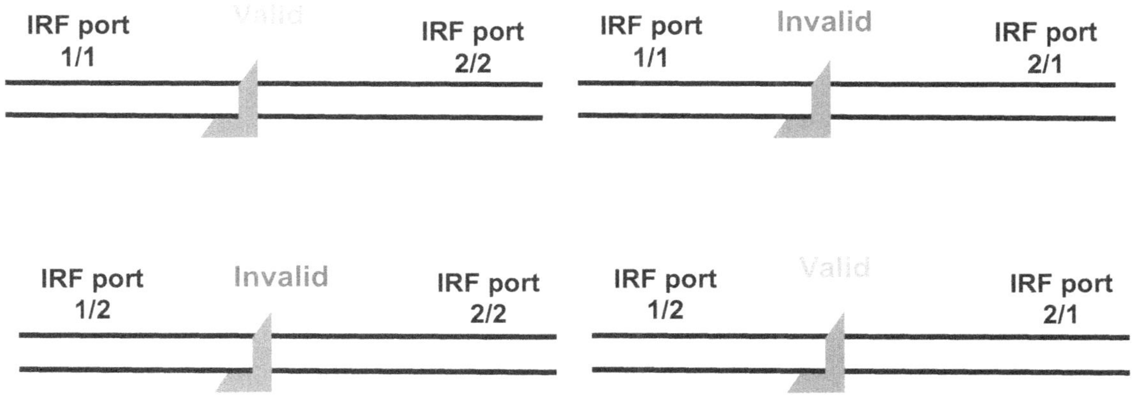


Figure 10-4: Valid and invalid IRF port connections

A logical IRF port is bound to one or more physical interfaces. You should often use more than one interface to enhance redundancy and to increase bandwidth on the IRF link. For modular switches, follow standard high-availability guidelines. For example, choose two links that are on different modules in case one of the modules fails. You can choose any physical interface on a Comware switch that meets the speed requirements.

On many Comware switches, the speed requirement is 10 GbE. But switches with 100 Mbps access ports only require 1 GbE for the links. And some data center switches with a high density of 10 GbE edge ports require 40 GbE interfaces for the IRF links. In fact, the HP 5900 Series requires you to dedicate several interfaces for the IRF links. These switches have interface modules with multiple ASICs, a quad of ports being allocated to each ASIC. You must dedicate all of the ports in an ASIC to the IRF links. In short, different requirements apply to different switch models, and you should check the requirements for your model.

After the physical interface is bound to the IRF port, you cannot configure settings on it. The port is now dedicated to IRF protocol traffic and to data traffic between IRF members. If you have bound multiple physical interfaces to a logical IRF port, the interfaces load balance this traffic much like a link aggregation does.

## IRF configuration process

Figure 10-5 outlines the IRF configuration process. Depending on your environment and switches, you can vary this process a bit, but this process generally works well.

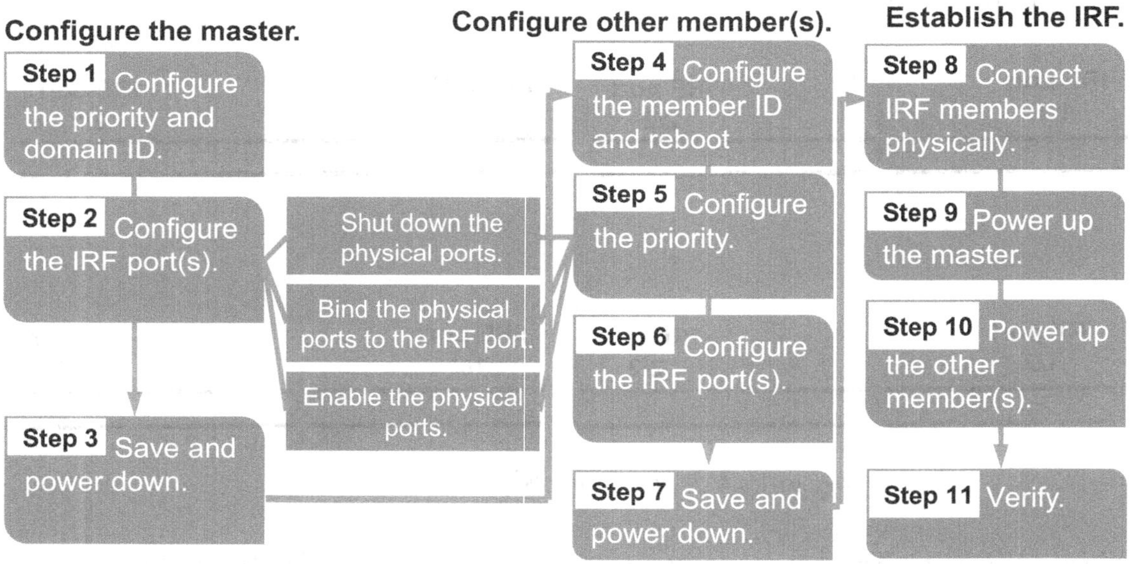


Figure 10-5: IRF configuration process

**Note**

The commands used to configure IRF vary slightly among the different switch models that support this feature. Please check the documentation for the switch model you are using or use the CLI help commands to discover the commands you need.

## Configure the master

You should always choose which switch you want to function as the master. This switch's configuration will be retained for the IRF virtual switch.

**Note**

For some modular switches, you must enable IRF `(chassis convert mode irf)`. The switch automatically reboots in IRF mode.

1. Configure the master's IRF settings:

    a. To ensure this switch is always elected master, you should set the IRF priority to its highest number, which is 32.

    ```
    [Switch] irf member 1 priority 32
    ```

    b. HP networking also recommends that you ensure this switch has a member ID of 1, which is the default setting. The master is not required to have ID 1, but using this convention will simplify the process of managing and troubleshooting an IRF virtual device.

    c. You should also configure the domain ID. The default ID is 1, and every IRF virtual device in your network should have a unique ID.

    ```
    [Switch] irf domain <ID>
    ```

2. Configure the ports used for IRF connectivity.

    a. Shut down the physical ports that you want to bind to the logical IRF ports.

    b. Assign the physical ports to an IRF logical port. If the physical ports are not shut down, the switch will not let you complete this step.

```
[Switch] irf-port memberid/port-number

[Switch-irf-port 1/1] port group interface <interface-type interface-number>
```

**Note**

In the example configuration included in this chapter, IRF members need only one IRF port because they connect to one other member. If the IRF virtual device included more members, you would repeat this step to create a second IRF port.

c. Enable the physical ports. To follow best practices, do *not* physically connect the ports yet.

3. Save the configuration and power the switch down.

Powering down the switch prevents the IRF virtual switch from establishing itself too soon. You will bring the switches up in an ordered fashion later.

## Configure other member(s)

Next you configure each of the remaining switches that will join the IRF virtual device. In the example configuration included in this chapter, you have only one other member, but a similar process applies to each member.

**Important**

Before you begin, make sure that the switches are *not* connected to each other or to the master.

4. Assign each of the other switches a unique member ID. If you do not assign each member a unique member ID, the IRF virtual device cannot be established.

You *must* reboot the switch for the change to take effect and for the IRF virtual device to form. After the reboot, the switch's physical interfaces are renumbered according to the new ID. For example, interface g1/0/1 becomes interface 2/0/1. On a modular switch, a member 2 interface might be g2/1/0/1. Note that you do not have to save the config after you change the ID and before you reboot. The member ID is stored in a separate file in the switch flash, *not* in its configuration. However, you might want to save to preserve other configurations.

```
[Switch] irf member <member-id> renumber <new-member-id>

[Switch] save

[Switch] quit

<Switch> reboot
```

**Note**

Modifying a member ID may change the switch configuration, resulting in the loss of some configuration settings. When configuring an IRF virtual device, however, it is best practice to return slaves to factory default settings. Only the master should have a configuration (and you should save a copy of the configuration to a safe location on your network).

5. Configure the member's priority. The priority can remain at 1, or you can set it to any value less than the master's. The member will take the domain ID from the master after it joins the IRF so that its ID matches the master's.

   ```
   [Switch] irf member 2 priority <value>
   ```

6. Configure the ports used for IRF connectivity:

   a. Shut down the physical ports that you want to bind to the logical IRF ports.

   b. Assign the physical ports to an IRF logical port.

      ```
      [Switch] irf-port memberid/port-number
      ```

      ```
      [Switch-irf-port 1/1] port group interface <interface-type interface-number>
      ```

   c. Enable the physical ports. To follow best practices, do *not* physically connect the ports yet.

7. Save the configuration and power down the switch. It is very important to save the configuration to preserve the IRF settings (even though any other configuration will be replaced by the master). Powering down the switch helps you to ensure that all switches are configured and have their IRF configurations saved before the IRF virtual device begins to establish.

## Establish the IRF virtual device

You are now ready to establish the IRF virtual device.

8. Physically connect the physical ports that you assigned to IRF ports. When connecting ports, remember to connect the physical ports that are bound to IRF port 1 on one switch to the physical ports bound to IRF port 2 on another switch.

9. Power on the switch you configured to be the master.

10. After the master powers up, power on another member connected to the master. The switch should automatically join the IRF virtual device recognize that this device has a master. It will automatically reboot as a slave.

11. When the IRF virtual device has finished forming, you should use display commands such as `display irf` to verify that all members are added.

### Activating the IRF port configuration

Many Comware switches have a command that activates the IRF port configuration:

```
[Switch] irf-port-configuration active
```

On these switches, the IRF virtual device does *not* begin to form automatically after the IRF ports are configured and the physical interfaces are connected. You must activate the port configuration first. Or you must power down and power on the switch, which also activates the ports and initiates a merge.

Using this command, you can take a different approach to the IRF configuration process if you want. As you configure the master and members, *save* their configuration but do not power them down. After you have configured and saved each member, connect the members physically. Then activate the IRF port configuration. The member with the highest priority becomes master, and the other members reboot.

## Example configuration: Establish an IRF virtual device

This section outlines the steps for establishing an IRF virtual device, using the example network shown in Figure 10-6. This network is running Open Shortest Path First (OSPF), as described in "Example configuration: Setting up a basic OSPF configuration," on page 339.

In the example network shown in Figure 10-6, Comware-1 and Comware-2 will be combined to form a single virtual device. You will learn how to:

- Establish an IRF daisy-chain topology between two Comware switches
- Implement IRF

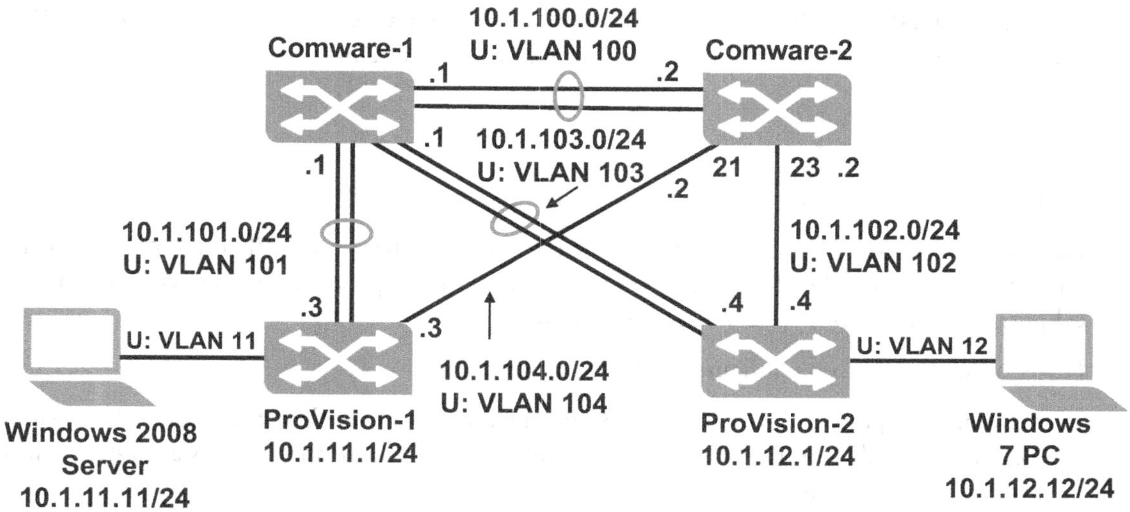


Figure 10-6: Example network

Figure 10-7 shows the network topology after Comware-1 and Comware-2 are combined into a single IRF device. The topology is physically similar to the topology in Figure 10-6. In Figure 10-7, however, Comware-1 and Comware-2 are connected through 10 GbE links, which are required for the IRF links. Also, after the IRF virtual device is formed, Comware-2 will be running a new configuration as part of the IRF virtual switch.

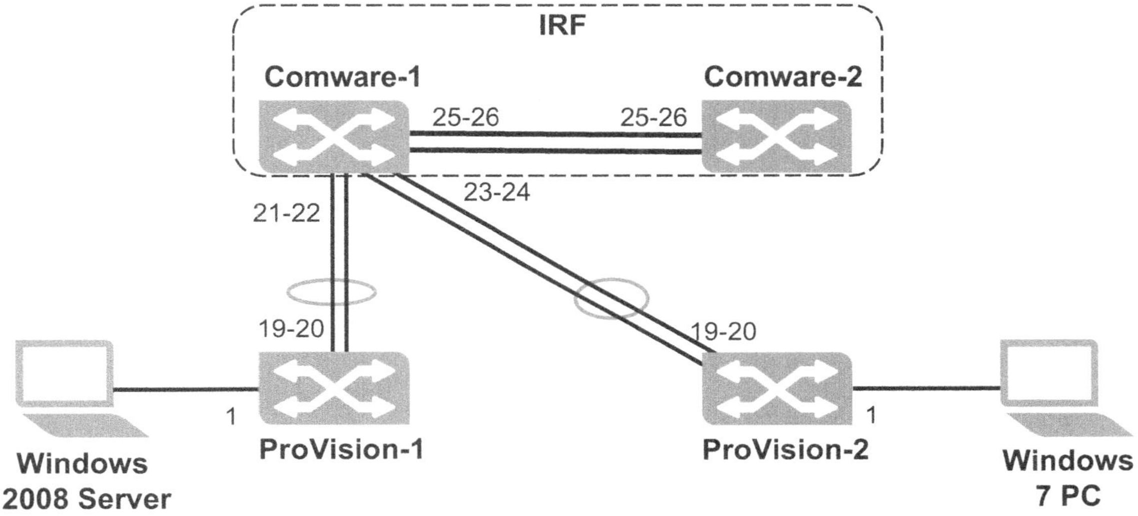


Figure 10-7: Example configuration: Establish an IRF virtual device

## Archive configurations

Before making a change like this, it is a good idea to archive the current configurations.

### Comware-1

1. Access the Comware-1 CLI.
2. Save the switch's current configuration to archive1.cfg.

   If you get a message stating that the file already exists, overwrite it.

   ```
   <Comware-1> save archive1.cfg
   ```

   ```
   The current configuration will be saved to flash:/ archive1.cfg.
   Continue? [Y/N]: y
   ```

3. View the contents of the saved configuration file. Make sure that the information in the file is correct. You should see OSPF settings near the end of the output.

   ```
   <Comware-1> more archive1.cfg
   ```

## Comware-2

4. Access the Comware-2 CLI. Back up the configuration.

```
<Comware-2> save archive1.cfg

The current configuration will be saved to flash:/archive1.cfg.
Continue? [Y/N]: y
```

5. View the contents of the saved configuration file on Comware-2. Make sure that the information in the file is correct. You should see OSPF settings near the end of the output.

```
<Comware-1> more archive1.cfg
```

## ProVision-1

6. Access the ProVision-1 CLI.

7. View the config files in the ProVision switch's flash. The output on your particular switch will differ.

```
ProVision-1# show config files

Configuration files:

 id | act pri sec | name

 ---+-------------+-----------

  1 |  *   *   *  | config1

  2 |             | config2

  3 |             |
```

8. Save the switch's configuration.

```
ProVision-1# write memory
```

9. Copy the startup-config file (based on the name of the file saved in slot 1) to a file called archive1.cfg. In this example, the name of the file is config1.

```
ProVision-1# copy config config1 config archive1.cfg
```

10. View the contents of the saved configuration file on ProVision-1. Make sure that the information in the file is correct. You should see a router ID near the beginning of the output.

```
ProVision-1# show config archive1.cfg
```

### ProVision-2

11. Access the ProVision-2 CLI and archive the startup-config.

## Configure the IRF master

You will set up Comware-1 as the master of the IRF virtual switch and Comware-2 as the slave.

1. Access the Comware-1 CLI and move to system view.

```
<Comware-1> system
```

2. Remove the connection between Comware-1 and Comware-2. The IRF link will take the place of this connection:

   - Shut down the physical interfaces.
   - Remove the bridge aggregation interface.
   - Remove VLAN 100, which is dedicated to this link.

```
[Comware-1] interface g1/0/19
[Comware-1-GigabitEthernet1/0/19] shutdown
[Comware-1-GigabitEthernet1/0/19] interface g1/0/20
[Comware-1-GigabitEthernet1/0/20] shutdown
[Comware-1-GigabitEthernet1/0/20] quit
[Comware-1] undo interface Bridge-Aggregation 1
[Comware-1] undo vlan 100
```

3. Verify the changes.

```
[Comware-1] display interface brief
The brief information of interface(s) under route mode:
Interface      Link      Protocol-link  Protocol type    Main IP
NULL0                UP         UP(spoofing)    NULL            --
Vlan1                DOWN       DOWN          ETHERNET   10.1.1.1
Vlan101              UP         UP            ETHERNET   10.1.101.1
Vlan103              UP         UP            ETHERNET   10.1.103.1
The brief information of interface(s) under bridge mode:
```

```
Link: ADM - administratively down; Stby - standby
Speed or Duplex: (a)/A - auto; H - half; F - full
Type: A - access; T - trunk; H - hybrid
Interface             Link Speed   Duplex Type PVID Description
BAGG2                 UP   2G(a)   F(a)   A    101
BAGG3                 UP   2G(a)   F(a)   A    103
<-output omitted->
GE1/0/19              ADM  auto    A      A    1
GE1/0/20              ADM  auto    A      A    1
<-output omitted->
```

4. Comware-1 will be the master in the IRF virtual device. Generally, the master should have ID 1, so you will keep the switch's current ID. Assign the switch a priority of 32 (highest).

```
[Comware-1] irf member 1 priority 32
```

5. Set the IRF domain ID. You will use 1.

```
[Comware-1] irf domain 1
```

6. Verify the changes.

```
 [Comware-1] display irf
Switch  Role   Priority  CPU-Mac         Description
 *+1   Master  32        d07e-28ce-c950  -----
--------------------------------------------------
 * indicates the device is the master.
 + indicates the device through which the user logs in.
 The Bridge MAC of the IRF is: d07e-28ce-c94f
 Auto upgrade                 : yes
 Mac persistent               : 6 min
 Domain ID                    : 1
```

7. Before you set up the IRF logical port, you *must* shut down the physical interfaces that you will bind to the port, Ten1/0/25 and Ten1/0/26. These ports should already be shutdown. Verify now.

```
[Comware-1] display interface ten1/0/25

Ten-GigabitEthernet1/0/25 current state: DOWN ( Administratively )

<-output omitted->

[Comware-1] display interface ten1/0/26

Ten-GigabitEthernet1/0/26 current state: DOWN ( Administratively )

<-output omitted->
```

8. If the ports were not shut down, do so now. Otherwise, move to the next step.

```
[Comware-1] interface Ten-GigabitEthernet 1/0/25

[Comware-1-Ten-GigabitEthernet1/0/25] shutdown

[Comware-1-Ten-GigabitEthernet1/0/25] interface Ten-GigabitEthernet
1/0/26

[Comware-1-Ten-GigabitEthernet1/0/26] shutdown

[Comware-1-Ten-GigabitEthernet1/0/26] quit
```

9. Create logical IRF port 1/1 and assign the two physical 10 GbE ports to it.

```
[Comware-1] irf-port 1/1

[Comware-1-irf-port 1/1] port group interface ten 1/0/25

Info: You are recommended to save the configuration after completing
your IRF configuration; otherwise, it may be lost after system reboot.

[Comware-1-irf-port 1/1] port group interface ten 1/0/26
```

10. Enable the two 10 GbE ports.

```
[Comware-1-irf-port 1/1] interface ten 1/0/25

[Comware-1-Ten-GigabitEthernet1/0/25] undo shutdown

[Comware-1-Ten-GigabitEthernet1/0/25] interface ten 1/0/26

[Comware-1-Ten-GigabitEthernet1/0/26] undo shutdown

[Comware-1-Ten-GigabitEthernet1/0/26] quit
```

11. Verify the IRF port. Both 10 GbE ports should be assigned to member 1 IRF port 1 (port 1/1).

```
[Comware-1] display irf configuration
 MemberID  NewID   IRF-Port1                     IRF-Port2
  1         1      Ten-GigabitEthernet1/0/25     disable
                   Ten-GigabitEthernet1/0/26
```

12. Verify the IRF topology. At this point, IRF port 1/1 is down because the other side of the link is not configured.

```
[Comware-1] display irf topology
                              Topology Info
 -------------------------------------------------------------
            IRF-Port1                IRF-Port2
 Switch   Link     neighbor   Link       neighbor    Belong To
 1        DOWN     --         DIS        --          d07e-28ce-c950
```

13. Save the configuration on Comware-1.

```
[Comware-1] save force
```

14. Power off Comware-1.

---

**Important**
It is *very* important to complete this step.

---

## Configure the other IRF member

You will now configure Comware-2, which will be a slave in the IRF virtual devices.

1. Access a terminal session with Comware-2.
2. Move to system view.
3. Although it is not strictly necessary, you can make sure that the connection between Comware-1 and Comware-2 is disabled on the Comware-2 side also.

```
[Comware-2] interface g1/0/19
[Comware-2-GigabitEthernet1/0/19] shutdown
[Comware-2-GigabitEthernet1/0/19] interface g1/0/20
```

```
[Comware-2-GigabitEthernet1/0/20] shutdown
[Comware-2-GigabitEthernet1/0/20] quit
[Comware-2] undo interface Bridge-Aggregation 1
[Comware-2] undo vlan 100
```

4. Verify the changes.

```
[Comware-2] display interface brief
The brief information of interface(s) under route mode:
Link: ADM - administratively down; Stby - standby
Protocol: (s) - spoofing
Interface            Link Protocol Main IP         Description
GE1/0/21             UP   UP       10.1.104.2
GE1/0/23             UP   UP       10.1.102.2
NULL0                UP   UP(s)    --
Vlan1                DOWN DOWN     10.1.1.2
The brief information of interface(s) under bridge mode:
Link: ADM - administratively down; Stby - standby
Speed or Duplex: (a)/A - auto; H - half; F - full
Type: A - access; T - trunk; H - hybrid
Interface            Link Speed   Duplex Type PVID Description
<-output omitted->
GE1/0/19             ADM  auto    A      A    1
GE1/0/20             ADM  auto    A      A    1
<-output omitted->
```

5. Save the changes on Comware-2.

```
[Comware-2] save force
```

6. Examine the default IRF information on the switch. As you see, the switch has member ID 1, by default, like Comware-1.

```
[Comware-2] display irf
```

```
Switch  Role    Priority  CPU-Mac         Description
 *+1    Master  1         d07e-28ce-d31f  -----
--------------------------------------------------
<-output omitted->
```

7. Each member of the IRF virtual switch needs a unique ID. Renumber the Comware-2 member number from 1 and 2.

```
[Comware-2] irf member 1 renumber 2
Warning: Renumbering the switch number may result in configuration
change or loss. Continue?[Y/N] y
```

8. Verify the change.

```
[Comware-2] display irf configuration
 MemberID  NewID  IRF-Port1                    IRF-Port2
  1         2     disable                      disable
```

9. Reboot the switch so that the new member ID takes effect. You do not need to save the configuration because the member ID is not held in the config file.

```
[Comware-2] quit
<Comware-2> reboot
Start to check configuration with next startup configuration file,
please wait.........DONE!
 This command will reboot the device. Continue? [Y/N]: y
After Comware-2 has rebooted, verify its member number.
<Comware-2> display irf
Switch  Role    Priority  CPU-Mac         Description
 *+2    Master  1         d07e-28ce-d31f  -----
--------------------------------------------------
<-output omitted->
```

10. Examine the interface nomenclature on Comware-2. Notice how all the interfaces begin with "2," the member number.

```
<Comware-2> display interface brief
<-output omitted->
```

```
Interface       Link     Speed       Duplex   Link-type  PVID
GE2/0/1         DOWN      auto         auto     access     1
GE2/0/2         DOWN      auto         auto     access     1
GE2/0/3         DOWN      auto         auto     access     1
<-output omitted->
```

11. When the interface IDs changed, their configuration was lost. Shut down all interfaces on the switch.

```
<Comware-2> system
[Comware-2] port-group manual allports
[Comware-2-port-group-manual-allports] group-member g2/0/1 to ten2/0/28
[Comware-2-port-group-manual-allports] shutdown
[Comware-2-port-group-manual-allports] quit
```

12. Set the IRF domain ID to match the ID on Comware-1.

```
[Comware-2] irf domain 1
```

13. When you shut down all the ports, you shut down the two 10 GbE interfaces for the IRF link. Therefore, you are ready to create the logical IRF port. Create logical IRF port, 2/2 and bind the two 10 Gb ports to it.

```
[Comware-2] irf-port 2/2
[Comware-2-irf-port 2/2] port group interface ten 2/0/25
 Info: You are recommended to save the configuration after completing
your IRF configuration; otherwise, it may be lost after system reboot.
[Comware-2-irf-port 2/2] port group interface ten 2/0/26
```

14. Enable the two 10 GbE ports.

```
[Comware-2-irf-port 2/2] interface ten 2/0/25
[Comware-2-Ten-GigabitEthernet2/0/25] undo shutdown
[Comware-2-Ten-GigabitEthernet2/0/25] interface ten 2/0/26
[Comware-2-Ten-GigabitEthernet2/0/26] undo shutdown
[Comware-2-Ten-GigabitEthernet2/0/26] quit
```

15. Verify the IRF configuration and topology. The physical interfaces should be assigned to IRF port 2, which enables them to connect to IRF port 1 on the other member.

```
[Comware-2] display irf configuration

 MemberID  NewID  IRF-Port1     IRF-Port2
   2        2     disable       Ten-GigabitEthernet2/0/25
                                Ten-GigabitEthernet2/0/26

[Comware-2] display irf topology

                         Topology Info
 ---------------------------------------------------------------
              IRF-Port1                    IRF-Port2
 Switch    Link       neighbor     Link     neighbor   Belong To
 2         DIS        --           DOWN     --         0023-89d9-c162
```

16. Save the configuration.

```
[Comware-2] save force
```

17. Power off Comware-2.

## Establish the IRF virtual switch

You will now connect the switches and power them on to create the IRF virtual switch.

18. Establish the physical connections between Comware-1 and Comware-2, as shown in Figure 10-8.

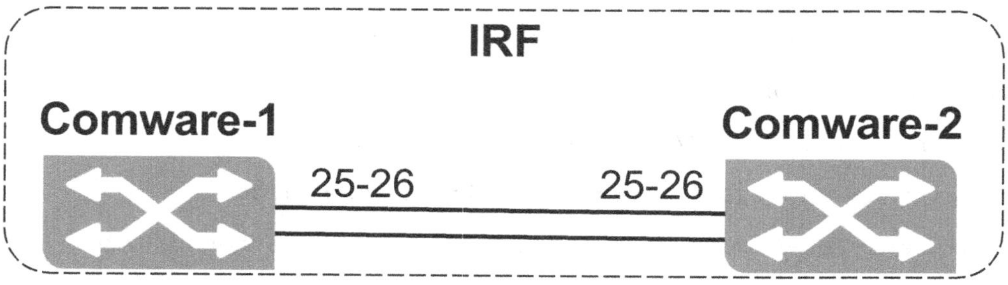


Figure 10-8: Physical connections between Comware-1 and Comware-2

**Note**
This example configuration was developed in a remote lab environment, which has permanent connections between Comware-1 and Comware-2. In such an environment, you must make sure that the IRF virtual switch does not start to form before you save the IRF settings on Comware-2. Otherwise, Comware-2 will lost its IRF settings when it joins the device and reboots. In this case, the IRF formation will fail.

## Comware-1

19. Power on Comware-1.

20. Access the Comware-1 CLI. When it has finished rebooting, verify the IRF configuration to ensure that the configuration booted correctly.

```
<Comware-1> display irf configuration

 MemberID  NewID  IRF-Port1                      IRF-Port2

  1         1      Ten-GigabitEthernet1/0/25      disable

                   Ten-GigabitEthernet1/0/26
```

## Comware-2

21. Power on Comware-2.

## Comware-1

22. Access the Comware-1 CLI. After 2 or 3 minutes, you will notice some messages concerning IRF. The following output is abbreviated with relevant messages highlighted. These messages indicate that Comware-2 is being added to the IRF virtual device. Note that the messages refer to Comware-2 as a "board"; Comware-2 is like a module being added to the virtual IRF switch.

```
%Apr 26 12:04:35:620 2000 Comware-1 IFNET/3/LINK_UPDOWN: Ten-GigabitEthernet1/0/25 link status is UP.

%Apr 26 12:04:35:741 2000 Comware-1 STM/6/STM_LINK_STATUS_UP:

 IRF port 1 is up.

#Apr 26 12:04:35:841 2000 Comware-1 STM/4/LINK STATUS CHANGE:

 Trap 1.3.6.1.4.1.25506.2.91.6.0.1: Physical index of the member is 2, member ID is 1. Link status of the IRF port with port index 1 and member ID 1 turned to 1.
```

```
%Apr 26 12:04:36:132 2000 Comware-1 IFNET/3/LINK_UPDOWN: Ten-GigabitEthernet1/0/26 link status is UP.

#Apr 26 12:04:55:586 2000 Comware-1 DEVM/1/BOARD INSERTED:

 Trap 1.3.6.1.4.1.25506.8.35.12.1.9: chassisIndex is 0, slotIndex 0.2

%Apr 26 12:04:55:746 2000 Comware-1 DEVM/2/BOARD_STATE_FAULT: Board state changes to FAULT on Chassis 0 Slot 2, type is unknown.

#Apr 26 12:05:57:544 2000 Comware-1 DEVM/1/FAN STATE CHANGES TO NORMAL: -Slot=2;

 Trap 1.3.6.1.4.1.25506.8.35.12.1.7: fan ID is 3

#Apr 26 12:05:57:704 2000 Comware-1 DEVM/1/BOARD STATE CHANGES TO NORMAL:

 Trap 1.3.6.1.4.1.25506.8.35.12.1.11: chassisIndex is 0, slotIndex 0.2

%Apr 26 12:05:57:885 2000 Comware-1 DEVM/5/FAN_RECOVERED: -Slot=2; Fan 3 recovered.

%Apr 26 12:05:57:985 2000 Comware-1 DEVM/5/BOARD_STATE_NORMAL: Board state changes to NORMAL on Chassis 0 Slot 2, type is MAIN_BOARD_TYPE_32C.

%Apr 26 12:06:03:487 2000 Comware-1 IFNET/3/LINK_UPDOWN: Ten-GigabitEthernet2/0/25 link status is UP.

%Apr 26 12:06:03:703 2000 Comware-1 IFNET/3/LINK_UPDOWN: Ten-GigabitEthernet2/0/26 link status is UP.

%Apr 26 12:06:04:053 2000 Comware-1 HA/5/HA_BATCHBACKUP_STARTED: Batch backup(s) of slave board(s) in slot 2 started.

%Apr 26 12:06:04:196 2000 Comware-1 IFNET/3/LINK_UPDOWN: -Slot=2; Aux0/0/0 link status is UP.

%Apr 26 12:06:10:784 2000 Comware-1 HA/5/HA_BATCHBACKUP_FINISHED: Batch backup of slave board in slot 2 is finished.

%Apr 26 12:06:12:735 2000 Comware-1 OPTMOD/4/MODULE_IN: -Slot=2;

 Ten-GigabitEthernet2/0/26: The transceiver is STACK_SFP_PLUS.
```

23. When Comware-2 joined the IRF virtual switch, it lost its interface configuration because Comware-1 does not have that configuration. Shut down all member 2 to prevent loops. Note that the 10 GbE ports bound to the IRF ports are automatically excluded from the range.

If you see an error when you specify the port range, wait until Comware-2 is fully joined.

```
[Comware-1] port-group manual all2ports
[Comware-1-port-group-manual-all2ports] group-member g2/0/1 to ten2/0/28
[Comware-1-port-group-manual-all2ports] shutdown
[Comware-1-port-group-manual-all2ports] quit
```

## Explore the IRF virtual device

You will now use display commands to explore the IRF configuration.

### Comware-1

24. On Comware-1, examine the IRF topology. Notice that IRF-Port2 of switch 2 (Comware-2) is now connected to IRF-Port 1 of switch 1 (Comware-1).

```
<Comware-1> display irf topology
                    Topology Info
 ---------------------------------------------------------------
               IRF-Port1                 IRF-Port2
 Switch  Link   neighbor     Link    neighbor    Belong To
 2       DIS    --           UP      1           d07e-28ce-c950
 1       UP     2            DIS     --          d07e-28ce-c950
```

25. View the IRF information. Make sure that Comware-1 is the master and Comware-2 is the slave.

```
<Comware-1> display irf
Switch  Role    Priority  CPU-Mac         Description
 *+1   Master   32        d07e-28ce-c950  -----
   2   Slave    1         d07e-28ce-d31f  -----
--------------------------------------------------
<-output omitted->
```

26. View interfaces on the IRF virtual switch. Notice that both 1/0 and 2/0 interfaces are listed. Scroll through the output until you reach the 10 GbE interfaces. Notice that the interfaces that are bound to the IRF ports no longer have PVID or port type (trunk/access) information. These interfaces now carry any traffic between members.

```
<Comware-1> display interface brief
<-output omitted->
XGE1/0/25           UP    --        --      --    --
XGE1/0/26           UP    --        --      --    --
XGE1/0/27           ADM   auto      A       A     1
XGE1/0/28           ADM   auto      A       A     1
XGE2/0/25           UP    --        --      --    --
XGE2/0/26           UP    --        --      --    --
<-output omitted->
```

## Comware-2

27. Access the Comware-2 CLI.

28. Notice that the prompt changed—Comware-1's configuration was applied to the slave (Comware-2). Be aware of this when you are working on the switches so that you are not confused.

```
<Comware-1>
```

## Comware-2

29. You can configure the IRF virtual device from the Comware-2 session, but return to the Comware-1 session to avoid confusion.

30. Save the configuration. Notice that the configuration is also saved to Comware-2 (slot 2).

```
<Comware-1> save force
Saved the current configuration to mainboard device successfully.
Slot 2:
 Save next configuration file successfully.
 Configuration is saved to device successfully.
```

## Learning check 1

Take a few minutes to test your understanding of IRF virtual devices by answering the following questions.

1. How did you make sure that Comware-1 would be elected master?

2. After you assigned Comware-2 a new member ID, how did you make the change take effect?

3. Why did you not save the configuration before rebooting Comware-2? Why did the change to the member ID persist?

4. After the IRF virtual device formed, what configuration was it running?

5. Why did Comware-2 lose all of its settings for its connections to the ProVision switches? Were any settings on Comware-2 preserved?

6. What did you observe when you saved the configuration on the IRF virtual device?

## Learning check 1 answers

1. How did you make sure that Comware-1 would be elected master?

   You assigned it the highest priority value (32).

2. After you assigned Comware-2 a new member ID, how did you make the change take effect?

   You had to reboot the switch.

3. Why did you not save the configuration before rebooting Comware-2. Why did the change to the member ID persist?

   The member ID is not stored in the typical configuration. This means that it persists across reboots and configuration resets. If you want to change the member ID, you must enter the command to change it and then reboot.

4. After the IRF virtual device formed, what configuration was it running?

   The IRF virtual device ran the master's configuration.

5. Why did Comware-2 lose all of its settings for its connections to the ProVision switches? Were any settings on Comware-2 preserved?

   Comware-2 joined the IRF virtual device as a slave. It lost all of its configurations and took the configurations from the IRF virtual device (master) instead. The IRF virtual device did not have any settings for member 2 physical interfaces because member 2 just joined. (You had to shut down the interfaces to prevent loops in your environment.)

   The only preserved settings on Comware-2 are its member ID and IRF port settings, which are added to the IRF virtual device settings.

6. What did you observe when you saved the configuration on the IRF virtual device?

   The configuration was also copied to slot 2, which means member 2. Slaves receive copies of the configuration because they might need to take over as master.

   In fact, the running-config is also synchronized to slaves when changes are made to it. The slaves apply global changes and changes to their own interface so that they behave correctly.

## Master election

In the example configuration, Comware-1 became master because it had the highest priority (and you booted it first).

As Figure 10-9 shows, master elections occur whenever the topology of the IRF virtual device changes:

- Multiple IRF virtual devices are merged
- An IRF virtual device is first formed
- The master fails or loses connectivity on its IRF links with the rest of the group
- A new member joins
- The IRF virtual device splits

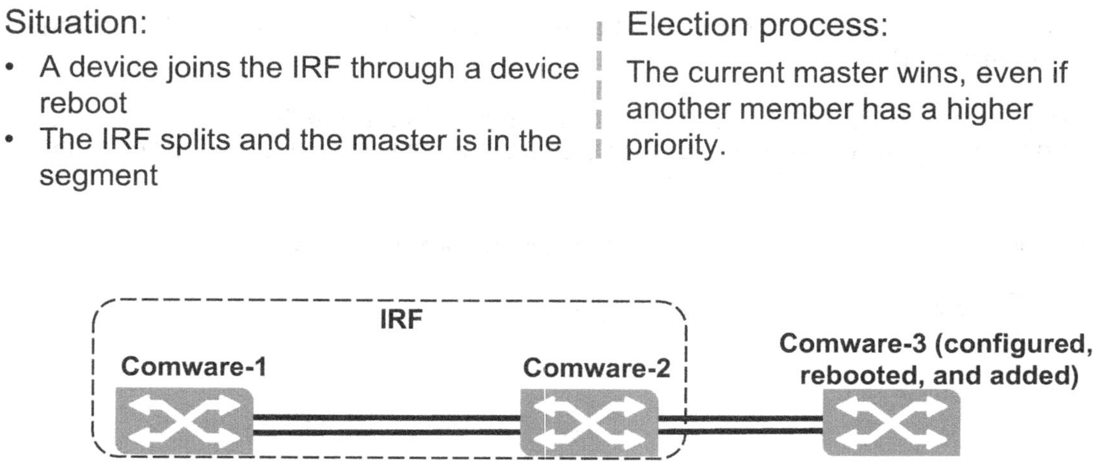


Figure 10-9: Master election

You should understand the master election process in a bit more detail. The precise process depends on the topology change and also whether new devices are joining the IRF as they boot or while they are online.

Begin by considering a situation in which a device is joining the IRF virtual device as that device boots. For example, you have an IRF virtual device with two members, as shown in Figure 10-9. Comware-1, with member ID 1, is currently the master. You set up IRF on Comware-3 and then shut down Comware-3, connect it to the existing IRF virtual switch, and boot Comware-3. Comware-3 joins the IRF.

In this scenario, the current master wins *even if the new member has a higher priority*. The new member reboots and takes its configuration from the existing IRF virtual device.

Refer to Figure 10-10 and consider several situations in which the rule about the current master staying master does not apply.

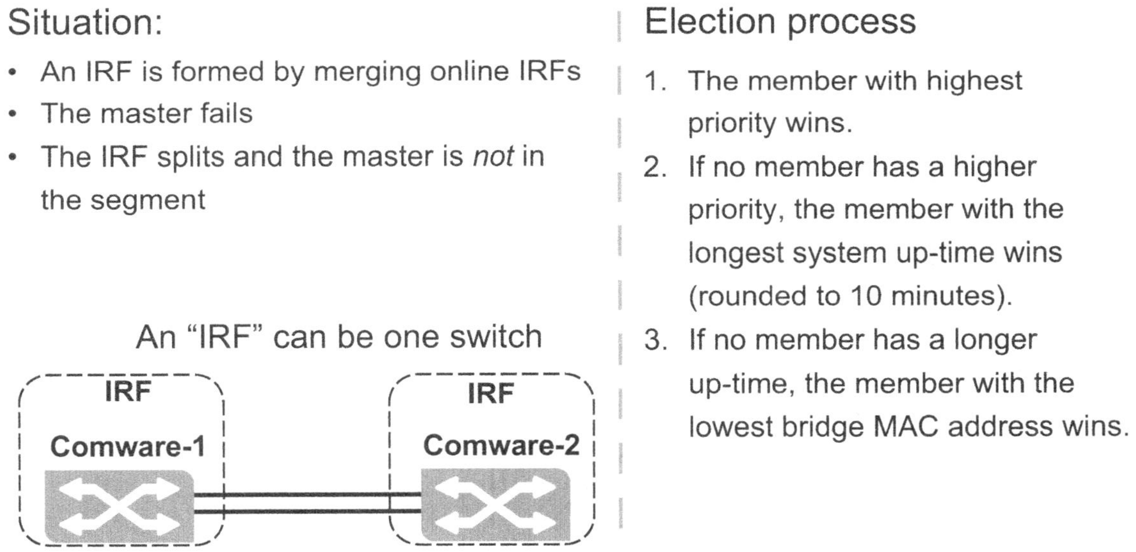


Figure 10-10: Master election

First consider a situation in which the IRF virtual device forms through a merge of online IRF virtual devices. A merge of online devices occurs when you use the `irf-port-configuration active` command to activate the IRF ports and initiate the merge rather than powering the switches down and the powering them up, one by one. Note that an "IRF virtual device" could be a single switch set up for IRF that has not yet merged with other devices.

In this case, the master election follows this process:

1. The member with highest priority wins.

   Be careful joining a new online member to an existing IRF virtual device (using the `irf-port-configuration active` command). If the new member has a higher priority it becomes the master. The devices in the new IRF virtual device reboot and take their configuration from the new master.

2. If no member has a higher priority, the member with the longest system uptime wins.

   The system uptime is rounded to 10 minutes. For example, two members are in the election. One switch has an uptime of 330 days, 5 hours, and 24 minutes and one has an uptime of 330 days, 5 hours, and 20 minutes. The switches have the same effective uptime.

3. If no member has a longer uptime, the member with the lowest bridge MAC address wins.

Other situations in which this process occurs include any topology change that removes the current master:

- The master fails
- The IRF virtual device splits, and the IRF segment does not include the current master

If you want a particular member to take over in the case of a master failure, give it a medium priority such as 25 or 30.

## Detailed look at IRF benefits

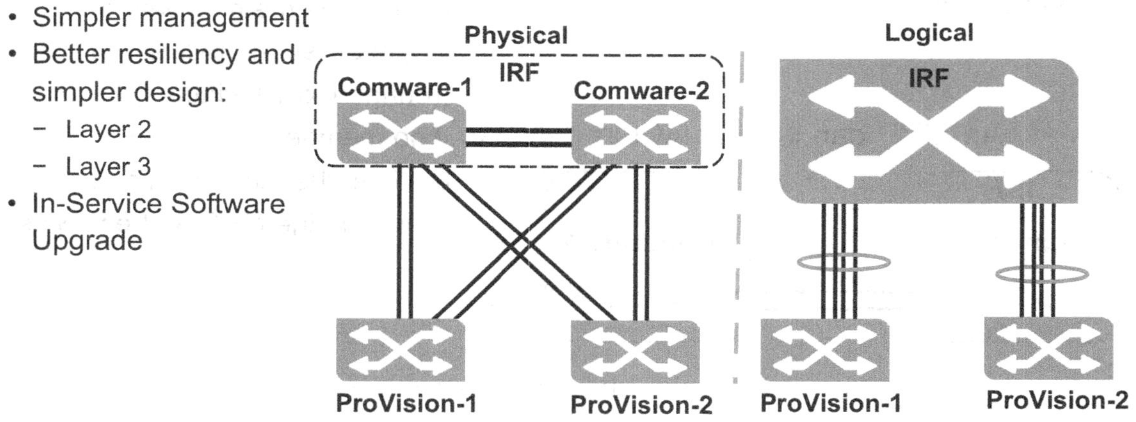


Figure 10-11: Detailed look at IRF benefits

IRF enables two or more Comware switches function as a single virtual switch, as illustrated in Figure 10-11. In the example network topology show in Figure 10-11, Comware-1 and Comware-2 are a single IRF virtual device. In effect, you now have a single modular switch at the core/distribution layer of your network; that "switch" is distributed physically in two chassis.

Begin to think about the benefits of this situation:

- **Management simplicity**—You no longer need to laboriously connect to, configure, and manage switches individually. You perform a configuration on the primary switch, and that configuration is distributed to all associated switches automatically, considerably simplifying network setup, operation, and maintenance.
- **Better resiliency and simpler design at Layer 2**—Chapter 7 introduced you to Rapid Spanning Tree Protocol (RSTP) and Multiple Spanning Tree Protocol (MSTP), demonstrating how important these protocols are for eliminating loops in redundant Layer 2 networks. But you also learned about some of the drawbacks:
    - **Relatively slow network convergence**—RSTP or MSTP convergence can take up to one second, which can be too long for modern applications.
    - **Management and troubleshooting complexity**—MSTP and RSTP take time to configure properly, especially in a large network. You must manage the switches individually, and you need to set up spanning-tree instances on each switch in turn, making sure that the parameters for one switch match those of its neighbor. The real problem emerges in maintenance, however. Troubleshooting spanning-tree-related issues is no easy task, usually requiring a great deal of time to locate the root cause of the failure.
    - **Blocked links**—RSTP blocks all but one set of redundant parallel paths. MSTP permits the use of some redundant links but does not load balance very granularly. Half (or more) of the available system bandwidth can be squandered in a backup role, off-limits to data traffic—not a very good use of the network infrastructure investment.

In Chapter 8, you saw how link aggregations compare favorably to RSTP and MSTP. However, traditionally, you can only create link aggregations between the same two devices. In a truly redundant Layer 2 network, a switch needs links to more than one other switch, so the network must use RSTP or MSTP in conjunction with link aggregations.

But with IRF, two or more physical switches can act as one virtual switch. Therefore, you can connect an access layer switch like ProVision-1 to both Comware-1 and Comware-2 in single link aggregation. Spanning tree is no longer necessary to eliminate loops on switch-to-switch links because at a logical level no loop exists. You still use spanning tree to eliminate accidental loops.

An IRF virtual device can also connect to servers that support link aggregation (sometimes called NIC teaming). The server and IRF virtual device treat the link like one logical link (as shown in Figure 10-12), but the server has the high availability that comes from connecting to multiple switches.

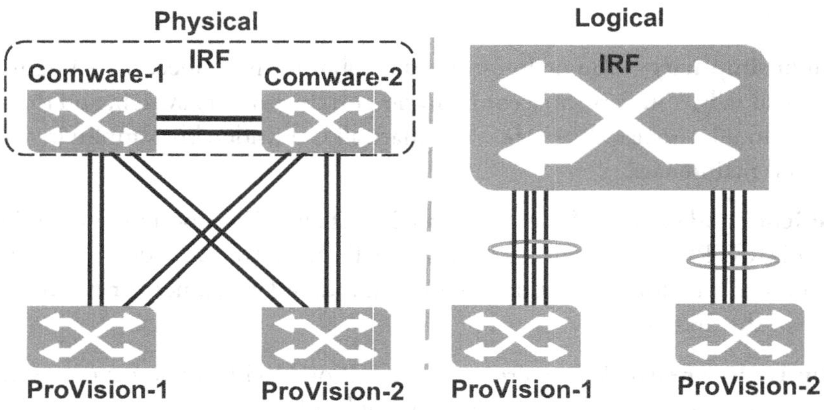

Figure 10-12: IRF virtual device connected to a server

IRF provides all the advantages of a completely link-aggregation-based Layer 2 design:

- Faster failover
- Load balancing
- Simple set up and design
- Better stability and less need for troubleshooting

- **Better resiliency and simpler design at Layer 3**—Companies require redundancy at Layer 3. In Module 8, you learned about using MSTP to provide redundancy at Layer 2, but this solution does not provide Layer 3 redundancy. Subnets have only a single default router. If a switch that acts as the clients' default routers fails and a backup is not available, clients and servers could no longer reach each other.

A standard protocol does exist for solving this problem—Virtual Router Redundancy Protocol (VRRP). (VRRP is not covered in this study guide.) IRF, however, provides an elegant solution for router redundancy. As an IRF virtual device, Comware-1 can be the default router. Because Comware-1 actually consists of two physical devices, it provides redundancy for router functions. If one of those devices fails, the other takes over seamlessly.

In this study guide, you have also learned that using static routing to create a redundant routing topology is not a trivial task. A dynamic routing protocol provides better redundancy, but still introduces complexity, particularly as routing devices and redundant links between them multiply. IRF helps to simplify the situation by creating fewer logical routing devices and simpler topologies between them. You do need to configure protocols to reduce the impact of an IRF master failure; however, these are beyond the scope of this study guide.

- **In-Service-Software-Upgrade:** IRF delivers a network-based In-Service-Software-Upgrade (ISSU) capability that allows an individual IRF-enabled switch to be taken offline for servicing or software upgrades without affecting traffic going to other switches in the IRF virtual switch.

---

**Note**

ISSU is beyond the scope of this study guide but is covered in the HP ASE-level networking courses.

---

## Example configuration: Set up distributed link aggregation with IRF

As you learned, a benefit of IRF is that you can create aggregated links between any members of the IRF virtual device and another switch. In this way, you can form a simple, but highly resilient, topology.

In this section, you will learn how to create these aggregated links in the example network shown in Figure 10-13. This is the same example network used in "Example configuration: Establish an IRF virtual device" on page 362.

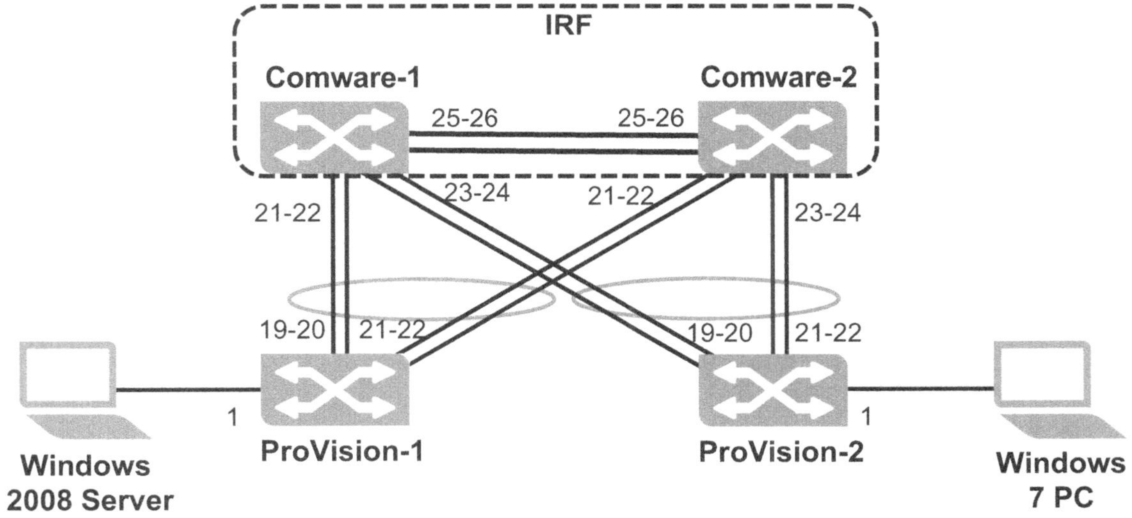


Figure 10-13: Configure distributed link aggregation with IRF

## Establish a distributed aggregated link between the IRF virtual switch and ProVision-1

In the example network shown in Figure 10-13, each ProVision switch can now connect to both Comware-1 and Comware-2 in the same link aggregation. This is because Comware-1 and Comware-2 are part of the same virtual switch.

ProVision-1 already connects to Comware-1 on a link aggregation. You will now simply add interfaces to these links.

### Comware-1

1. Access the Comware-1 CLI and move to the system view.
2. Add interfaces 2/0/21 and 2/0/22 (which are on Comware-2) to the BAGG interface 2. Also enable the interfaces.

```
[Comware-1] interface g2/0/21
[Comware-1-GigabitEthernet2/0/21] port link-aggregation group 2
[Comware-1-GigabitEthernet2/0/21] undo shutdown
[Comware-1-GigabitEthernet2/0/21] interface g2/0/22
[Comware-1-GigabitEthernet2/0/22] port link-aggregation group 2
[Comware-1-GigabitEthernet2/0/22] undo shutdown
[Comware-1-GigabitEthernet2/0/22] quit
```

3. You must configure the BAGG interface VLAN assignment again so that it applies to the new ports.

```
[Comware-1] interface bridge-aggregation 2
[Comware-1-Bridge-Aggregation2] port access vlan 101
[Comware-1-Bridge-Aggregation2] quit
```

### ProVision-1

You are now ready to configure the ProVision-1 side of the link aggregation.

4. Access a terminal session with ProVision-1.
5. Examine the existing aggregated link.

```
ProVision-1(config)# show trunk
 Load Balancing Method: L3-based (Default), L2-based if non-IP traffic
```

```
  Port    | Name                 Type       | Group  Type
  ------ + ------------------- --------- + ------ --------
  19      |                      100/1000T | Trk1   Trunk
  20      |                      100/1000T | Trk1   Trunk
```

6. Add the two interfaces that connect to Comware-2 to this trunk. (The VLAN configuration already applied to the trunk automatically takes effect.)

```
ProVision-1(config)# trunk 21-22 trk1 trunk
```

7. Enable interface 22 on ProVision-1.

```
ProVision-1(config)# interface 22 enable
```

## Comware-1

8. Return to the Comware-1 terminal session. Verify the status of the link aggregation.

```
[Comware-1] display link-aggregation verbose Bridge-Aggregation 2
Loadsharing Type: Shar -- Loadsharing, NonS -- Non-Loadsharing
Port Status: S -- Selected, U -- Unselected
Flags:  A -- LACP_Activity, B -- LACP_Timeout, C -- Aggregation,
        D -- Synchronization, E -- Collecting, F -- Distributing,
        G -- Defaulted, H -- Expired
Aggregation Interface: Bridge-Aggregation2
Aggregation Mode: Static
Loadsharing Type: Shar
  Port              Status    Oper-Key
-----------------------------------------------------
  GE1/0/21          S         1
  GE1/0/22          S         1
  GE2/0/21          S         1
  GE2/0/22          S         1
```

## Establish a distributed link aggregation between the IRF virtual switch and ProVision-2

This task is similar to Task 1. You will add links between ProVision-2 and Comware-2 to the link aggregation that already exists between ProVision-2 and Comware-1. This link aggregation uses Link Aggregation Control Protocol (LACP).

### Comware-1

9. You should be in a terminal session with Comware-1.
10. Add interfaces 2/0/23 and 2/0/24 (physically on Comware-2) to BAGG interface 3. Remember to enable the interfaces.

```
[Comware-1] interface g2/0/23
[Comware-1-GigabitEthernet2/0/23] port link-aggregation group 3
[Comware-1-GigabitEthernet2/0/23] undo shutdown
[Comware-1-GigabitEthernet2/0/23] interface g2/0/24
[Comware-1-GigabitEthernet2/0/24] port link-aggregation group 3
[Comware-1-GigabitEthernet2/0/24] undo shutdown
[Comware-1-GigabitEthernet2/0/24] quit
```

11. You must configure the BAGG interface VLAN assignment again so that it applies to the new ports.

```
[Comware-1] interface bridge-aggregation 3
[Comware-1-Bridge-Aggregation3] port access vlan 103
[Comware-1-Bridge-Aggregation3] quit
```

### ProVision-2

12. Access a terminal session with ProVision-2. Move to global configuration mode.
13. Examine the existing aggregated link.

```
ProVision-2(config)# show trunk
 Load Balancing Method: L3-based (Default), L2-based if non-IP traffic
  Port   | Name                Type       | Group  Type
  ------ + ------------------ --------- + ------ --------
```

```
  19     |                        100/1000T | Trk1    LACP
  |                        100/1000T | Trk1    LACP
```

14. Add the two interfaces that connect Comware-2 to this trunk.

```
ProVision-2(config)# trunk 21-22 trk1 lacp
```

15. Enable interface 22 on ProVision-2.

```
ProVision-2(config)# interface 22 enable
```

16. Examine LACP peer information for the link aggregation on ProVision-2. Note that the peer system ID and operational key are the same on all interfaces whether they connect to the physical Comware-1 or physical Comware-2.

```
ProVision-2(config)# show lacp peer

LACP Peer Information.

System ID: d4c9ef-84fe80

Local Local                        Port     Oper LACP   Tx
Port  Trunk System ID      Port Priority Key  Mode   Timer
----- ----- ------------- ---- -------- ---- ------ -----
19    Trk1  d07e28-cec94f 23   32768    3    Active Slow
20    Trk1  d07e28-cec94f 24   32768    3    Active Slow
21    Trk1  d07e28-cec94f 59   32768    3    Active Slow
22    Trk1  d07e28-cec94f 60   32768    3    Active Slow
```

## Comware-1

17. Return to the Comware-1 terminal session.

18. Verify the status of the link aggregation. Remember that you can verify that LACP link aggregation is functioning correctly by looking for the ACDEF flags.

```
[Comware-1] display link-aggregation verbose Bridge-Aggregation 3

Loadsharing Type: Shar -- Loadsharing, NonS -- Non-Loadsharing

Port Status: S -- Selected, U -- Unselected

Flags:  A -- LACP_Activity, B -- LACP_Timeout, C -- Aggregation,

        D -- Synchronization, E -- Collecting, F -- Distributing,
```

```
        G -- Defaulted, H -- Expired
Aggregation Interface: Bridge-Aggregation3
Aggregation Mode: Dynamic
Loadsharing Type: Shar
System ID: 0x8000, d07e-28ce-c94f
Local:
  Port              Status  Priority Oper-Key  Flag
--------------------------------------------------------------
  GE1/0/23          S       32768    3         {ACDEF}
  GE1/0/24          S       32768    3         {ACDEF}
  GE2/0/23          S       32768    3         {ACDEF}
  GE2/0/24          S       32768    3         {ACDEF}
Remote:
Actor    Partner Priority Oper-Key  SystemID               Flag
--------------------------------------------------------------------
GE1/0/23   19     0        562     0xfe80, d4c9-ef84-fe80 {ACDEF}
GE1/0/24   20     0        562     0xfe80, d4c9-ef84-fe80 {ACDEF}
GE2/0/23   21     0        562     0xfe80, d4c9-ef84-fe80 {ACDEF}
GE2/0/24   22     0        562     0xfe80, d4c9-ef84-fe80 {ACDEF}
```

## Task 3: Verify connectivity

You will now test connectivity in your environment.

### ProVision-1

19. Access the terminal session with ProVision-1.
20. View routes on ProVision-1. You should see one route to VLAN 12.

```
ProVision-1(config)# show ip route
                          IP Route Entries
```

```
Destination   Gateway     VLAN Type       Sub-Type Metric Dist.
------------ ----------- ---- ---------- -------- ------ -----
10.1.11.0/24  VLAN11      11   connected           1      0
10.1.12.0/24  10.1.101.1 101  ospf       IntraArea 3      110
10.1.101.0/24 VLAN101     101  connected           1      0
10.1.103.0/24 10.1.101.1 101  ospf       IntraArea 2      110
127.0.0.0/8   reject           static              0      0
127.0.0.1/32  lo0              connected           1      0
```

## Windows server

21. Access the Windows 2008 server and open a command prompt.
22. Ping the Windows 7 PC (10.1.12.12) and verify success.

```
ping 10.1.12.12
```

23. Trace the route.

```
tracert -d 10.1.12.12
Tracing route to 10.1.12.12 over a maximum of 30 hops
  1    <1 ms    <1 ms    <1 ms  10.1.11.1
  2     1 ms     1 ms     1 ms  10.1.101.1
  3    <1 ms    <1 ms    <1 ms  10.1.103.4
  4    <1 ms    <1 ms    <1 ms  10.1.12.12
Trace complete.
```

# Tracing traffic flow

You should understand how IRF virtual devices handle traffic forwarding. Although the master runs protocols and builds forwarding tables, all traffic does *not* need to pass through it. Instead traffic is forwarded much more efficiently.

IRF uses a distributed forwarding and routing model. The members in the IRF virtual device synchronize forwarding and routing tables, but each member in the stack forwards frames or routes packets independently.

## Switching at Layer 2

Each member in the IRF system learns MAC addresses on its own ports, which it forwards to the master. The master then distributes learned MAC addresses to all members so that each member can then handle traffic immediately no matter where the traffic arrives, minimizing traffic flooding.

When a member device receives a frame to be forwarded at Layer 2, it finds the outbound interface of the frame by searching its Layer 2 forwarding table. It then forwards the packet out the outbound interface, which might be on the local member or on another member device. If the packet is on another member device, the source member forwards the packet on an IRF link, encapsulating the frame in an IRF header that indicates the proper outbound interface without further processing by other members.

Note that an IRF virtual device learns MAC addresses on the link aggregation as a whole. When an IRF member needs to forward traffic on a link aggregation, it prefers its own ports in that link aggregation. In a design like this one—in which all members in the IRF virtual device have connections in aggregated links to connected switches—members can forward traffic very efficiently with a minimum of traffic that must travel over IRF links.

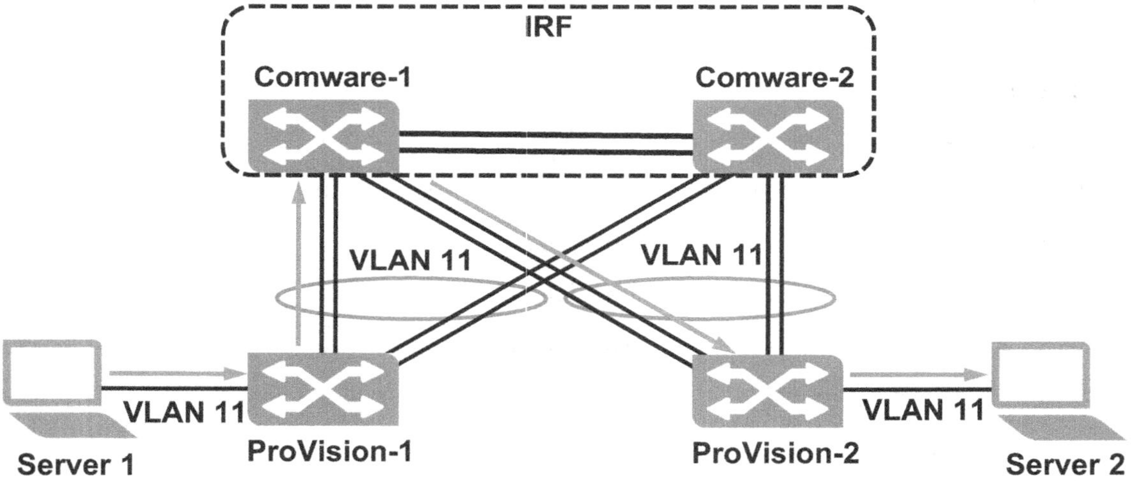


Figure 10-14: Tracing traffic flow

For example, consider a topology like the example network shown in Figure 10-14 but using an IP topology in which VLANs extended across the network. Assume that a server connected to ProVision-1 attempts to reach a server connected to ProVision-2, as shown in Figure 10-14. For this example, the network has been active for some time, and the switches have learned the proper ports for all devices.

1. The server sends the traffic on its link to ProVision-1.
2. ProVision-1 knows that the forwarding port for server 2 is trk1. It chooses one of its four links for the traffic; for the sake of the example, it chooses one of the links to Comware-1.

3. The IRF virtual device has learned bridge interface 3 as the forwarding port for server 2. So Comware-1 chooses one of its local links in this aggregation and forwards the traffic to ProVision-2.

4. ProVision-2 forwards the traffic to server 2.

## Routing at Layer 3

The master builds IP routing tables. It programs the tables into the forwarding information based on members.

When a member receives packets to route at Layer 3, it scans this table to identify the next hop and the forwarding egress port. It assembles the correct new Layer 2 header, as routers do, and forwards the packet out the correct egress port, which could be local or on another member. Just as with Layer 2 forwarding, a packet that needs to egress on another member travels across the IRF links.

In this way, all members of the device can actively route traffic. However, to the outside network, the packets have traveled just one hop—as if one device routed the packets.

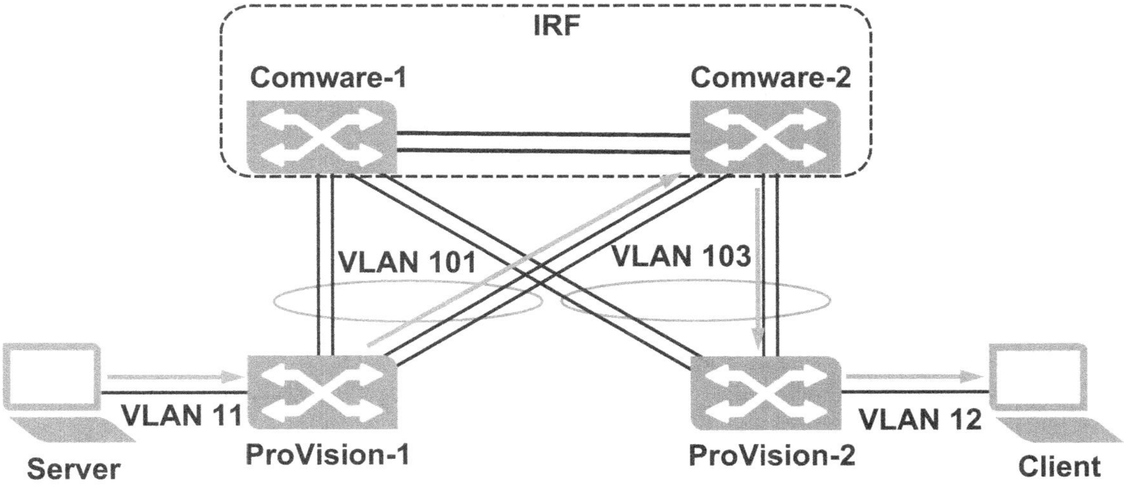


Figure 10-15: Tracing the traffic flow (routed)

Now consider the example network shown in Figure 10-15. Suppose the server in Figure 10-13 attempts to send traffic to the client.

1. The server sends the packet to ProVision-1, its default router.

2. Using its default route, ProVision-1 routes the traffic to 10.1.101.1 in VLAN 101. This VLAN is on the link aggregation to the IRF virtual device, so ProVision-1 chooses one of the links in this aggregation for this traffic. For this example, ProVision-1 selects a link to Comware-2.

3. The IRF virtual device has a route to 10.1.12.0/24 through 10.1.104.4 in VLAN 104. Comware-2 routes the traffic accordingly, assigning it for forwarding on bridge interface 3 (which is the only interface in VLAN 104). Comware-2 can determine the correct next hop and forwarding interface even though it is not the master because the master has programmed it to do so. Comware-2 selects one of its local links in bridge interface 3 for forwarding the traffic.
4. ProVision-2 receives the packet and routes it to the client.

## IRF failover

IRF provides very fast failover for a number of situations. As you can see in Figure 10-16, if a single link fails, traffic fails over to other links in the same link aggregation. If an entire device fails, traffic fails over to the other links—links connected to another member. In addition, the other members can continue to handle Layer 3 functions. In this way, IRF gives good redundancy for default router functions and routing protocols.

Table 10-1 indicates some of the failover times that you can expect for various situations. At the millisecond level, these times are suitable for data centers and modern applications.

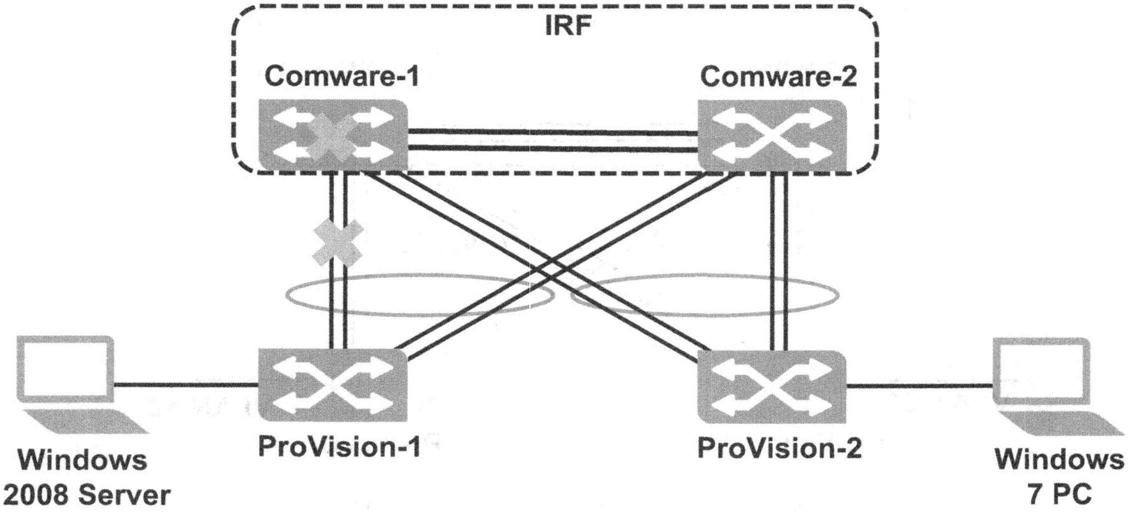


Figure 10-16: IRF failover

Table 10-1: IRF failover time

| Scenario | Failover time |
|---|---|
| A port is removed from/added to a link aggregation | 2 ms/0.7 ms |
| An entire interface is removed/added (removing or adding links to a link aggregation) | 2 ms/1 ms |
| An entire physical device fails/recovers | 2 ms/0.14 ms |
| In-Service Software Upgrade occurs | 2 ms |

### IRF link failure

When you bind multiple physical interfaces to an IRF port, one of the links can also fail without affecting the IRF functionality. (Of course, bandwidth for the link decreases.)

If all of the physical interfaces fail, then the IRF link fails, which can lead to a split stack or "split brain" situation. It is important to avoid such a situation because it can interfere with connectivity. Attend the *Deploying HP FlexNetwork Core Technologies* course to learn how to configure Multi-Active Detection (MAD) to prevent issues with split stacks.

## Add IRF members

In the example configurations in this chapter, the IRF virtual switch includes two members. In some circumstances you can create virtual switches with more members. The setup and functionality are similar, but you have more choices for the topology.

Each member has two logical IRF ports so that it can connect to one or two other members. You can create a chain topology, such as in Figure 10-17.

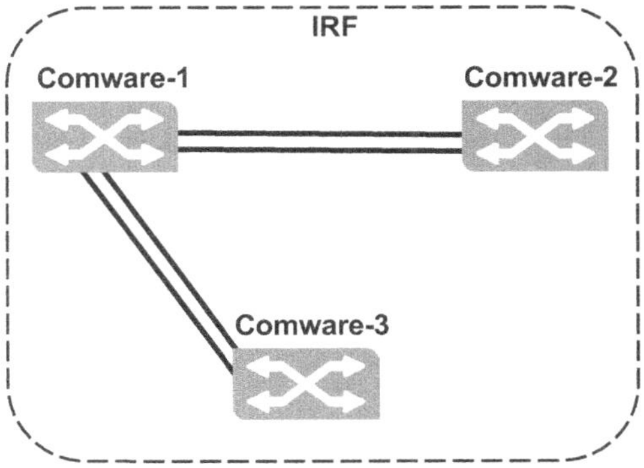


Figure 10-17: Adding IRF members

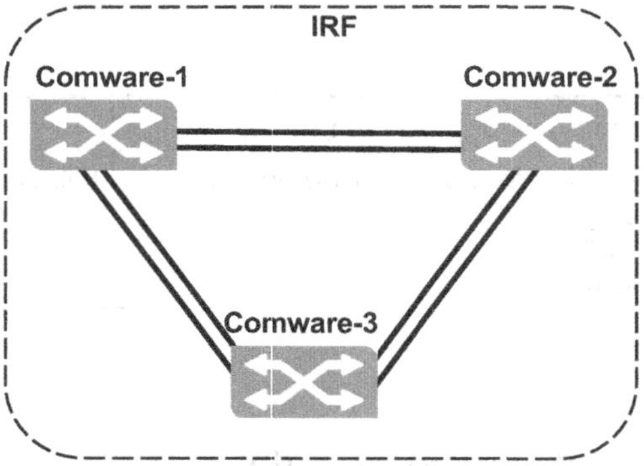


Figure 10-18: Recommended ring topology

As shown in Figure 10-18, you can also create a ring topology, which is preferred; it offers better resiliency in case an IRF link fails. The failure turns the ring topology into a chain, but the device does not split.

The number of members supported in an IRF virtual device depends on the switch model and sometimes the software version. Refer to the details about your switch models on the HP networking web site or in the switch documentation.

## Summary

You now understand how IRF combines multiple devices into a single virtual device, and you can set up your own IRF virtual device. You also understand how the IRF virtual device fits within a network topology and the benefits that it brings.

## Learning check 2

1. Describe the link aggregation to the IRF virtual device from the point of view of ProVision-2.

2. Describe what you observed when you traced a route from the server to the client. How many hops is the IRF virtual device?

3. How many default routes is ProVision-1 using? Does it still have redundancy in case Comware-1 (the physical switch) fails?

4. For many Comware switches, including HP 5800 Series switches, what is the lowest speed interface that can be part of an IRF link?
   a. 10 Mbps
   b. 100 Mbps
   c. 1 Gbps
   d. 10 Gbps

5. Order the criteria for electing the master in an IRF virtual device. If an option is not used, do not include it in the list. For example, only highest priority or lowest priority is used; not both.
   a. The member with the longest system uptime
   b. The member with the highest priority
   c. The member with the lowest priority
   d. The member with the highest MAC address
   e. The member with the lowest MAC address
   f. The current master of an existing IRF (assuming new members join during their boot)

6. What must you do before you can create the logical IRF port?
   a. Shut down the physical interface(s) that will be used for the logical IRF port.
   b. Reboot the switch.
   c. Save the switch configuration.
   d. Establish the physical connections for the logical IRF port.

7. Which of the following commands creates a logical IRF port?
   a. interface irf-port <member-id/port-number> port group interface <interface-id>
   b. interface bridge <member-id/port-number> port group interface <interface-id>
   c. irf-port <member-id/port-number> port group interface <interface-id>
   d. interface irf-port <member-id/port-number> member-interface <interface-id>

8. Which command activates the IRF port configuration?

## Learning check 2 answers

1. Describe the link aggregation to the IRF virtual device from the point of view of ProVision-2.

   To ProVision-2, the link aggregation appears exactly like a link aggregation to a single device. LACP detects that all the links connect to the same partner whether a link is physically connected to Comware-1 or Comware-2. Comware-1 and Comware-2 are a single logical device running just one LACP protocol.

2. Describe what you observed when you traced a route from the server to the client. How many hops is the IRF virtual device?

   The packet was routed to ProVision-1, to the IRF virtual device, to ProVision-2, to the client. The IRF virtual device is one hop.

3. How many default routes is ProVision-1 using? Does it still have redundancy in case Comware-1 (the physical switch) fails?

   ProVision-1 is using just one default route, through 10.1.101.1 (the IRF virtual device's IP address on the VLAN between ProVision-1 and the IRF virtual device). Similarly, ProVision-2 is using just one default route.

   But ProVision-1 still has redundancy in case Comware-1 (the physical switch) fails. Comware-2 would take over as the master of the IRF virtual device and continue routing traffic.

4. For many Comware switches, including HP 5800 Series switches, what is the lowest speed interface that can be part of an IRF link?

   a. 10 Mbps

   b. 100 Mbps

   c. 1 Gbps

   d. 10 Gbps

5. Order the criteria for electing the master in an IRF virtual device. If an option is not used, do not include it in the list. For example, only highest priority or lowest priority is used; not both.

   f, b, a, e

6. What must you do before you can create the logical IRF port?

   a. Shut down the physical interface(s) that will be used for the logical IRF port.

   b. Reboot the switch.

   c. Save the switch configuration.

   d. Establish the physical connections for the logical IRF port.

7. Which of the following commands creates a logical IRF port?

   a. `interface irf-port <member-id/port-number> port group interface <interface-id>`

   b. `interface bridge <member-id/port-number> port group interface <interface-id>`

   c. `irf-port <member-id/port-number> port group interface <interface-id>`

   d. `interface irf-port <member-id/port-number> member-interface <interface-id>`

8. Which command activates the IRF port configuration?

   `irf-port-configuration active`

# 11 Wireless Networks for Small Offices

## EXAM OBJECTIVES

✓ Describe the basics of wireless communications and 802.11 standards

✓ Define a Wireless LAN (WLAN)

✓ Differentiate between wireless security options

✓ Configure basic settings on autonomous HP MultiService Mobility (MSM) Access Points (APs)

## INTRODUCTION

This chapter introduces you to wireless networks, focusing on setting up a wireless network for a small office environment using an HP MSM AP.

### Assumed knowledge

Before reading this chapter, you should have a basic understanding of:

- Wired access
- Broadcast domain
- Virtual LANs (VLANs)

### Wireless communications

This study guide is not designed to make you an expert in wireless communications. If you want to learn more about wireless technologies, you should attend other training such as *Introduction to Aruba WLANs (IAW)* and *Scalable WLAN Design and Implementation (SWDI)*. This chapter does teach you enough about wireless technologies, though, for you understand basically how devices use wireless technology to communicate.

As Figure 11-1 shows, IEEE 802.11 defines the standard for wireless communications. At the Data Link layer, devices use Ethernet frames for data. They encapsulate these frames with 802.11 headers for delivery in the wireless network. The 802.11 standard also defines various management and control frames for maintaining the network and controlling data transmissions, but you do not need to learn about these for this study guide.

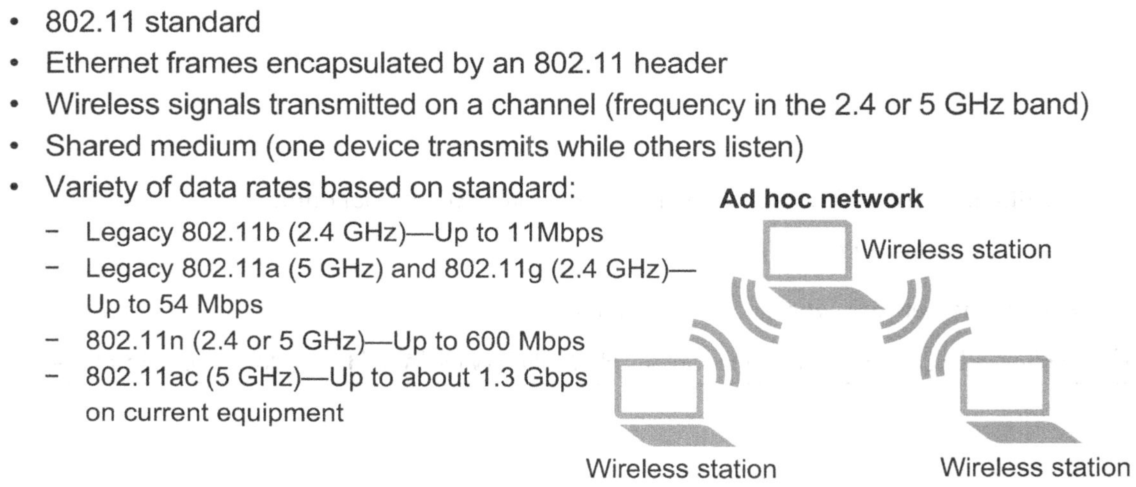


Figure 11-1: Wireless communications

At the physical layer, communications use radio frequency (RF) technology. Each 802.11 standard uses a specific frequency band or bands. (A job aid at the end of this chapter indicates the bands for each channel.) A frequency band is further divided into narrower frequencies called channels. A wireless device sends and transmits its traffic as a modulated signal on a channel.

The wireless medium is shared, much like Ethernet was historically shared when devices connected through hubs. Only one device can transmit on a channel at once. Other devices must listen and wait their turn. Devices that support 802.11 use Carrier Sense Multi-Access Collision Avoidance (CSMA/CA) to listen for other devices that are transmitting, to back off if necessary, and to take a chance to transmit themselves.

Data is sent at a specific data rate. Each 802.11 standard specifies multiple data rates. A job aid at the end of this chapter shows you the precise data rates. Highlights include:

- 802.11b, 802.11g, and 802.11a are legacy standards that support data rates up to 11 Mbps (b) or 54 Mbps (g and a)
- 802.11n supports theoretical rates up to 600 Mbps
- 802.11ac supports theoretical rates up to about 1.3 Gbps in current equipment (and higher with support for more features in the future)

The data rate determines the rate of transmission for a frame. But a device's data rate does not predict the device's overall throughput. Because the wireless medium is shared, each device only receives a fraction of the chances to transmit or "airtime." For example, consider ten devices are sharing the wireless medium, all are using 135 Mbps for their data rate, and all are trying to use the network equally. Each device will receive at most 13.5 Mbps throughput. In fact, due to overhead, they will receive less than that. An exception is that 802.11ac introduces multiuser multiple input-multiple output MIMO (MU-MIMO), which allows an AP to transmit traffic to up to four stations at once.

The 802.11 standard defines several different types of networks, all of which use these basic communication rules. Figure 11-1 shows several wireless-capable laptops communicating directly with each other. This type of network is called an ad-hoc wireless network. On the next page, you will learn about other types of networks.

## Infrastructure mode communications

Enterprise environments usually feature infrastructure mode networks, which are the focus for this study guide. An infrastructure mode network (which is illustrated in Figure 11-2) uses the same basic communication rules that you already learned, but in this mode, an AP establishes the wireless network. It identifies the network with a wireless network name or Service Set Identifier (SSID), which it advertises in periodic beacons. When you see a list of wireless networks on your wireless devices, you are seeing SSIDs beaconed by APs.

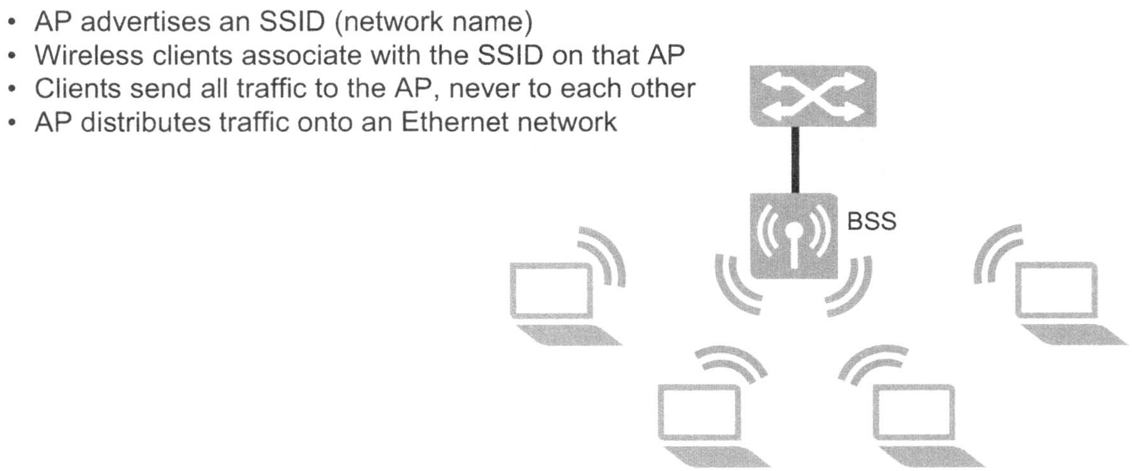


Figure 11-2: Infrastructure mode communications

A wireless station (a device such as a laptop, smartphone, tablet, and so on) joins the wireless network by associating with the AP.

An AP and all of the stations associated with it compose a basic service set (BSS). You probably will not hear the BSS term very often. An AP and its station are also often called a wireless cell.

Each station sends all of its traffic to the AP, and the AP is responsible for sending the traffic toward its destination. Similarly, the AP receives traffic destined to the station and forwards this traffic on to the station.

The AP manages the settings for the wireless network, including among others:

- The channel used—If you have multiple APs in the same general area, each AP should operate on a different channel so that the cells' communications do not interfere with each other. For the 2.4 GHz range, the channels also must be non overlapping. Other HP training provides more details about channel planning.
- The wireless standard or standards that can be used—A wireless cell might use just one standard, but some standards provide backward compatibility with others. For example, 802.11g can be backward compatible with 802.11b, and 802.11n can be backward compatible with 802.11b, 802.11g and 802.11a.
- The data rates that devices must support to connect to the network (basic set)
- The data rates that devices can use in the network (supported set)
- Security settings, discussed later

The AP advertises many of these settings in a beacon so that stations can agree to them when they associate.

In addition to connecting to wireless stations, the AP is connected to a wired network. As the interface between the wired and the wireless networks, the AP receives wireless traffic from stations, decapsulates the 802.11 header, and forwards an Ethernet frame on to the wired network. Likewise, the AP receives and forwards traffic from the wired network to correct wireless stations. In this way, wireless devices can communicate on an Ethernet LAN.

## AP types

As shown in Figure 11-3, APs generally come in two basic types: autonomous or controlled. HP MSM APs can operate as either type, but this study guide focuses on autonomous APs.

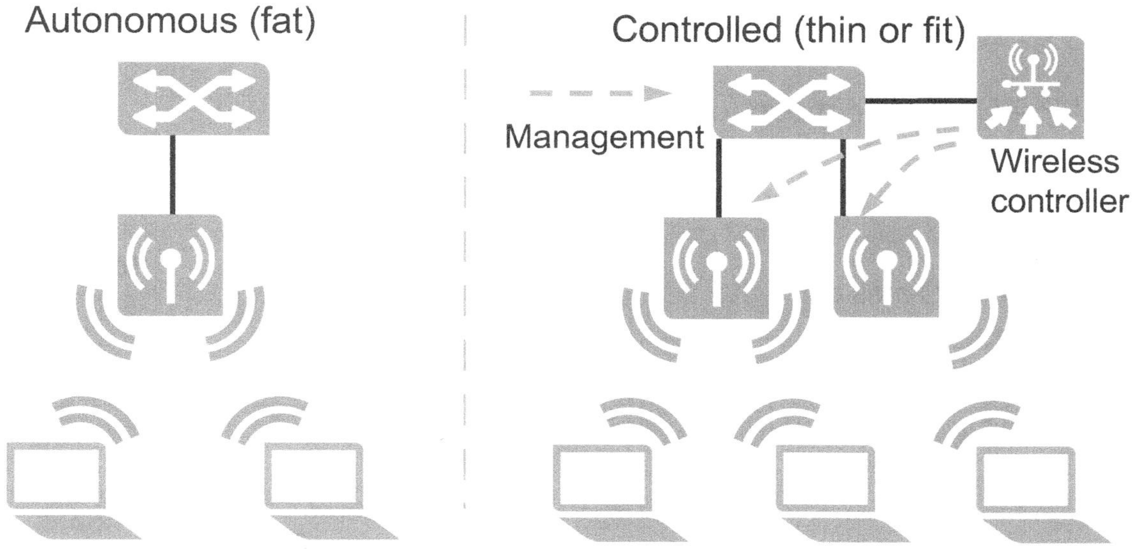


Figure 11-3: AP types

## Autonomous

Autonomous APs operate on their own. The APs might be configured to support the same wireless network, but each of the APs is configured and functions on its own. These APs are more common in smaller networks with few APs that need to be managed.

Autonomous APs are sometimes also called "stand-alone" or "fat" APs.

## Controlled

For larger networks, you usually deploy controlled APs. Controlled APs establish the wireless network like autonomous APs. However, all of their settings are managed by a centralized wireless access controller.

Some controlled APs, sometimes called "thin" APs, encapsulate all traffic from wireless users and tunnel it to the wireless controller. The controller can then apply certain security and QoS settings and forward the traffic on the Ethernet LAN.

Other controlled APs, sometimes called "fit" APs, receive their settings from a controller but continue to forward wireless traffic directly onto the Ethernet LAN themselves.

HP controlled APs can operate in either way, depending on how you configure them.

If you want to learn more about HP wireless products, refer to www.hp.com/networking.

## Accessing an HP MSM AP

- AP = Controlled mode (default)
- Access the AP's Web browser interface to change to autonomous.
- AP's initial IP address:
  - DHCP
  - If DHCP times out:
    - 192.168.1.1/24
    - No default gateway (Layer 2 connectivity required)

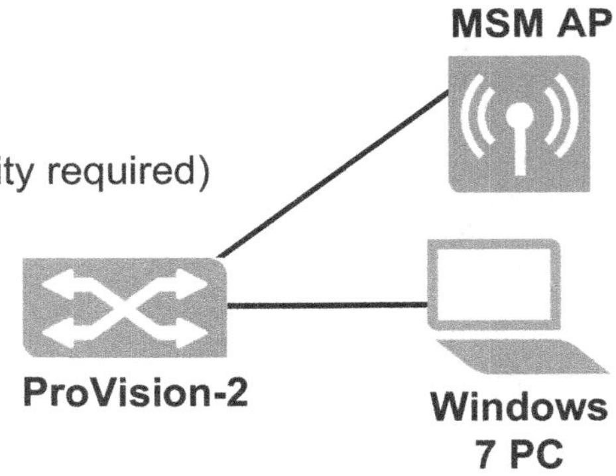


Figure 11-4: Accessing an HP MSM AP

In this chapter, you will learn how to access an HP MSM AP that is operating at factory default settings and to begin configuring it. The AP operates in controlled mode, by default, so you will learn how to access its Web browser interface and change its mode to autonomous.

---

**Note**

You can use any compatible Web browser such as IE or Firefox to access and manage the MSM AP.

---

To reach the AP's interface, you must:

- Establish connectivity between the AP and your management station
- Know the AP's IP address

### AP connected to a switch

You can access an MSM AP that is operating at factory default settings in a couple of ways. Figure 11-4 outlines the approach that is used in the "Example configuration: Complete initial setup on an autonomous MSM AP" on page 406. You connect the AP to a switch. If the switch supports Power over Ethernet (PoE), you will not need to power up the AP with an external power source. The AP can receive its power over its Ethernet connection instead. In fact, many APs require PoE and do not have external power supplies.

Your management station does not necessarily need to connect to the same switch as the AP, but it must be able to reach the AP's IP address. To ensure that it can, you must understand how the AP obtains its IP address.

First, the AP attempts to receive a DHCP address. If you connect the AP to a switch and the AP's VLAN is DHCP-enabled (has a DHCP server or DHCP relay), the AP will receive an IP address and default gateway address. Work with the DHCP administrator to determine the AP's address. You should then be able to contact the AP from your management station even if the station is on a different subnet (provided that IP routing and access control lists [ACLs] do not interfere).

If, however, the AP cannot receive an IP address, it starts to use IP address 192.168.1.1/24. It does not have a default gateway, so you must connect your management station in the same VLAN as the AP. You must also give your management station an IP address on 192.168.1.0/24 so that the station has Layer 2 connectivity to the AP.

After you access the AP at its initial IP address and log in with the default credentials (**admin** and **admin**), you can change the AP to autonomous mode. Changing to autonomous mode causes the AP to reboot.

You can then give the AP an IP address that is valid on your network.

### AP connected directly to the management station

A few APs have external power supplies. In this case, you can connect your management station directly to the AP's Ethernet port (port 1 if the AP has multiple ports). This approach is illustrated in Figure 11-5 below.

Figure 11-5: Connecting directly to an MSM AP

In this case, the AP will not receive a DHCP address. Assign your management station an IP address in 192.168.1.0/24 and contact the AP at 192.168.1.1. Log in with the default credentials (admin and admin).

You can then change the AP to autonomous mode and assign it an IP address that is valid on your network. You will then need to connect the AP to your network and contact it at its new IP address.

## Example configuration: Complete initial setup on an autonomous MSM AP

This section provides step-by-step instructions for adding an MSM AP to the example network shown in Figure 11-6. This is the same example network used in Module 9. (See "Example configuration: Setting up a basic OSPF configuration" on page 339.)

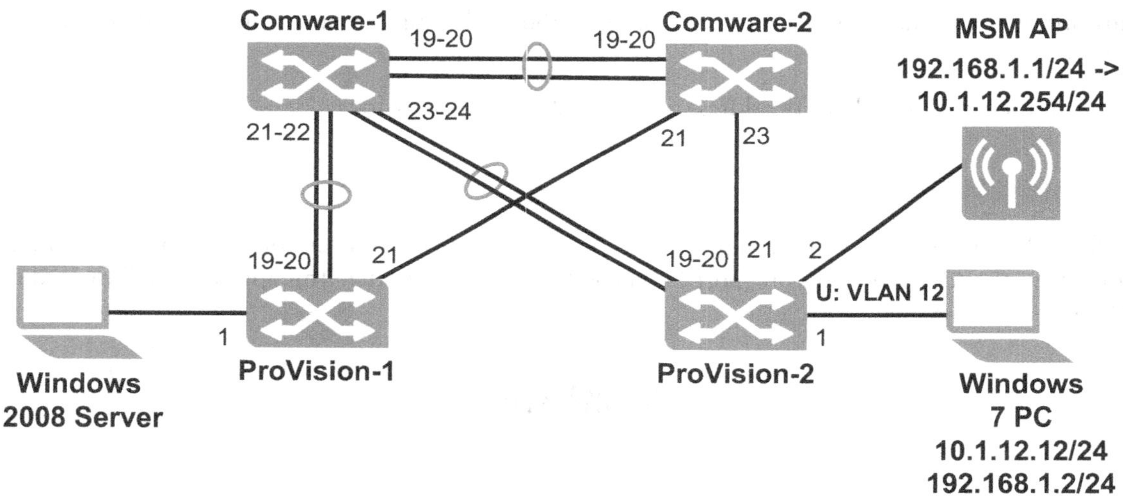


Figure 11-6: Example configuration: Complete initial setup on an autonomous MSM AP

The MSM AP should be operating at factor default settings. If the MSM AP you are using is not running at factory default settings, check the MSM AP documentation for instructions on how to return it to factory default settings.

### Prepare access to the MSM AP

In the example network, the MSM AP will be managed from the wired Windows 7 PC.

1. Connect the AP to port 2 on the ProVision-2 switch, as shown in Figure 11-7. If the MSM AP has more than one port, use the first port.

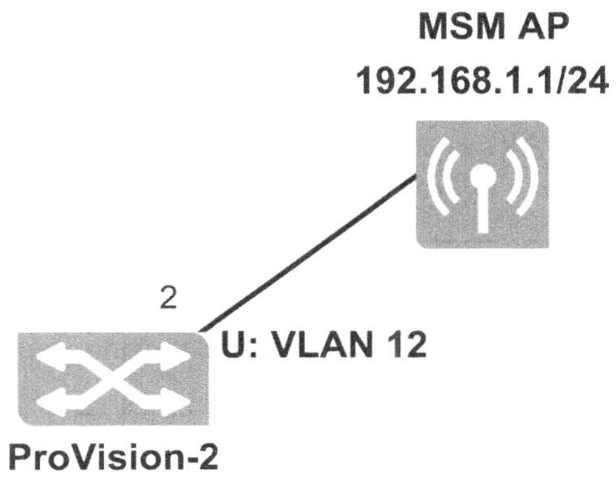


Figure 11-7: Connect the MSM AP to ProVision-2

## ProVision-2

2. Access a terminal session with the ProVision-2 switch and move to global configuration mode context.
3. Enable interface 2 and assign it as an untagged member of VLAN 12.

```
ProVision-2(config)# interface 2 enable
ProVision-2(config)# vlan 12 untagged 2
```

4. Verify that the PC's and AP's ports are both in VLAN 12.

```
ProVision-2(config)# show vlan 12
 Status and Counters - VLAN Information - VLAN 12
  VLAN ID : 12
  Name : VLAN12
  Status : Port-based
  Voice : No
  Jumbo : No
Port Information Mode     Unknown VLAN Status
  ---------------- -------- ------------ ----------
1                Untagged Learn        Up
2                Untagged Learn        Up
```

## Windows client

The example network is not set up for DHCP services on VLAN 12. Therefore, the PC needs an IP address in the same subnet as the AP's default IP address. You will add this IP address as a secondary address on the PC's NIC so that the PC can continue to use its normal VLAN 12 address as well.

---

**Important**
Make sure to add the 192.168.1.2/24 IP address and keep the 10.1.12.12/24 IP address as well. Use the steps below.

---

5. Access the desktop of the wired Windows 7 PC.
6. Use the shortcut to access the network properties. Then select IP Protocol Version 4 and click **Properties**.
7. Click **Advanced**.

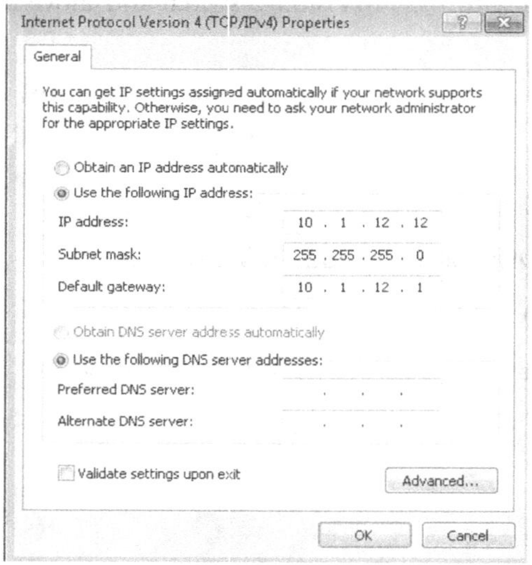


Figure 11-8: Assign a second IP address to the PC

8. In the IP Address section (which is shown in Figure 11-8), click **Add**.
9. Add a *secondary* IP address of 192.168.1.2/24 and click **Add**, as shown in Figure 11-9.

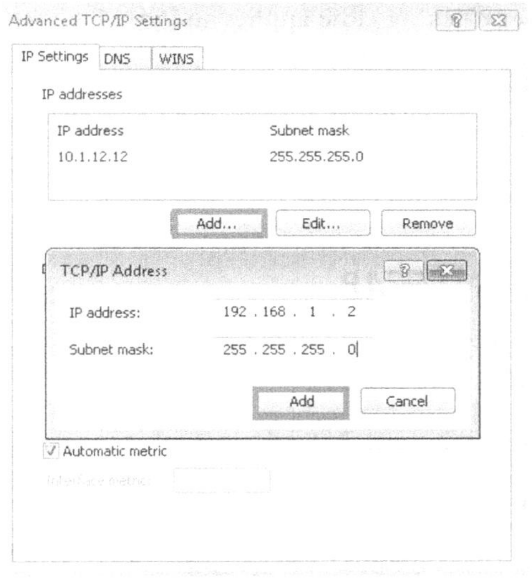


Figure 11-9: Assign a second IP address to the PC

10. Your window should look like the one shown in Figure 11-10 when you have finished.

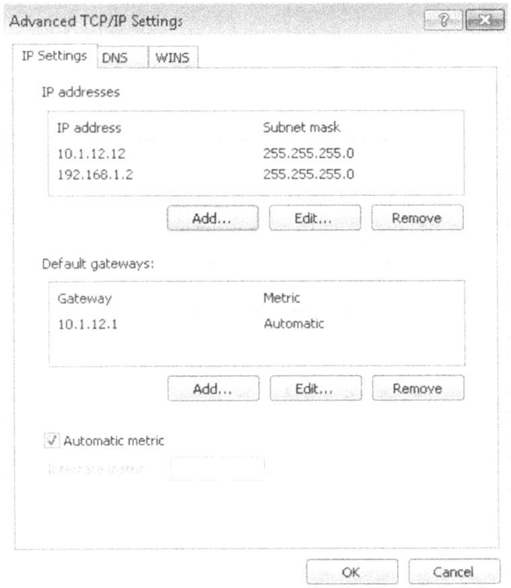


Figure 11-10: Assign a second IP address to the PC

11. Click **OK** and then **Close**. Clicking close applies and saves the settings.
12. Open a command prompt on the client.
13. Ping the MSM AP's default IP address (192.168.1.1) and verify success.

    ping 192.168.1.1

## Access and initialize the MSM AP

You will now access the MSM AP, change the AP to autonomous mode, and change its IP address to 10.1.12.254/24, a valid address on VLAN 12.

1. Open a web browser on the Windows 7 PC desktop.
2. Navigate to http://192.168.1.1.
3. You will see a certificate error because the AP redirects you to HTTPS. The AP uses a self-signed certificate to secure the HTTP session, and your PC does not trust that certificate. Choose to ignore the error.
4. You should be presented with the Welcome screen. Log in using a username and password of **admin** and **admin**.
5. As Figure 11-11 shows, you will see a yellow text bar indicating that the AP is running in "controlled mode." At the bottom of the window, click **Switch to Autonomous Mode**.

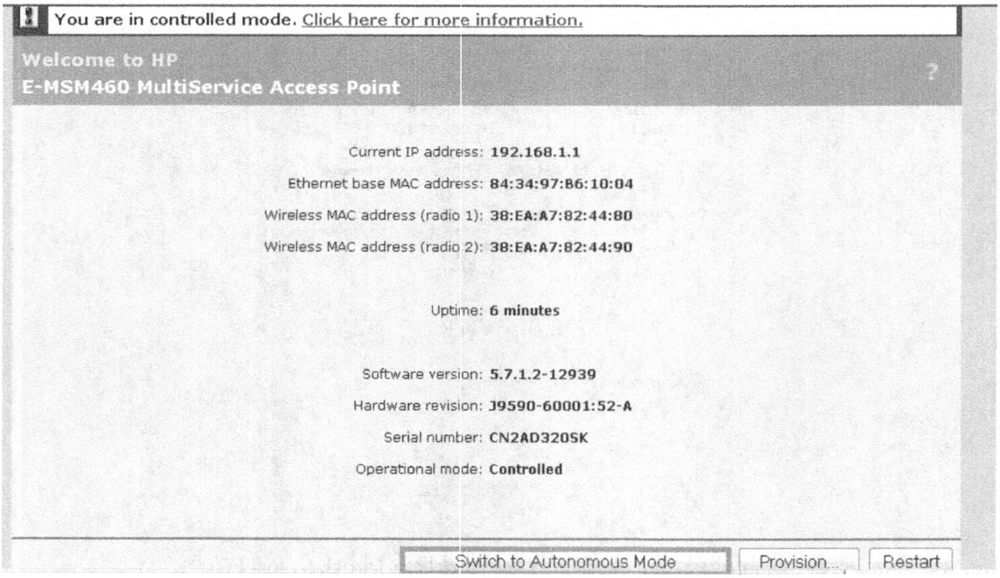


Figure 11-11: Switch to Autonomous Mode on the MSM AP

6. The pop-up window shown in Figure 11-12 displays a warning. Click **OK**.

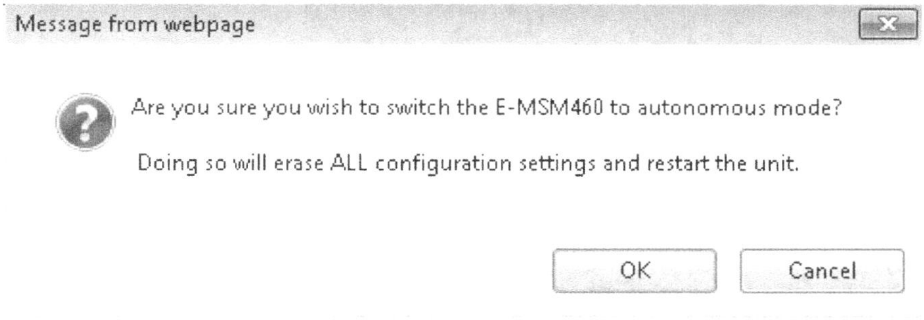


Figure 11-12: Switch to Autonomous Mode on the MSM AP

7. It will take about a minute for the AP to reboot. After the reboot, log back in using a username and password of **admin** and **admin**.
8. On the License window, click the **Accept HP End User License Agreement** button to accept the license.
9. On the Registration window, which is shown in Figure 11-13, click the **Register Later** button to skip the registration.

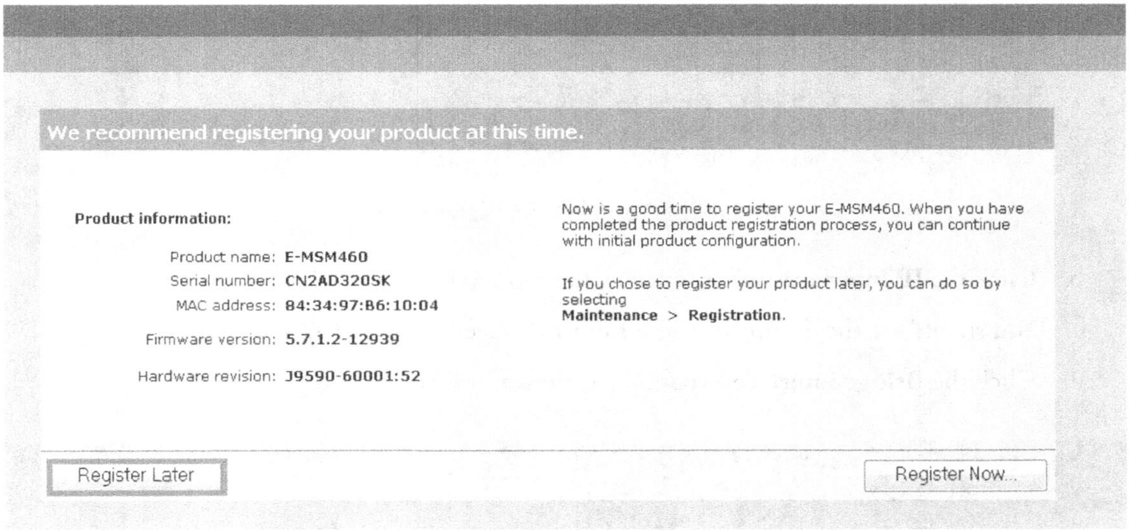


Figure 11-13: Choose to Register Later

10. In the Country window, which is shown in Figure 11-14, select the country in which the AP will be installed and click **Save**.

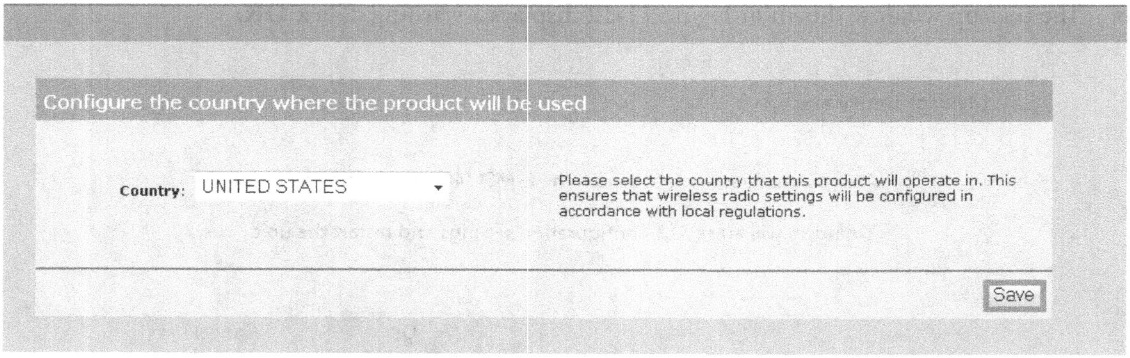


Figure 11-14: Choose to Register Later

11. You are taken to the initial configuration wizard, which begins by asking you to change the password. This wizard is not used in this example configuration. Click **Cancel**.
12. Change the IP address of the AP to 10.1.12.254/24, with a default gateway of the ProVision-2 switch (10.1.12.1).
    a. Click the **Network** tab, as shown in Figure 11-15.

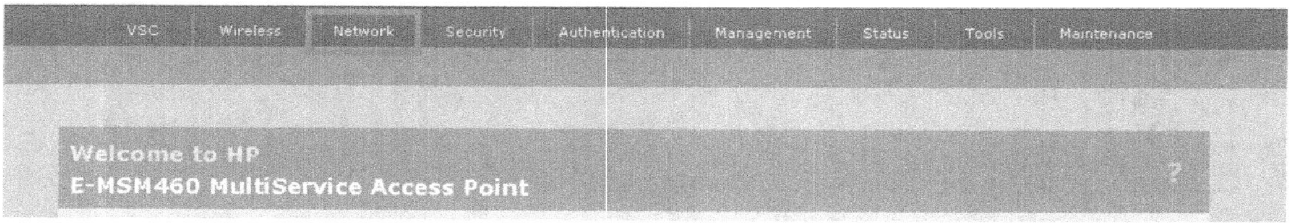


Figure 11-15: Click the Network tab

    b. Click the **IP interfaces** sub-tab (below the top-level tabs).
    c. You should see the default IP address of the bridge interface: 192.168.1.1/24.
    d. Click the **Bridge interface** hyperlink, as shown in Figure 11-16.

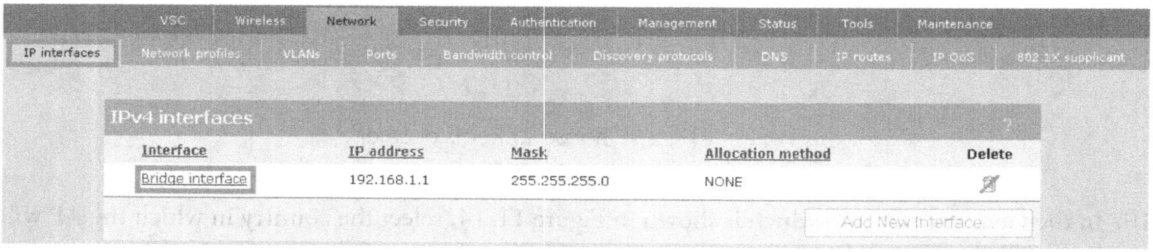


Figure 11-16: Click the Network tab

e. Select the **Static** radio button as shown in Figure 11-17and click the associated **Configure** button.

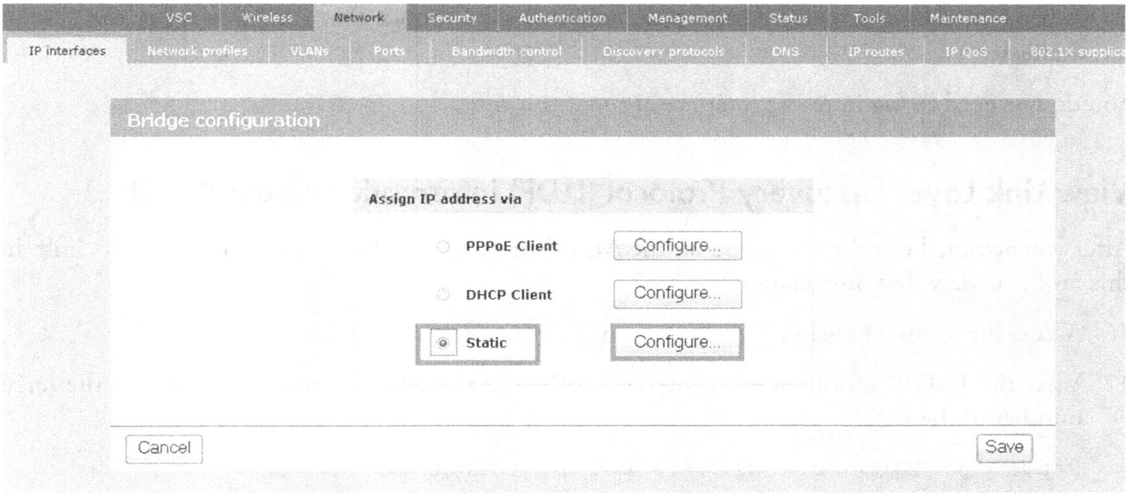


Figure 11-17: Click the Network tab

f. In the IP Address textbox, enter **10.1.12.254**.

g. In the Address mask textbox, enter **255.255.255.0**

h. In the Default gateway textbox, enter **10.1.12.1**.

i. Click the **Save** button at the bottom on the window as shown in Figure 11-18.

Figure 11-18: Click the Network tab

13. At this point you will lose connectivity to the MSM AP.
14. Navigate the Web browser to the new address: **http://10.1.12.254**.
15. Accept the self-signed certificate. In the login window, notice that the AP is using the new IP address.

You do not need to log in to the interface again at this time.

## View Link Layer Discovery Protocol (LLDP) information about the AP

After you accept the end-user license on the AP, it begins to send LLDP information on its link. In this task you view that information.

16. Access the terminal session with ProVision-2.
17. View the LLDP information on interface 2. Notice that the system name it uses is the serial number of the AP.

```
ProVision-2(config)# show lldp info remote-device 2

LLDP Remote Device Information Detail

  Local Port    : 2

  ChassisType   : mac-address

  ChassisId     : 843497-b61004

  PortType      : mac-...

  PortId        : 84 34 97 b6 10 04

  SysName       : APDynamicName

               System       Descr      :        HP        AP
Autonomous,CN2AD320SK,J9590-60001:52-A,5.7.1.2-12939

  PortDescr     : Port 1

  Pvid          :

  System Capabilities Supported  : wlan-access-point

  System Capabilities Enabled    : wlan-access-point

  Remote Management Address

     Type    : ipv4

     Address : 10.1.12.254

------------------------------------------------------------

  Local Port    : 2
```

```
ChassisType   : local
ChassisId     : 010105-000202
PortType      : local
PortId        : eth0
SysName       :
System Descr  : 5.7.1.2-12939E-MSM460
PortDescr     :
Pvid          :
System Capabilities Supported  :
System Capabilities Enabled    :
Remote Management Address
   Type    : ipv4
   Address : 10.1.12.254
```

Save the configuration on ProVision-2 (**write memory**).

## Learning check 1

Answer the following questions to test your understanding of the concepts outlined in this section of the chapter.

1. Does an MSM AP operate in autonomous mode, by default? If not, how do you change it?

2. Your AP was not able to receive a DHCP address. What IP address did it use? Why did you need to change your Windows PC's IP settings to reach the AP?

3. You were asked to set a country code on the AP. What purpose do you think this setting serves?

4. What information were you able to obtain about the AP by looking at LLDP information on ProVision-2?

## Learning check 1 answers

1. Does an MSM AP operate in autonomous mode, by default? If not, how do you change it?

   MSM APs operate in controlled mode, by default. You change the AP to autonomous mode by contacting its Web browser interface and clicking the Switch to Autonomous Mode button on the first window after you log in.

2. Your AP was not able to receive a DHCP address. What IP address did it use? Why did you need to change your Windows PC's IP settings to reach the AP?

   It used IP address 192.168.1.1/24. You had to change your Windows PC's IP settings so that the PC was in the same subnet as the AP. The AP does not have a default gateway, and your network is not set up to use the 192.168.1.0/24 subnet, so routing between the PC and AP does not work at this point.

3. You were asked to set a country code on the AP. What purpose do you think this setting serves?

   Different countries have different regulations for using RF. For example, different countries allow APs to use different channels. Setting the country code ensures that the AP follows the regulations for its region.

4. What information were you able to obtain about the AP by looking at LLDP information on ProVision-2?

   You can collect a great deal of information. You can see the AP's management IP address, its system name, its MAC address, its firmware version, and so on.

## Wireless LAN

You will now learn how to set up a wireless network on an MSM AP. Figure 11-19 summarizes the settings you will need to configure.

A wireless network is often called a wireless LAN (WLAN). A WLAN is defined by the SSID.

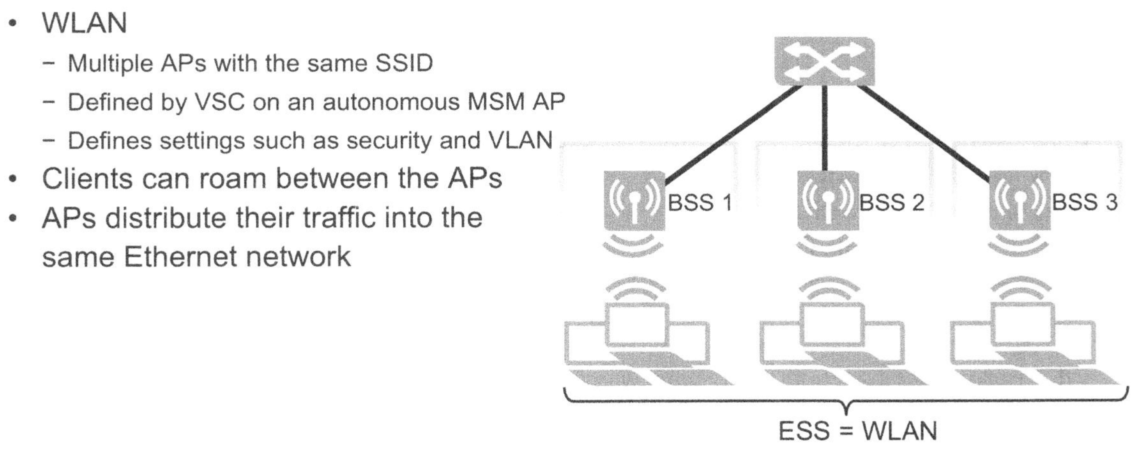


Figure 11-19: Wireless LAN

As well as the SSID, the WLAN defines various other settings, including security settings and the VLAN on which user traffic is forwarded. You will look at these settings in a moment. On MSM APs, you configure the WLAN settings through a Virtual Service Community (VSC).

Multiple APs can support the same SSID. Earlier you looked at an infrastructure mode cell, or BSS. When multiple APs support the same SSID, they form an extended service set (ESS), which is equivalent to the WLAN. Wireless devices can roam from one AP to another that within the same WLAN; that is, as a station moves, it can disassociate from one AP and reassociate to a new AP with a better signal without disrupting the connection or the user noticing. The APs should be forwarding the stations' traffic into the same Ethernet network, so the devices can continue to receive traffic without disruption.

---

**Important**

In theory, wireless devices can roam between APs that support the same WLAN without any additional setup. However, in practice, many factors come into play. The RF coverage must not have dead zones. Depending on the WLAN's security settings, you might need to configure fast roaming features. Finally, the backend Ethernet infrastructure can disrupt the roam if APs are deployed on different network segments and forward user traffic into different VLANs. To learn more about these considerations, attend other training such as *Introduction to Aruba WLANs (IAW)* and *Scalable WLAN Design and Implementation (SWDI)*.

---

## WLAN security settings

Figure 11-20 provides an overview of WLAN security settings. Review the detailed information that follows for a more thorough understanding.

Basic security options*

- None
- WPA-PSK
- WPA2-PSK**
- WPA with 802.1X (RADIUS)
- WPA2 with 802.1X (RADIUS)***
- 802.1X without encryption

Options that can be added to any of the others:

- MAC-Auth to a RADIUS server
- Local MAC filters

*WEP options are insecure, do not work with 802.11n and 802.11ac, and are not recommended.
**Focus for this course
***Typical recommendation for an enterprise network

Figure 11-20: WLAN security settings

The shared wireless medium has a serious security drawback. Any device can eavesdrop on any other device's communications. For this reason, most wireless networks use a combination of:

- **Encryption**—Encrypts traffic so that only the correct wireless station and the AP can read it. Encryption (data privacy) is typically joined with a data integrity algorithm, which ensures that data is not intercepted and changed in transit.
- **Authentication**—Ensures that only authorized users can connect to the WLAN.

It is recommended that you use WPA2 with AES for encryption. WPA with TKIP for encryption is a less secure alternative.

Both WPA and WPA2 let you choose between two forms of authentication.

WPA2 with preshared keys (PSK) provides basic authentication for smaller networks without back-end authentication servers. The PSK is essentially a password. You configure the password in the WLAN (VSC) settings, and users must enter the password to connect to the network.

In an enterprise environment, you should usually combine WPA2 with 802.1X RADIUS authentication instead. This form of authentication requires users to log in to a network RADIUS server with individual credentials such as a username and password or a digital certificate. (A wireless controller can sometimes fulfill the RADIUS server role.)

MSM APs also support MAC-Auth to a RADIUS server. When a station attempts to associate, the AP submits the station's MAC address to the RADIUS server. The server decides whether the station is permitted to connect. Or you can create MAC filters locally on the AP, which allow or block specific MAC address. You can add these options to a network without security or on top of one of the other security options. See Table 11-1 for details on the MSM wireless protection options.

**Note**

Controlled AP solutions support an additional option for authentication: Web-Auth (sometimes called guest or portal authentication). You could also create a wireless network without any security on autonomous MSM APs and then use another solution (such as your switches) for implementing Web-Auth.

Table 11-1: MSM wireless protection options

| Option | Mode | Key source | Encryption and data integrity | Authentication |
|---|---|---|---|---|
| WPA | WPA | Preshared | TKIP | Password (preshared key) |
| | | RADIUS | | 802.1X individual credentials |
| | WPA2 | Preshared | | Password (preshared key) |
| | | RADIUS | | 802.1X individual credentials |
| | WPA or WPA2 | Preshared | AES-CCMP or TKIP (based on stations' abilities) | Password (preshared key) |
| | | RADIUS | | 802.1X individual credentials |
| 802.1X | | | None | 802.1X individual credentials |

**Note**

MSM APs also support, Wired Equivalent Privacy (WEP), the original security standard for 802.11 networks. This study guide does not cover this standard because it is insecure, and HP does not recommend using it. In addition, WEP does not work with the latest 802.11 standards, 802.11n and 802.11ac.

## Data rates and 802.11n and 802.11ac features

Earlier you learned that various 802.11 standards offer different sets of data rates. If a standard supports multiple rates, what makes a client and AP use one rate rather than another? Factors that affect the choice include:

- The strength of the signal between the transmitting and receiving devices (the stronger the signal, the higher the data rate)
    - For 802.11a, 802.11b, and 802.11g, signal strength is the primary factor for the data rate. Different forms of signal modulation can transmit more information within the signal, but the signal must be stronger for these forms of modulation to work.
    - For 802.11n and 802.11ac, signal strength is the primary factor for determining the modulation and coding scheme (MCS). Different MCSs are associated with different sets of data rates.
- The ability of the transmitting and receiving devices to use special 802.11n and 802.11ac features

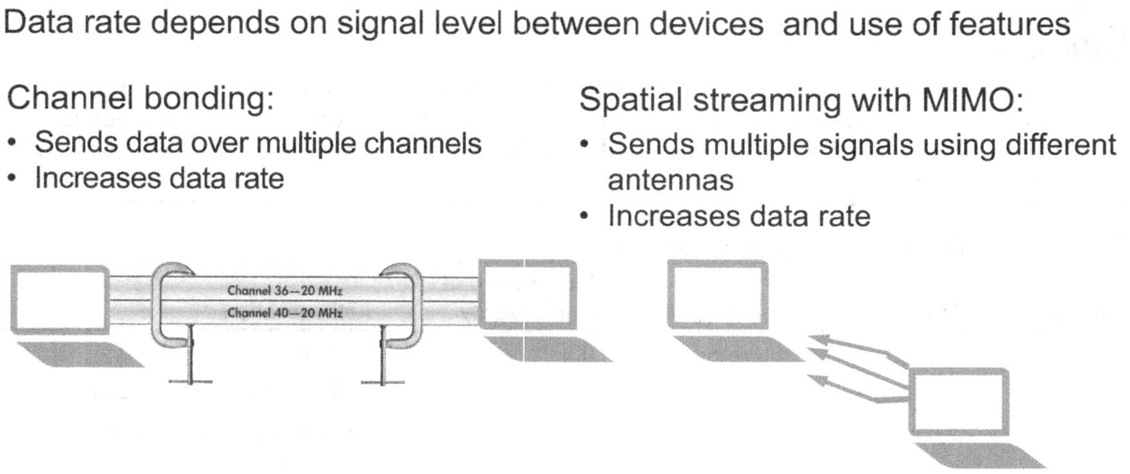


Figure 11-21: Data rates and 802.11n and 802.11ac features

Some of the most important 802.11n and 802.11ac features are illustrated in Figure 11-21:

- Channel bonding—Wireless devices send wireless signals over a frequency of a certain width called a channel. A normal 802.11 channel has a width of 20 Hz. Channel bonding allows the wireless device to send signals over a wider frequency, leading to higher bandwidth for the communications. With 802.11n channel bonding, devices use two 802.11n channels for their communications, which doubles the data rate (or actually slightly more than doubles the rate because devices can use a small area between channels as well). 802.11ac permits even wider channels. Channel bonding increases data rate, but does come at the cost of fewer channels available for neighboring APs.

- Spatial streaming—802.11n and 802.11ac use Multi-Input Multi-Output (MIMO) technology. MIMO allows a wireless device to send and receive different data streams over different antennas. Each of these data streams is called a spatial stream because it uses a different antenna and follows a slightly different path to the receiver. Using two spatial streams doubles the data rate. Using three spatial streams triples the data rate, and so on.

Refer to Job Aid 11-1 for a list of the 802.11 standards, their frequencies, and their data rates. Not all data rates for 802.11n and 802.11ac are shown for simplicity. Instead, for each set of features, the range of available data rates are shown from the lowest data rate (the lowest MCS using the long guard interval [GI]) to the highest data rate (the highest available MCS using the short GI). Also note that some 802.11ac features will not be available in the first products that come to market; for example, APs and clients will initially support a maximum of four spatial streams. In addition, enterprises will typically only use up to four bonded channels.

# Example configuration: Create a WAN with WPA2-PSK security

This section provides step-by-step instructions for creating a WLAN on the autonomous MSM AP in the example network shown in Figure 11-22. (This is the same example network featured in "Example configuration: Complete initial setup on an autonomous MSM AP" on page 406.

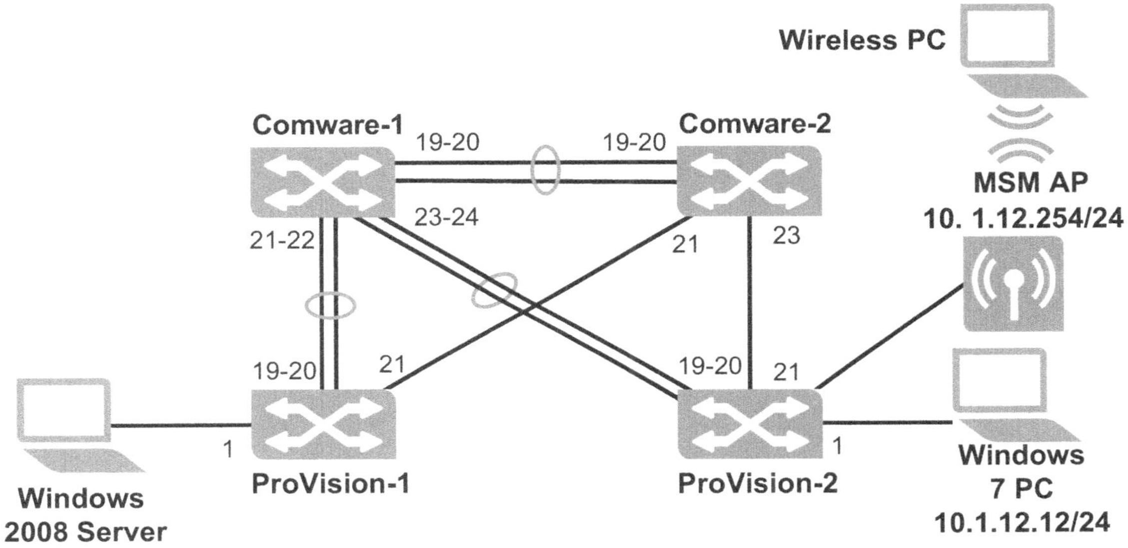


Figure 11-22: Configure a WAN with WPA2-PSK security

## Create a VSC for the WLAN

You will now create a VSC, which defines the settings for the WLAN. You should assign the WLAN a unique SSID so that you can connect your client to *your* AP.

1. Access the desktop of the Windows 7 PC, the wired client that you are using to manage the AP.
2. Open a Web browser and navigate to the AP's IP address, 10.1.12.254.
3. Log in with the default credentials, **admin** and **admin**.
4. Configure your VSC.
   a. In the top-level tabs, click **VSC**.
   b. Notice that a default VSC, called "HP," already exists. You will edit this VSC to use the settings that you want.
   c. Click **HP**, as shown in Figure 11-23.

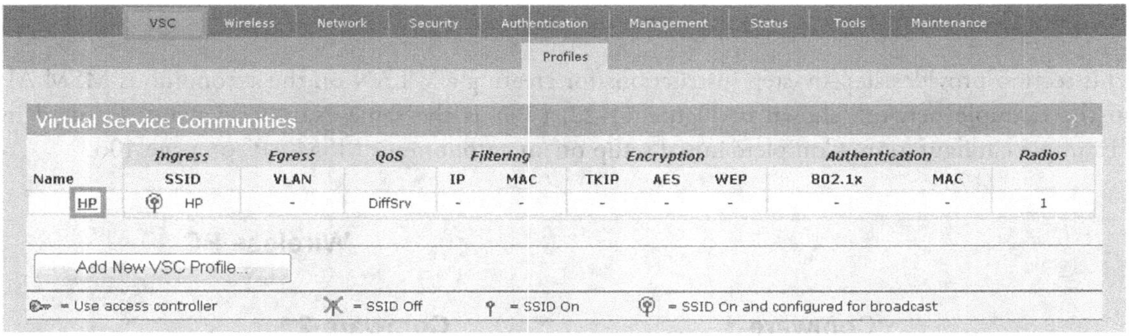


Figure 11-23: Click the HP VSC

5. Change the name of the VSC to **Wireless Users,** as shown in Figure 11-24. This name is locally significant only.

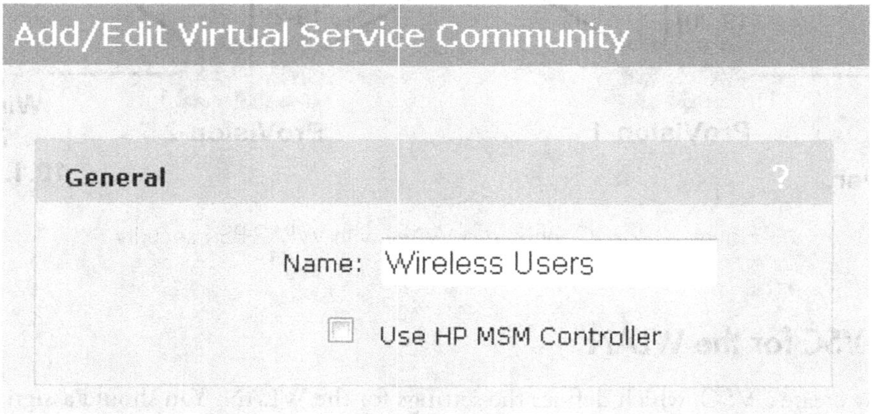


Figure 11-24: Change the name of the VSC

6. Make sure that the Virtual AP checkbox is selected. This lets the AP support a WLAN for this VSC.
7. In the Virtual AP section, you specify the wireless settings:
   a. For Name (SSID), enter your unique SSID value.
   b. For Transmit/receive on, choose Radio 1 and 2.
   c. Make sure that the Broadcast name (SSID) check box is selected, as shown in Figure 11-25.

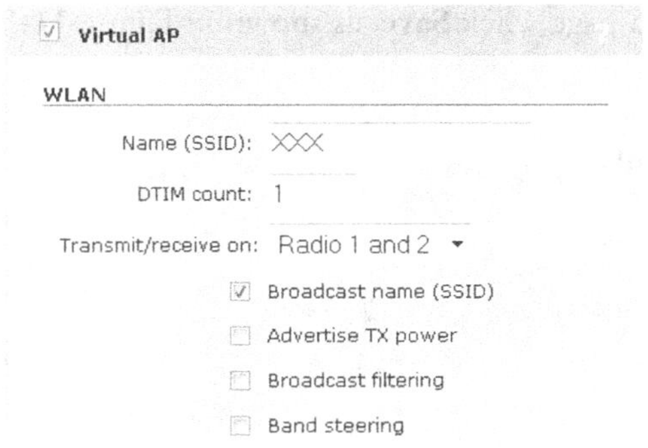


Figure 11-25: Configure the SSID

8. In the Wireless security filters section on the bottom left of the web page, make sure that the checkbox is *clear*.
9. You are now ready to configure the wireless settings. Scroll to the Wireless protection section in the top right of the page.
   a. Select the checkbox for Wireless protection and make sure that **WPA** is selected, as shown in Figure 11-26.
   b. For Mode, select **WPA2 (AES)**.
   c. For Key source, choose **Preshared Key**.
   d. For Key, enter **hp12345678**.
   e. For Confirm key, enter **hp12345678**.

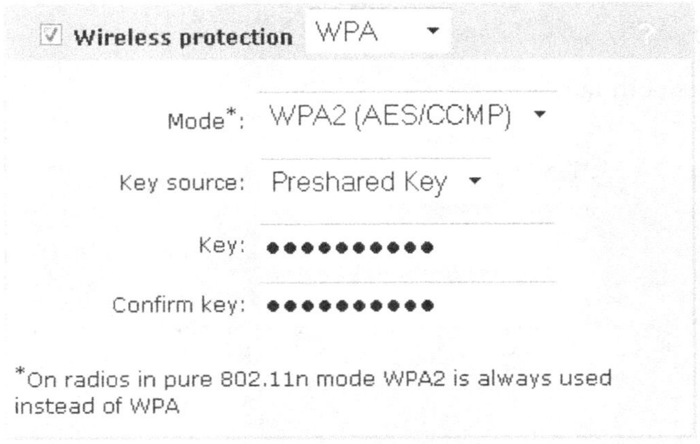


Figure 11-26: Configure WPA2-PSK security

10. At the bottom of the web page, click **Save**, as shown in Figure 11-27.

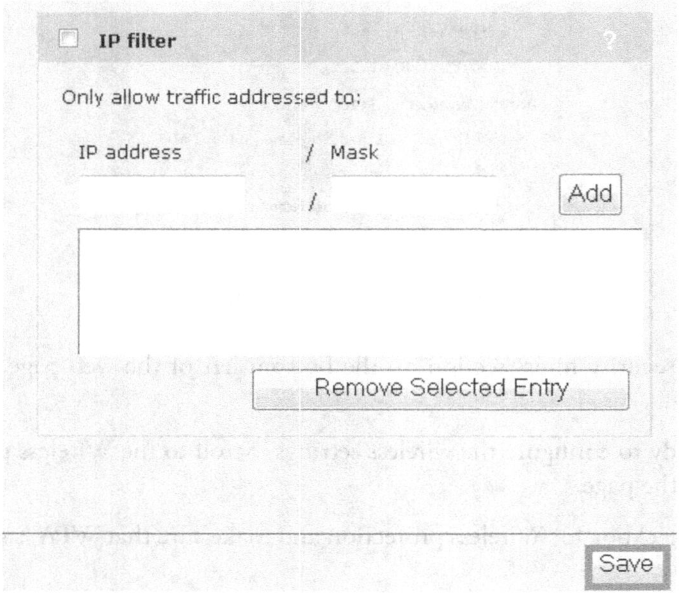


Figure 11-27: Configure a WAN with WPA2-PSK security

11. You are taken back to the VSC window, where you can see your new profile, as shown in Figure 11-28. Verify your settings:
    - SSID is your unique SSID.
    - Encryption is AES.
    - The VSC uses both radios.

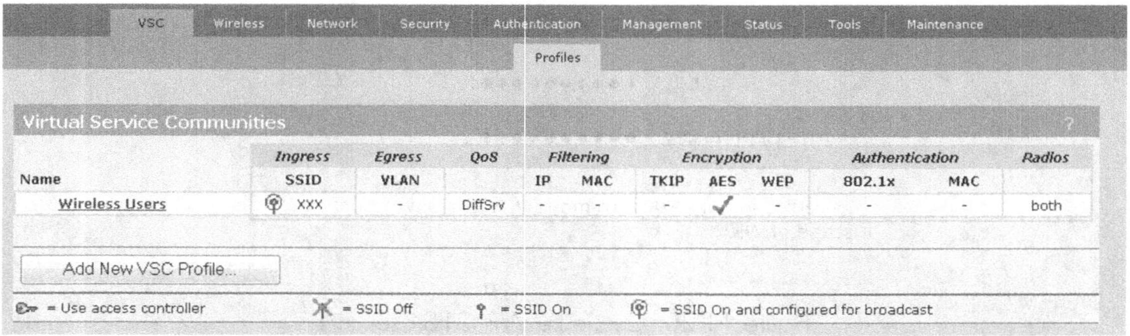


Figure 11-28: Configure a WAN with WPA2-PSK security

## Enable DHCP for VLAN 12

Wireless clients need to be able to receive IP addresses. In this configuration, your AP will forward wireless client traffic in VLAN 12, so you will enable DHCP on this VLAN. Comware-2 is already acting as a DHCP server for this VLAN, so you just need to configure an IP helper address on ProVision-2.

1. Access a terminal session with ProVision-2 and move to global configuration mode.
2. Configure VLAN 12 with an IP helper address to Comware-2, 10.1.102.2.

```
ProVision-2(config)# vlan 12 ip helper-address 10.1.102.2
```

## Test wireless access to the AP

You will now test your set up using the wireless Windows 7 PC.

1. Access the desktop of the *wireless* Windows 7 PC.
2. Click the network icon in the system tray.

> **Note**
> If your client's system tray does not include this icon, you can follow these alternative steps. Open the Network and Sharing Center (**Start > Control Panel > Network and Sharing Center**). Then click **Connect to a network**.

3. In the list of networks that is displayed, select your SSID.

> **Note**
> If you do not see your SSID, the AP might be broadcasting the SSID on a radio with a standard not supported by your client. Log back into the MSM AP, move to the **VSC>VCS profiles** window, and verify that the VSC is using both radios.

4. Click **Connect**.
5. When prompted for the Security key (as shown in Figure 11-29), enter **hp12345678** and click **OK**.

> **Note**
> Clear the Hide characters check box to see the password. Select the check box to hide them.

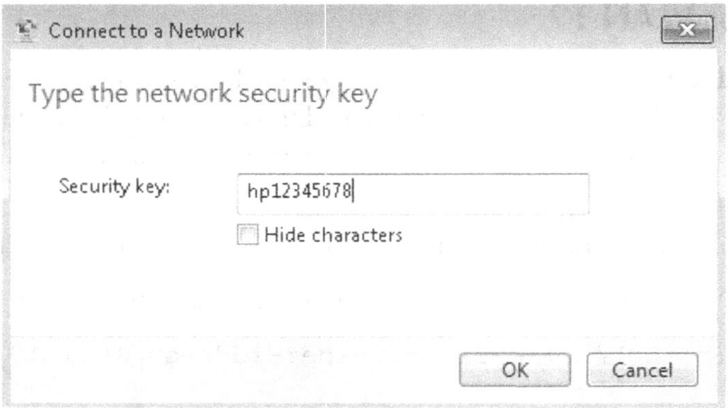


Figure 11-29: Connect to the VSC you created

You should see the status change to connected.

6. Open a command prompt and check the client's IP address on its Wireless Network Connection.

```
ipconfig

Windows IP Configuration

Wireless LAN adapter Wireless Network Connection:

   Connection-specific DNS Suffix  . :
   Link-local IPv6 Address . . . . . : fe80::3450:766d:aceb:2913%21
   IPv4 Address. . . . . . . . . . . : 10.1.12.22
   Subnet Mask . . . . . . . . . . . : 255.255.255.0
   Default Gateway . . . . . . . . . : 10.1.12.1
<-output omitted->
```

**Note**

If the client has not received an IP address, check the helper address on ProVision-2. Then check the DHCP configuration on Comware-2.

You should also check the IPv4 properties for the client's wireless adapter and verify that automatic configuration is enabled.

7. Try pinging the client's default gateway and the server. Both pings should be successful.

```
ping 10.1.12.1
ping 10.1.11.11
```

You have verified that the wireless client is fully connected to the network.

Save the configuration on ProVision-2 (**write memory**) and Comware-2 if you changed its configuration (**save force**).

## Summary

This chapter has introduced you to basic 802.11 concepts that help you to set up an autonomous MSM AP correctly. You know how to use a VSC to create a WLAN, configure an SSID for that WLAN, and secure the WLAN with WPA2-PSK.

## Learning check 2

Answer the following questions to test your understanding of wireless technologies:

1. Where do you set the WPA2-PSK in the AP settings?

2. In the VSC settings, you made sure that the Broadcast name (SSID) check box was selected. What do you think would happen if you cleared this check box?

3. Match the following IEEE standards with the appropriate description. Each option can only be used once. (Use the job aid on the next page for help.)

| | | |
|---|---|---|
| a. | 802.11a | Supports speeds up to 54 Mbps and uses the 5 GHz frequency |
| b. | 802.11b | Supports MIMO for theoretical data rates up to 600 Mbps |
| c. | 802.11g | Supports speeds up to 54 Mbps and uses the 2.4 frequency |
| d. | 802.11n | Supports MIMO for theoretical data rates in the Gbps range |
| e. | 802.11ac | Was the first IEEE wireless standard to achieve widespread adoption and uses the 2.4 frequency |

4. A VSC defines all the following settings for an MSM AP, except which?
   a. SSID
   b. Power settings for the AP radios
   c. Wireless protection, like WEP or WPA2
   d. MAC authentication

5. What is the default IP address of an MSM AP, if one cannot be obtained via DHCP?

6. What are the default username and password of an MSM AP?

## Learning check 2 answers

1. Where did you set the WPA2-PSK in the AP settings?

   You configure these settings in the VSC settings in the **Wireless protection** section.

2. In the VSC settings, you made sure that the Broadcast name (SSID) check box was selected. What do you think would happen if you cleared this check box?

   If you do not know, make your best guess.

   If you clear the check box, the AP does not broadcast the SSID in its beacons. This type of operation is called closed mode. In the past, closed mode was used for a degree of security. Users were still able to connect to the network, but they had to configure their wireless client with the SSID in advance. However, many wireless devices can detect SSIDs even when the AP does not broadcast it (because data frames from already-associated devices include the SSID). So closed operation does not provide real security and can make it harder for legitimate users to connect.

   In the example configuration, you set up an open mode system in which the AP *does* broadcast the SSID.

3. Match the following IEEE standards with the appropriate description. Each option can only be used once.

| | | | |
|---|---|---|---|
| a. | 802.11a | a | Supports speeds up to 54 Mbps and uses the 5 GHz frequency |
| b. | 802.11b | d | Supports MIMO for theoretical data rates up to 600 Mbps |
| c. | 802.11g | c | Supports speeds up to 54 Mbps and uses the 2.4 frequency |
| d. | 802.11n | e | Supports MIMO for theoretical data rates in the Gbps range |
| e. | 802.11ac | b | Was the first IEEE wireless standard to achieve widespread adoption and uses the 2.4 frequency |

4. A VSC defines all the following settings for an MSM AP, except which?

   a. SSID

   b. **Power settings for the AP radios**

   c. Wireless protection, like WEP or WPA2

   d. MAC authentication

5. What is the default IP address of an MSM AP, if one cannot be obtained via DHCP?

   192.168.1.1/24

6. What are the default username and password of an MSM AP?

   Both the username and the password are **admin**.

Job Aid 11-1: 802.11 standards

| Standard | RF frequency band | Number of spatial streams | Use of channel bonding | Data rates (Mbps) |
|---|---|---|---|---|
| 802.11b | 2.4 GHz | N/A | N/A | 1, 2, 5.5, 11 |
| 802.11g | 2.4 GHz | N/A | N/A | 6,9,12,18,24,36,48,54 |
| 802.11a | 5 Ghz | N/A | N/A | 6,9,12,18,24,36,48,54 |
| 802.11n | 2.4 GHz or 5 GHz | One spatial stream | **MSC 0–MSC 7** | |
| | | | No channel bonding | 6.5–72.2 |
| | | | Channel bonding (2) | 13.5–150 |
| | | Two spatial streams | **MSC 8 – MSC 15** | |
| | | | No channel bonding | 13–144.4 |
| | | | Channel bonding (2) | 27–300 |
| | | Three spatial streams | **MSC 16–MSC 23** | |
| | | | No channel bonding | 19.5–216.7 |
| | | | Channel bonding (2) | 40.5–450 |
| | | Four spatial streams | **MSC 24–MSC 31** | |
| | | | No channel bonding | 26–288.9 |
| | | | Channel bonding (2) | 54–600 |
| 802.11ac | 5 GHz | One spatial stream | **MSC 0 – MSC 9** | |
| | | | No channel bonding* | 6.5–86.7 |
| | | | Channel bonding (2) | 13.5–200 |
| | | | Channel bonding (4) | 29.3–433.3 |
| | | | Channel bonding (4+4) | 58.5–866.7 |
| | | Two spatial streams | **MSC 0–MSC 9** | |
| | | | No channel bonding* | 13–173.3 |
| | | | Channel bonding (2) | 27–400 |
| | | | Channel bonding (4) | 58.5–866.7 |
| | | | Channel bonding (4+4) | 117–1733.3 |
| | | Three spatial streams | **MSC 0–MSC 9** | |
| | | | No channel bonding | 19.5–288.9 |
| | | | Channel bonding (2) | 40.5–600 |
| | | | Channel bonding (4)** | 87.8–1300 |
| | | | Channel bonding (4+4)* | 175.5–2340 |

*Continued*

Job Aid 11-1: Continued

| Standard | RF frequency band | Number of spatial streams | Use of channel bonding | Data rates (Mbps) |
|---|---|---|---|---|
| | | Four spatial streams | **MSC 0–MSC 9** | |
| | | | No channel bonding* | 26–346.7 |
| | | | Channel bonding (2) | 54–800 |
| | | | Channel bonding (4) | 117–1733.3 |
| | | | Channel bonding (4+4) | 234–3466.7 |
| | | Five spatial streams | **MSC 0–MSC 9** | |
| | | | No channel bonding* | 32.5–433.3 |
| | | | Channel bonding (2) | 67.5–1000 |
| | | | Channel bonding (4) | 146.3–2166.7 |
| | | | Channel bonding (4+4) | 292.5–4333.3 |
| | | Six spatial streams | **MSC 0–MSC 9** | |
| | | | No channel bonding | 39–577.8 |
| | | | Channel bonding (2) | 81–1200 |
| | | | Channel bonding (4)* | 175.5–2340 |
| | | | Channel bonding (4+4) | 351–5200 |
| | | Seven spatial streams | **MSC 0–MSC 9** | |
| | | | No channel bonding* | 45.5–606.7 |
| | | | Channel bonding (2) | 94.5–1400 |
| | | | Channel bonding (4)** | 204.8–3033.3 |
| | | | Channel bonding (4+4) | 409.5–6066.7 |
| | | Eight spatial streams | **MSC 0–MSC 9** | |
| | | | No channel bonding* | 52–693.3 |
| | | | Channel bonding (2) | 108–1600 |
| | | | Channel bonding (4) | 234–3466.7 |
| | | | Channel bonding (4+4) | 468–6933.3 |

* MSC9 is not available.
** MCS6 is not available

# 12 HP Intelligent Management Center (IMC)

## EXAM OBJECTIVES

- ✓ Access Intelligent Management Center (IMC) and set up basic discovery
- ✓ Configure HP network devices to be discovered by IMC

## INTRODUCTION

In this study guide, you have focused on configuring and managing HP ProVision and Comware devices from the command line interface (CLI). In larger networks, though, you will find it helpful to have a centralized tool for monitoring, managing, configuring, and troubleshooting all of your network infrastructure devices.

This chapter introduces you to HP IMC, a powerful network management tool. You will learn about IMC capabilities and how IMC uses Simple Network Management Protocol (SNMP) to manage devices. You will also learn how to discover network infrastructure devices with IMC so that IMC can manage them.

**Note**

This study guide focuses on HP IMC version 7.

### Assumed knowledge

Before reading this chapter, you should have a basic understanding of the following:

- Simple Network Management Protocol (SNMP)
- Telnet and Secure Shell (SSH)
- IP addressing

## Introduction to IMC

IMC is a centralized network management platform that allows you to manage both your physical and virtual networks. It integrates fault management, element configuration, and network monitoring from a central location. From a single interface, you can monitor and manage network traffic and devices. HP IMC supports both HP and third-party network devices. In fact, IMC supports thousands of network devices from dozens of vendors, including Cisco.

IMC includes an auto discovery feature, which you can invoke to locate all the devices on the network, categorize the devices into types of network devices (such as switches, routers, servers, and desktops), and map them on a network topology.

You can view the network devices based on IP address or device type (such as routers, switches, servers, desktops, and so on). You can also create custom views to make it easier to view and manage devices. IMC provides device inventories and detailed topologies. Supported topology views include Layer 2 and Layer 3 topologies, as well as VLAN topology. You can also create custom views. Customization enables administrators to organize and control the network infrastructure based on their preferred organizational model. IMC can even manage and map virtual machines (VMs) in your data center.

In addition, IMC allows you to establish baseline configurations and software images. You can compare configurations, track versions, and establish alerts if configuration changes are made. VLAN management services help you to monitor your existing VLANs and set up new ones more quickly while Access Control List (ACL) management services help you to secure the network.

IMC also features flexible, centralized reporting. Centralized report management simplifies an organization's report administration. IMC software's flexible historical reports provide the information necessary for network trend analysis and capacity planning and offer predefined reports or customization options to define parameters. Reports can be viewed in a number of formats, including .pdf and .xls, and can be sent automatically via email, or be set to run on a particular schedule.

IMC has a modular architecture, which allows you to add new modules to enrich a network's management capabilities. Modules for user access management, VPN management, and traffic analysis can be quickly added and provide instant benefits. The architecture allows modules to share information and provide collaborative policy creation and reports. This study guide, however, focuses on the base platform.

## IMC editions

As Figure 12-1 shows, IMC comes in several editions.

IMC Standard is the basic offering for midsized enterprises.

IMC Enterprise is intended for larger enterprises. It supports more nodes with the base license, and it also scales to support more nodes with additional licenses because it supports a hierarchical deployment.

That is, you can deploy multiple IMC servers beneath the master server, and each server manages part of the larger network. (Note that the master must use IMC Enterprise edition, but other servers below it in the hierarchy can use IMC Standard edition.)

| Good for: | IMC Standard | IMC Enterprise | IMC Basic or IMC Basic WLAN |
|---|---|---|---|
| Network size | • Medium networks | • Large networks | • Small networks |
| Network requirements | • Simple architecture | • Hierarchical architecture deployment | • Basic monitoring, basic configuration |
| Additional requirements | • Suitable for BYOD, unified wired and WLAN management | • Same as IMC Standard | • WLAN management optional. Limited solution management. |

Figure 12-1: HP IMC editions

Other enhancements in IMC Enterprise include:

- Network Traffic Analyzer (NTA) module with two licenses
- Access to extensible APIs

You can either extend an existing function to support third-party devices by compiling interactive scripts and XML files, or you can customize a function by compiling interactive scripts, XML files, and UI configuration files.

IMC Basic and IMC Basic WLAN are designed for ease of use for small-to-medium-sized businesses (SMBs). It has a fixed 50-node license and a limited feature set.

This study guide focuses on the Standard and Enterprise editions, both of which support modules that enhance their functionality and provide features such as Bring Your Own Device (BYOD) and unified wired and wireless management.

## IMC deployment options

IMC supports two types of installations, as shown in Figure 12-2:

- **Centralized:** All IMC components and add-on modules are installed and deployed on the same server.
- **Distributed:** All IMC components and add-on modules are installed on the master IMC server, but some components for modules are deployed on slave servers. The master server is the

management center for IMC; slave servers are responsible for specific management tasks. For example, User Access Manager (UAM), an IMC module, provides user authentication and includes a RADIUS server. A slave server might run the UAM and RADIUS processes. You access the master server to complete all management tasks, and it interacts with slave servers as needed to manage the network.

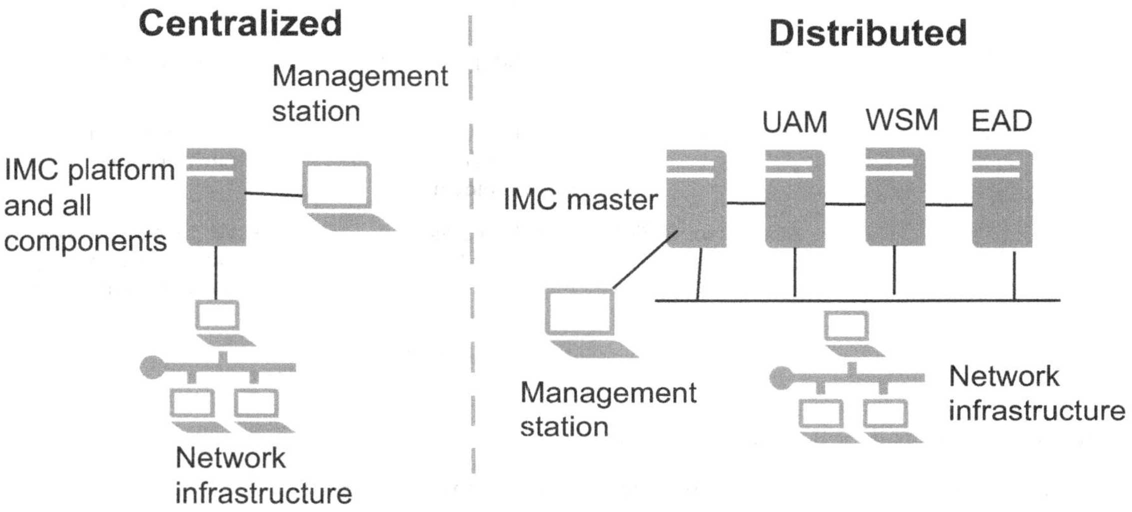


Figure 12-2: IMC deployment options

If you are using the IMC platform, but no other add-on modules, HP recommends you use the centralized deployment. Otherwise, use the distributed installation.

For either centralized or distributed installations, IMC requires a database to store information about managed devices and users. The IMC Standard Edition ships with Microsoft SQL Server Express, which you can install at the same time you install IMC.

Note that hierarchical deployment is different from distributed deployment. Both IMC Standard and IMC Enterprise support distributed deployment, in which different servers support different IMC components. Only IMC Enterprise supports hierarchical deployments, in which different servers manage different network segments.

**Note**

Refer to the latest *HP IMC Installation and Configuration Guides* to learn about systems that support IMC installation requirements, and installation processes.

# IMC management protocols

IMC uses several different protocols and mechanisms to monitor and manage network devices, as illustrated in Figure 12-3. The sections below describe the most important ones for you to understand. (IMC supports others as well.)

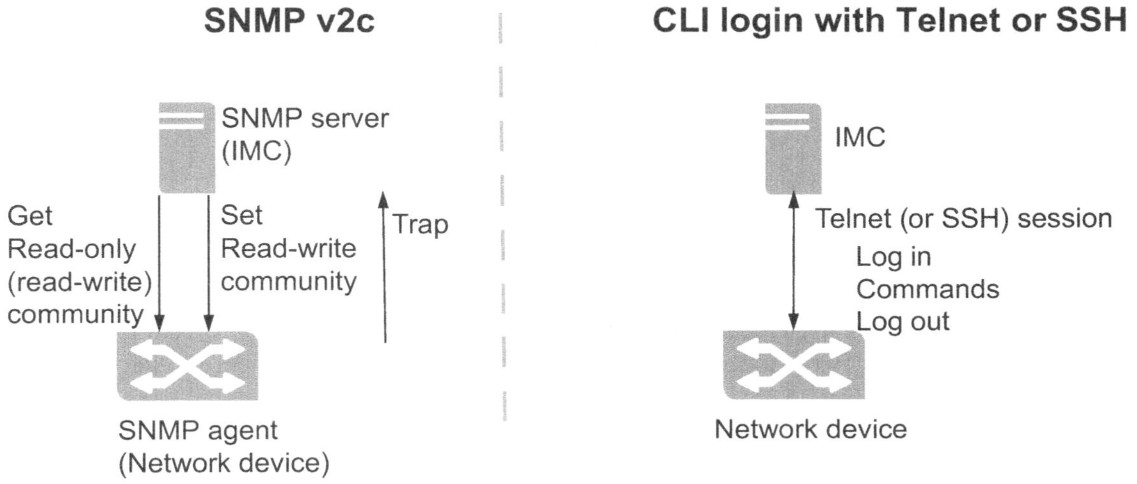


Figure 12-3: IMC management protocols

## Simple Network Management Protocol (SNMP)

SNMP is an industry-standard protocol that allows you to manage and monitor a variety of network devices from a central location.

SNMP works in a client–server relationship. SNMP agents, which function as clients, run on managed devices such as switches. The SNMP server is a management application that requests, handles, and analyzes the information from the managed devices. IMC is an SNMP server. IMC, as an SNMP server, sends Get messages to the agents to query them about their settings, their logs, their status, and so on. It also sends Set messages to alter the managed devices' settings—for example, to shut down a port.

Agents mostly only reply to server messages, but they can send SNMP traps to their trap server to update the server when an alarm or an event, such as a failed link, occurs. IMC configures itself as a trap server when it discovers devices; trap messages also have a community, and IMC configures the device to use the correct one.

When your solution uses SNMP v2c, the SNMP servers and agents include a community string in the messages. SNMP agents will only accept GET messages if the community matches their read-only or read-write community string. They will only accept WRITE messages if the community matches their read-write community string.

SNMPv2 is easy to set up but is not secure because the community is included in messages in plaintext. Therefore, the community string cannot function as a true password.

---

**Note**

SNMPv3 provides data privacy (encryption) and data integrity. Instead of community strings, SNMPv3 uses user credentials. The server credentials (*not* transmitted in plaintext) must match user credentials on the managed device, and those credentials are associated with various read and write privileges. For more information on SNMPv3, attend the *Deploying HP FlexNetwork Core Technologies* course.

---

### Logging in to the CLI

To complete some functions, IMC logs into a network device's CLI and sends commands over a terminal session, just like you do as an admin. The process happens in the background when you use a service such as VLAN management just like SNMP GET and WRITE messages occur in the background. You simply see that you make a change through the IMC management interface, and the change is applied.

IMC supports Telnet or SSH for logging into switches.

## Preparing HP devices for discovery

You must configure SNMP and Telnet or SSH settings on your switches to prepare them to be discovered by IMC. Figure 12-4 summarizes the settings for Comware switches, ProVision switches, and autonomous MSM Access Points (APs).

ProVision

- Define SNMP settings
- Configure manager password
- Set up SSH if desired

Comware

- Define SNMP settings
- Enable Telnet or SSH
- Configure VTY interface authentication

Autonomous MSM AP

- Define SNMP settings

Figure 12-4: Preparing HP devices for discovery

The example configuration provided in this study guide focuses on SNMP v2c. You should be aware that SNMP v3 is more secure and should, therefore, be used for real-world environments.

### HP ProVision

You must enable SNMP on ProVision switches, which support v2c by default. You must then set the SNMP read-only and read-write communities to match the settings configured in IMC.

HP ProVision switches support the "public" community, by default, and they assign unrestricted (write) privileges to it. In the real world, you should typically delete this community for security purposes. Then you should create your own communities.

Telnet is enabled by default on ProVision switches. (If you want to use SSH, set it up.) Generally, you would set a manager password and give IMC the manager username and correct password. Again, authentication is not used in the example configuration to simplify the setup. In a real-world environment, you should secure switches with passwords.

### HP Comware

You must enable SNMP for the correct version. For the example configuration, which uses SNMP v2, you must set the SNMP read-only and read-write communities to match the ones that you will use in IMC.

You must also enable Telnet or set up SSH on Comware switches. You must then configure the VTY interface authentication to match the credentials that you will set on IMC. You must make sure that IMC will receive level 3 privileges. Generally, you would set a password on the VTY interface, or you would use the scheme authentication method and set up management local users (not covered in this study guide).

### Autonomous HP MSM APs

You must also configure SNMP settings on the MSM AP so that IMC can discover and manage it. You do not typically use the CLI to manage MSM products, so you will not configure those settings.

**Note**

In a controlled wireless deployment, you configure the SNMP settings on the wireless controller, rather than on the AP. In addition to configuring SNMP settings on an MSM controller, you must also configure IMC's IP address as the controller's PMM Server Console.

## Device discovery

Once you configure these settings, you can use IMC to discover the devices. As Figure 12-5 illustrates, IMC involves the following:

- Contacting the devices
- Checking devices' management settings

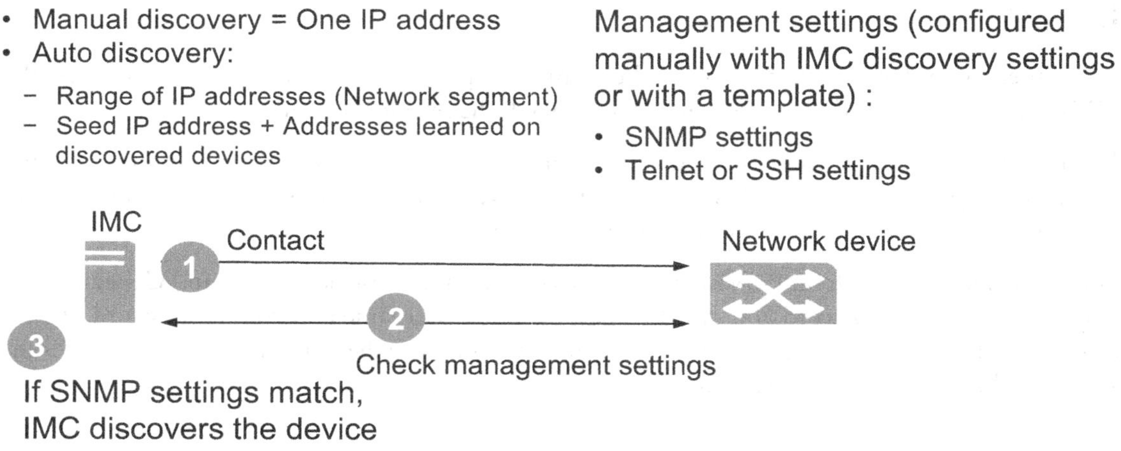


Figure 12-5: Device discovery

## Discovery methods

You will first consider some of IMC's discovery methods—its methods for learning the IP addresses at which to contact devices.

IMC supports manual discovery in which you manually enter the IP address of a device that you want to discover. However, manual discovery is not feasible for a large network.

IMC also supports multiple modes of auto discovery, each of which takes a different approach toward informing IMC how to find new devices to discover. For example, the basic auto discovery mode has you specify a range of IP addresses for IMC to discover. Other modes specify a seed IP address for a device which IMC discovers, and IMC then uses information on that device to discover more devices. Table 12-1 summarizes these auto-discovery modes.

Table 12-1: IMC auto discovery modes

| Mode | Description | Advantages | Disadvantages | Use case |
|---|---|---|---|---|
| Routing-based | You specify a seed device. IMC discovers the device and reads its routing table. IMC attempts to discover all nodes in all IP subnets in the routing table. | You achieve full automatic discovery of network devices. | If the routing table is large, discovery could take hours or even days. If some networks are missing from the routing table, you must add them manually. | Use with networks that have few subnets. |

*Continued*

Table 12-1: Continued

| Mode | Description | Advantages | Disadvantages | Use case |
|---|---|---|---|---|
| ARP(ND)-based | You specify a seed device, typically one of your network's default routers.<br>IMC discovers the device and reads its ARP table (or ND table for IPv6).<br>IMC attempts to discover IP addresses in the table.<br>IMC repeats the process on other discovered devices. | The search is restricted to active devices, so the search is quicker than a routing-based search. It will still typically capture most devices. | Inactive devices will not be discovered, so you might need to run discovery multiple times. Discovery can still be rather slow in a large network. | Use for a quicker discovery process when you do not want to specify ranges for managed devices' IP addresses. |
| IPsec VPN-based | IMC scans IP addresses on the remote end of an IPsec VPN. | The search focuses on remote devices related to IPsec VPNs. | The process might be time-consuming if you have a large hub-and-spoke site-to-site topology. | Use to discover devices in remote sites connected through IPsec tunnels. |
| Network segment-based | You enter specific ranges of IP addresses. IMC discovers just devices in those ranges. | You can target specific IP addresses that are likely to have network devices:<br>• Management VLAN IP subnets<br>• IP address ranges that you have reserved for managed devices | You must manually configure the ranges for all devices. | Use for larger networks when you know the ranges for managed devices and want to reduce discovery time. |
| PPP-Protocol- based | You specify a seed device, which has a PPP connection. IMC discovers devices on the other side of the connection. | You can target dialup devices and discover their remote peers. | This form of discovery is limited to a relatively rare use case. (Not many networks use dialup or PPP.) | Use to discover remote devices connected through PPP. |

## Management settings

After IMC contacts a device, it checks the device's management settings to ensure that it can manage the device. Only then will it complete the discovery process. You specify the management settings while you set up discovery, either configuring the settings then or selecting a preconfigured template. The management settings include:

- SNMP settings—You can use SNMPv2c or SNMPv3 settings. It is easier to discover devices using SNMPv2c. However, SNMPv2c is not secure, so in a real network, you should then configure IMC and the discovered devices to use SNMPv3 instead. For SNMPv2c, you must set the read-only and read-write communities to match those on your network devices.
- Login settings—You can choose either Telnet or SSH. You must then specify the credentials that IMC uses to log in under Telnet or SSH settings.

If the SNMP settings that you specify on IMC match the device SNMP settings, IMC finishes discovering the device and adds it as a managed device. The login settings do not have to match, but, if they do not, IMC might not be able to manage the device properly.

---

**Note**

You can also configure discovery settings such that IMC will discover the device if it receives a response from it even if the SNMP settings do not match. You might choose this approach when you want to discover desktops and servers.

---

# Example configuration: Discover devices with HP IMC

This section outlines the steps for configuring network devices for IMC discovery. The steps are provided in the context of configuring devices in the example network shown in Figure 12-6.

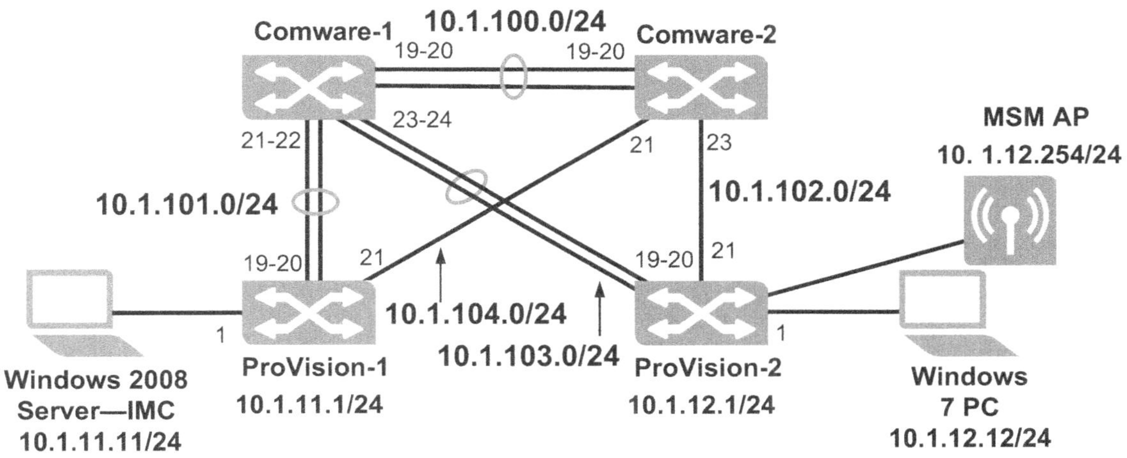


Figure 12-6: Example configuration: Discover devices with HP IMC

Specifically, you will learn how to:

- Configure SNMP v2 and login settings on ProVision and Comware switches
- Define SNMP settings on an MSM AP
- Discover the network devices in IMC

Figure 12-6 provides a detailed illustration of the example network.

## Prepare HP Comware switches for discovery by IMC

You first define SNMP v2 and login settings on the Comware switches. You will match the settings that you plan to use for IMC.

### HP Comware-1

1. Access the Comware-1 CLI and move to system view.
2. Enable SNMP version 2c.

```
[Comware-1] snmp-agent sys-info version v2c
```

3. Configure the read-only community string.

```
[Comware-1] snmp-agent community read public
```

4. Configure the read-write community string.

```
[Comware-1] snmp-agent community write private
```

5. Configure the IMC server IP address as the trap server. Use the read-only community string for the v2c security name.

```
[Comware-1] snmp-agent target-host trap address udp-domain 10.1.11.11
params securityname public v2c
```

6. Verify the SNMP settings.

```
[Comware-1] display snmp-agent community
   Community name: public
       Group name: public
       Storage-type: nonVolatile
   Community name: private
       Group name: private
       Storage-type: nonvolatile
```

7. IMC also needs to log in to the switches, so you must enable Telnet (if not already enabled).

```
[Comware-1] telnet server enable
```

8. For simplicity in example configuration, allow Telnet access without credentials. Configure the VTY interfaces to disable authentication. In a real-world environment, of course, you would enforce tighter security measures.

```
[Comware-1] user-interface vty 0 15
[Comware-1-ui-vty0-15] authentication-mode none
```

9. IMC requires full privileges.

```
[Comware-1-ui-vty0-15] user privilege level 3
[Comware-1-ui-vty0-15] quit
```

## HP Comware-2

10. Access a terminal session with Comware-2 and move to the system view.

11. Define the same SNMP settings.

```
[Comware-2] snmp-agent sys-info version v2c

[Comware-2] snmp-agent community read public

[Comware-2] snmp-agent community write private

[Comware-2] snmp-agent target-host trap address udp-domain 10.1.11.11
params securityname public v2c
```

12. Verify Comware-2's SNMP settings.

```
[Comware-2] display snmp-agent community

   Community name: public

       Group name: public

       Storage-type: nonVolatile

   Community name: private

       Group name: private

       Storage-type: nonvolatile
```

13. Set up Telnet access with no authentication and full privileges. Again, this configuration is used only for a test environment. For a real-world environment, you would secure Telnet access.

```
[Comware-2] telnet server enable

% Telnet server has been started

[Comware-2] user-interface vty 0 15

[Comware-2-ui-vty0-15] authentication-mode none

[Comware-2-ui-vty0-15] user privilege level 3

[Comware-2-ui-vty0-15] quit
```

## HP ProVision-1

Next, define SNMPv2 settings on the ProVision switches. ProVision switches have Telnet enabled and no passwords by default, so you do not need to configure additional login settings. Again, you would tighten security for a real-world environment.

1. Access a terminal session with ProVision-1 and move to global configuration mode.
2. Define the SNMP read-only community.

```
ProVision-1(config)# snmp-server community public operator restricted
```

**Note**
HP ProVision switches support the "public" community, by default. However, they assign manager unrestricted (write) privileges to it. You want to change the privileges for "public" to operator restricted (read).

3. Define the SNMP read-write community.

```
ProVision-1(config)# snmp-server community private manager unrestricted
```

4. Configure the IMC server as the trap server.

```
ProVision-1(config)# snmp-server host 10.1.11.11 community public
```

5. Enable SNMP, which operates in v2c by default.

```
ProVision-1(config)# snmp-server enable
```

6. Verify the SNMP settings.

```
ProVision-1(config)# show snmp-server

SNMP Communities

 Community Name                   MIB View Write Access
 -------------------------------- -------- ------------
 public                           Operator Restricted
 private                          Manager  Unrestricted

 Trap Receivers

  Link-Change Traps Enabled on Ports [All] : All

  Traps Category                       Current Status
  _____________________________        __________________

  SNMP Authentication               :  Extended
  Password change                   :  Enabled
  Login failures                    :  Enabled
  Port-Security                     :  Enabled
  Authorization Server Contact      :  Enabled
  DHCP-Snooping                     :  Enabled
  Dynamic ARP Protection            :  Enabled
```

```
 Dynamic IP Lockdown            : Enabled
 Startup Config change          : Disabled
 Running Config Change          : Disabled
 MAC address table changes      : Disabled
 MAC Address Count              : Disabled
 Address      Community   Events   Type   Retry   Timeout
 ------------ ----------- -------- ------ ------- -------
 10.1.11.11   public      None     trap   3       15
Excluded MIBs
Snmp Response Pdu Source-IP Information
 Selection Policy   : rfc1517
Trap Pdu Source-IP Information
 Selection Policy   : rfc1517
```

### HP ProVision-2

7. Access a terminal session with ProVision-2 and move to global configuration mode.
8. Define the same SNMP settings for this switch.

```
ProVision-2(config)# snmp-server community public operator restricted
ProVision-2(config)# snmp-server community private manager unrestricted
ProVision-2(config)# snmp-server host 10.1.11.11 community public
ProVision-2(config)# snmp-server enable
```

9. Verify the settings.

```
ProVision-2(config)# show snmp-server
 SNMP Communities
  Community Name         MIB View Write Access
  ---------------------- -------- ------------
  public                 Operator Restricted
  private                Manager  Unrestricted
<-output omitted->
```

## Prepare the HP MSM AP for discovery by IMC

You then configure SNMPv2 settings on the MSM AP.

1. Access the desktop on the wired Windows 7 PC.
2. Open a web browser and navigate to the MSM AP (**http://10.1.12.254**).
3. Log in using a username and password of **admin** and **admin**.
4. Click **Management** in the top navigation bar.
5. Click **SNMP**.
6. In this window, you configure the SNMP settings. Select the **Notifications** checkbox, as shown in Figure 12-7.

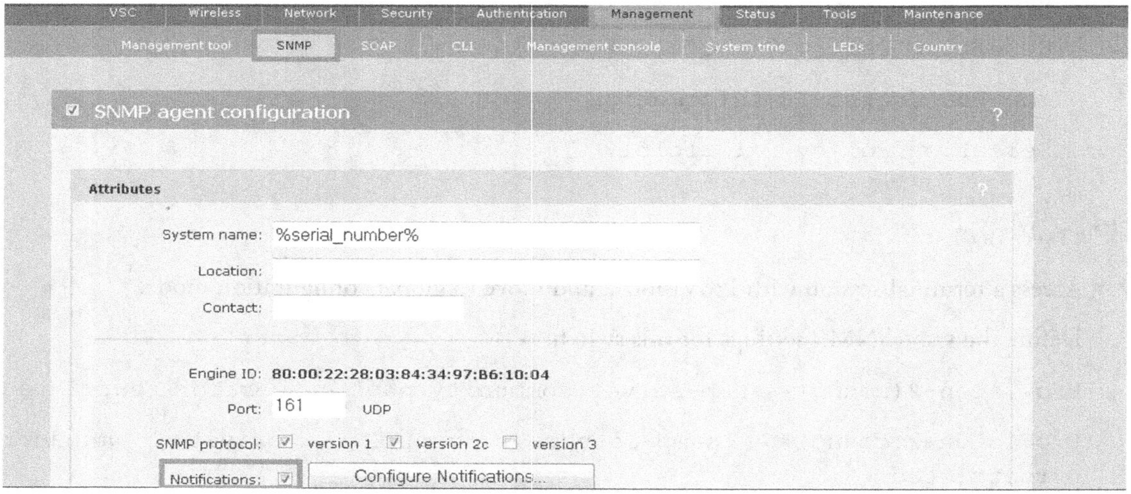


Figure 12-7: Configure SNMP settings on the MSM AP

7. Figure 12-8 shows the SNMP v2 configuration fields. For Community name, enter **private**.
8. For Confirm community name, enter **private**.
9. For Read-only name, enter **public**.
10. For Confirm read-only name, enter **public**.

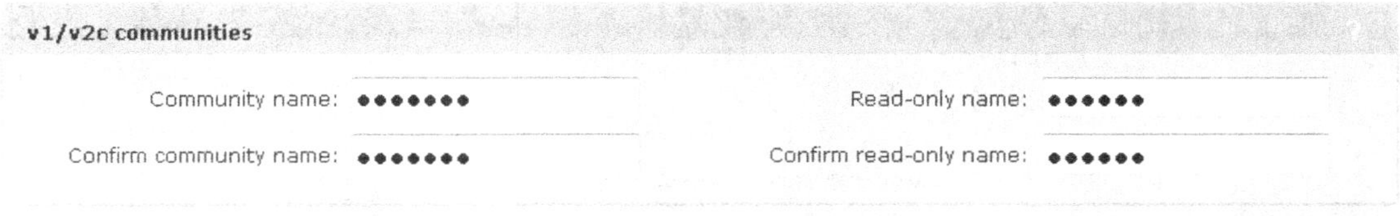


Figure 12-8: Configure SNMP v2 community settings on the MSM AP

11. Scroll to the bottom of the window and click **Save.**

## Verify that IMC is ready

For this example, IMC has been installed successfully. The IMC installation process is not covered.

### Windows server

1. Access the desktop of the Windows 2008 server.
2. Open a command prompt and enter `ipconfig`. Verify the server's IP settings. In the example network, the server should have the IP address 10.1.11.11/24.
3. Open the IMC Deployment Monitoring Agent from the Start Menu, as shown in Figure 12-9.

Figure 12-9: IMC Deployment Monitoring Agent

4. Verify that IMC is running. The Start IMC button should be grayed out, as shown in Figure 12-10.

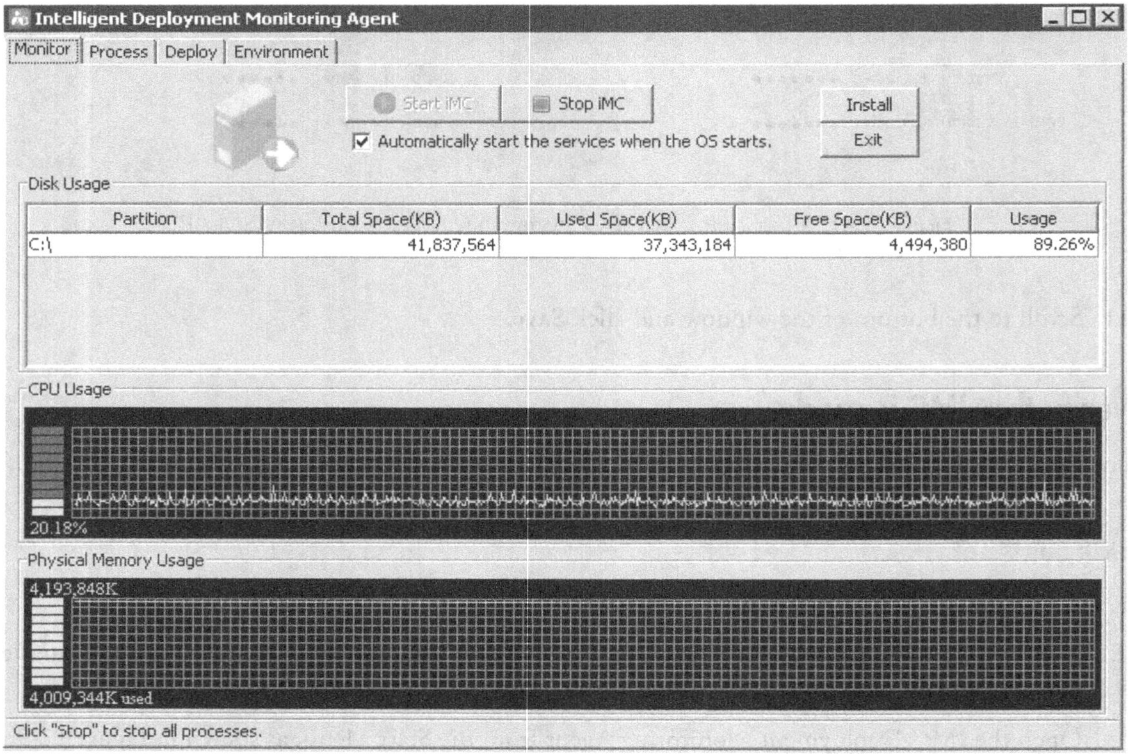


Figure 12-10: IMC Deployment Monitoring Agent

5. Click the **Process** tab, as shown in Figure 12-11. Verify that the IMC processes are running. In particular, look for the jserver process, which is often the last process to start and which is required for you to reach the IMC Web interface. If the portalserver and mschapv2 server are not running, however, you can still complete the configuration and discovery process.

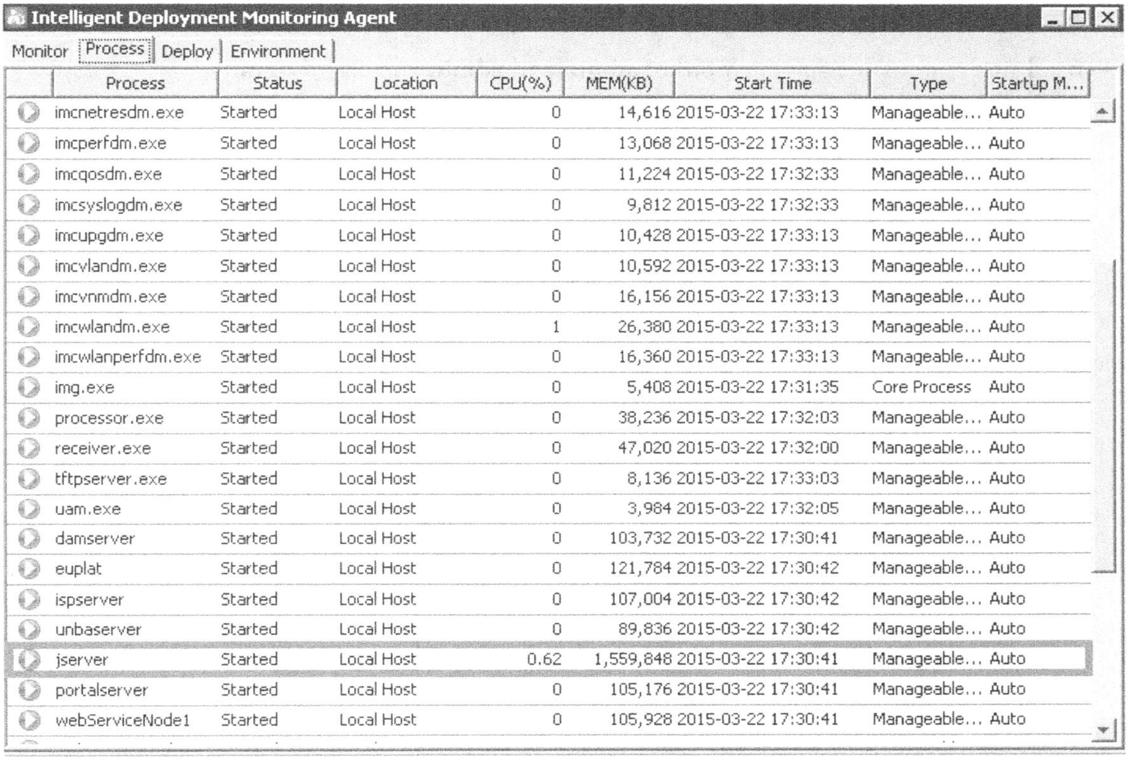


| Process | Status | Location | CPU(%) | MEM(KB) | Start Time | Type | Startup M... |
|---|---|---|---|---|---|---|---|
| imcnetresdm.exe | Started | Local Host | 0 | 14,616 | 2015-03-22 17:33:13 | Manageable... | Auto |
| imcperfdm.exe | Started | Local Host | 0 | 13,068 | 2015-03-22 17:33:13 | Manageable... | Auto |
| imcqosdm.exe | Started | Local Host | 0 | 11,224 | 2015-03-22 17:32:33 | Manageable... | Auto |
| imcsyslogdm.exe | Started | Local Host | 0 | 9,812 | 2015-03-22 17:32:33 | Manageable... | Auto |
| imcupgdm.exe | Started | Local Host | 0 | 10,428 | 2015-03-22 17:33:13 | Manageable... | Auto |
| imcvlandm.exe | Started | Local Host | 0 | 10,592 | 2015-03-22 17:33:13 | Manageable... | Auto |
| imcvnmdm.exe | Started | Local Host | 0 | 16,156 | 2015-03-22 17:33:13 | Manageable... | Auto |
| imcwlandm.exe | Started | Local Host | 1 | 26,380 | 2015-03-22 17:33:13 | Manageable... | Auto |
| imcwlanperfdm.exe | Started | Local Host | 0 | 16,360 | 2015-03-22 17:33:13 | Manageable... | Auto |
| img.exe | Started | Local Host | 0 | 5,408 | 2015-03-22 17:31:35 | Core Process | Auto |
| processor.exe | Started | Local Host | 0 | 38,236 | 2015-03-22 17:32:03 | Manageable... | Auto |
| receiver.exe | Started | Local Host | 0 | 47,020 | 2015-03-22 17:32:00 | Manageable... | Auto |
| tftpserver.exe | Started | Local Host | 0 | 8,136 | 2015-03-22 17:33:03 | Manageable... | Auto |
| uam.exe | Started | Local Host | 0 | 3,984 | 2015-03-22 17:32:05 | Manageable... | Auto |
| damserver | Started | Local Host | 0 | 103,732 | 2015-03-22 17:30:41 | Manageable... | Auto |
| euplat | Started | Local Host | 0 | 121,784 | 2015-03-22 17:30:42 | Manageable... | Auto |
| ispserver | Started | Local Host | 0 | 107,004 | 2015-03-22 17:30:42 | Manageable... | Auto |
| unbaserver | Started | Local Host | 0 | 89,836 | 2015-03-22 17:30:42 | Manageable... | Auto |
| jserver | Started | Local Host | 0.62 | 1,559,848 | 2015-03-22 17:30:41 | Manageable... | Auto |
| portalserver | Started | Local Host | 0 | 105,176 | 2015-03-22 17:30:41 | Manageable... | Auto |
| webServiceNode1 | Started | Local Host | 0 | 105,928 | 2015-03-22 17:30:41 | Manageable... | Auto |

Figure 12-11: IMC Deployment Monitoring Agent

## Discover devices in IMC

You should still be in the desktop of the Windows server.

1. Log in to IMC.
    a. Launch Internet Explorer (or Firefox).

**Note**

IMC 7.0 and later requires IE9/10, Firefox 26 or later, or Chrome 26 or later because it requires HTML5 support.

b. Navigate to http://localhost:8080/imc.

c. Enter a username and password of **admin** and **admin** and click **Login**.

2. In the top menu, navigate to **Resource** > **Auto Discovery**, as shown in Figure 12-12. (You might also be prompted to go to discovery settings.)

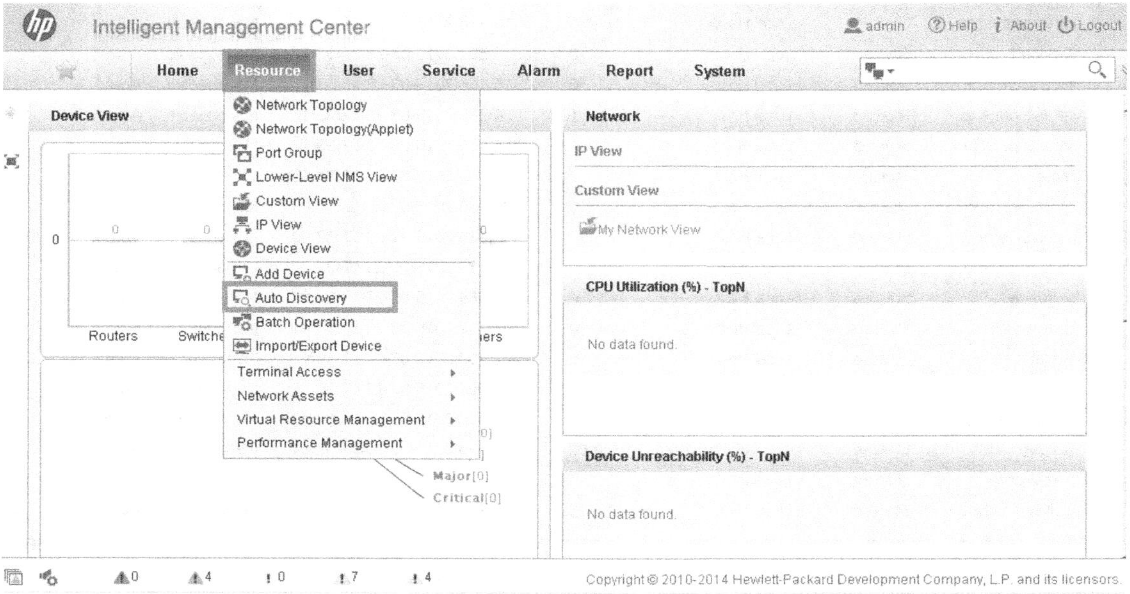


Figure 12-12: Select Auto Discovery

3. The switches in the example network do not have contiguous IP addresses, so advanced discovery will work better. Click **Go to Advanced**, as shown in Figure 12-13.

ARP discovery lets you give IMC the IP address of a single network device as a seed. IMC discovers that device and uses its ARP table to find more devices to discover. This method will work well for your network in which your switches have IP addresses on several different subnets.

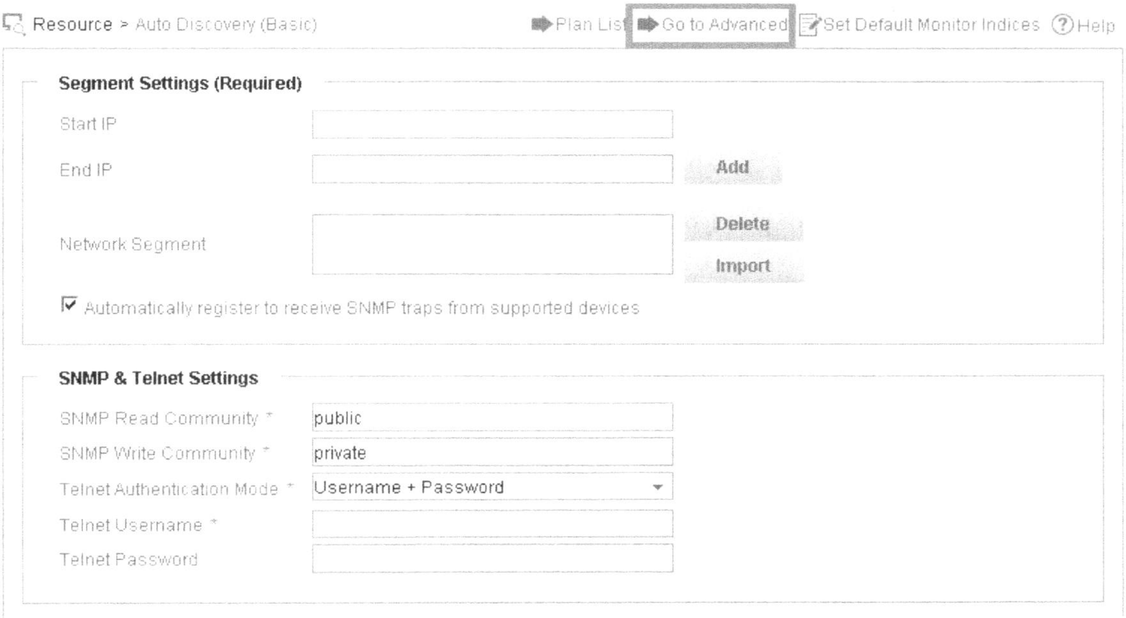


Figure 12-13: Select Go to Advanced

4. Select **ARP(ND)-Based**, as shown in Figure 12-14.
5. Click **Next**.

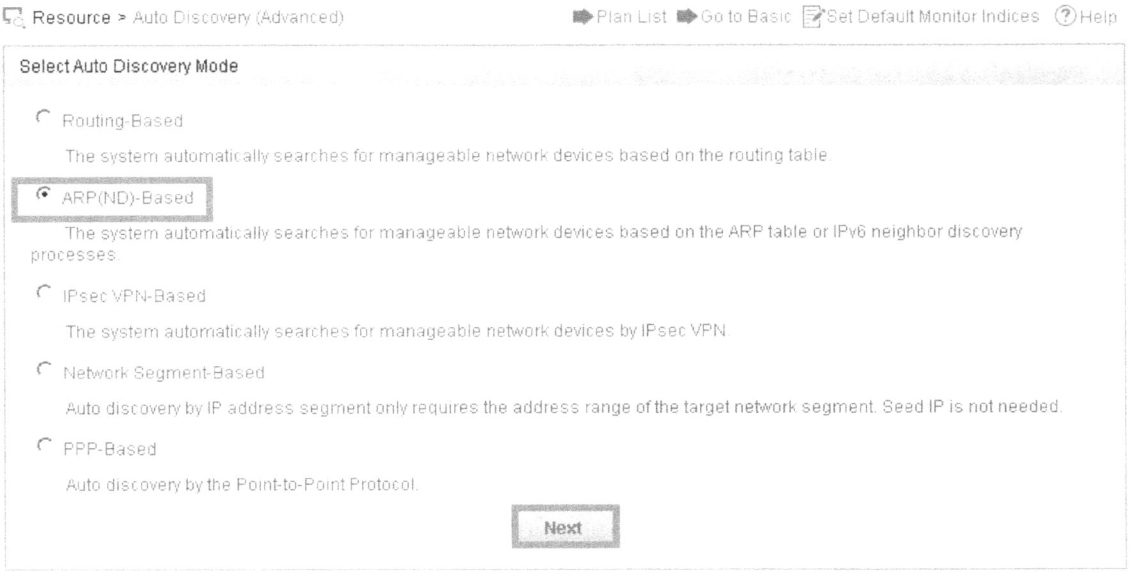


Figure 12-14: Select ARP (ND)-Based

6. You can accept the default settings, shown below, for the basic settings.
7. For Seed IP, enter **10.1.11.1** (which is the IP address ProVision-1).
8. Click **Add**.
9. You will now configure the SNMP settings to match the ones that you set on the switches.
   a. Click the small tool or gear icon for SNMP settings, as shown in Figure 12-15.

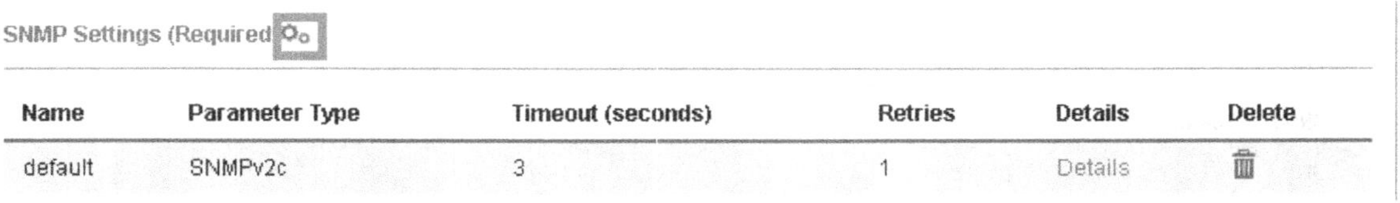


Figure 12-15: Select the small tool icon

   b. Make sure that SNMPv2c is selected for Parameter Type, as shown in Figure 12-16.
   c. For the Read-Only Community String, enter **public**.
   d. For the Read-Write Community String, enter **private**.
   e. Click **OK**.

Figure 12-16: Configure SNMP settings

10. You must also configure the Telnet settings.
    a. Click the tool or gear icon for Telnet parameters.
    b. For Authentication mode, choose No Username + No Password, as shown in Figure 12-17. In a real-world environment, of course, you would configure more secure settings.
    c. Click **OK**.

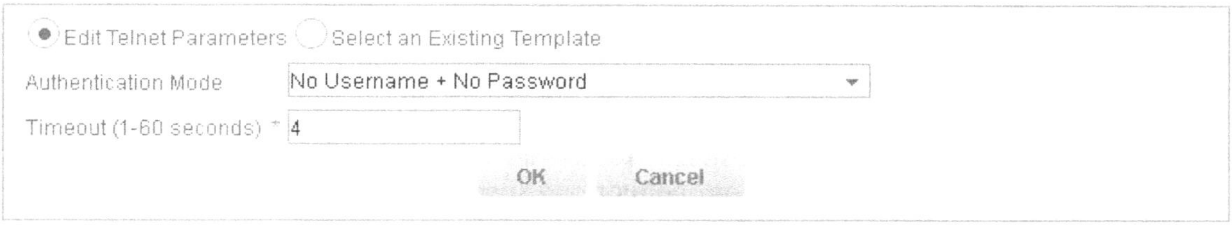


Figure 12-17: Configure Telnet settings

11. You have finished configuring the discovery settings. Scroll to the bottom of the window, and click **Auto Discovery**.
12. Be patient as IMC discovers the devices. You will see messages as the devices are discovered. You should eventually see all four HP switches and the MSM AP (which is using its serial number for its hostname), as shown in Figure 12-18.

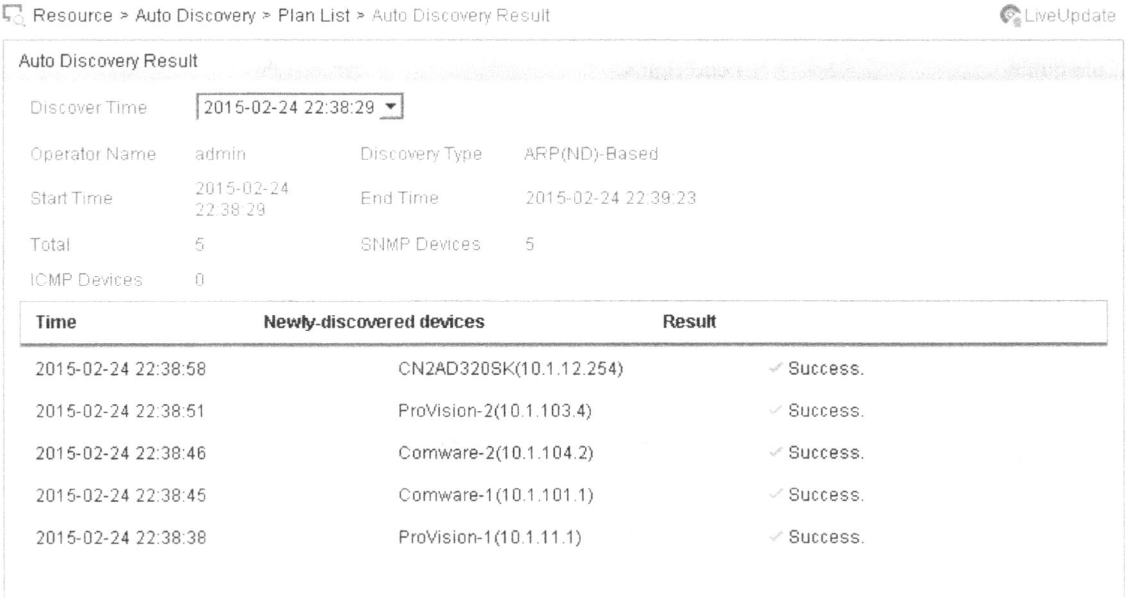


Figure 12-18: IMC Deployment Monitoring Agent

**Note**

If IMC did not discover any of the devices, verify the SNMP and Telnet settings on that device. Then run auto discovery again.

## Save

Save the current configuration on each of the four switches. For ProVision, use the `write memory` command. For Comware, use the `save [force]` command.

# Learning check 1

You will now answer some questions about the concepts you have learned in this chapter.

1. For ProVision switches, you did not specify read-write or read-only for the community. Instead you set the community to manager or operator and specified the restricted or unrestricted option

   The manager unrestricted community provides read-write access to all MIBs, manager restricted provides read-only access to all MIBs, operator unrestricted provides read-write access to all MIBs except the CONFIG MIB, and operator restricted provides read-only access to all MIBs except the CONFIG MIB.

   - Based on the information in Table 12-2, which manager option is associated with read-only access and which with read-write access?
   - When might you want to use the operator-restricted options for the SNMP read-only community?

Table 12-2: ProVision SNMPv2 communities

| Community | Read rights | Write rights |
|---|---|---|
| manager unrestricted | **ManagerReadView**<br>All managed objects | **ManagerWriteView**<br>All managed objects except a few related to setting SNMP rights (vacmContextTable, vacmAccessTable, vacmViewTreeFamilyTable) |
| manager restricted | **ManagerReadView**<br>All managed objects | **DiscoveryView**<br>No managed objects except a sampling probe |
| operator unrestricted | **OperatorReadView**<br>All managed objects except some related to SNMP security and TFTP settings (icfSecurityMIB, hpSwitchIpTftp-Mode, vacmContextTable, vacmAccessTable, vacmViewTreeFamily-Table, usmUserTable, snmpCommunityTable) | **OperatorReadView**<br>All managed objects except some related to SNMP security and TFTP settings (icfSecurityMIB, hpSwitchIpTftp-Mode, vacmContextTable, vacmAccessTable, vacmViewTreeFamily-Table, usmUserTable, snmpCommunityTable) |
| operator restricted | **OperatorReadView**<br>All managed objects except some related to SNMP security and TFTP settings (icfSecurityMIB, hpSwitchIpTftp-Mode, vacmContextTable, vacmAccessTable, vacmViewTreeFamily-Table, usmUserTable, snmpCommunityTable) | **DiscoveryView**<br>No managed objects except a sampling probe |

**Note**

By default, the Comware read-only community permits read access to ViewDefault objects: all managed objects except to a few related to SNMP security settings (snmpUsmMIB, snmpVacmMIB, snmpModules.18). The read-write community permits write access to ViewDefault objects.

2. What steps did you have to complete on the Comware switches to set up login access for IMC?

3. For the example configuration, you allowed IMC Telnet access without authentication. How would you change the set up for the real world?

4. The basic auto discovery process has you enter a range of IP addresses to discover. When would this type of discovery be appropriate?

5. Which form of discovery did you use? Why was this type of discovery appropriate for your network? How does it work?

6. The switches in the example network have multiple IP addresses. Which does IMC use to manage the switches?

## Learning check 1 answers

1. For ProVision switches, you did not specify read-write or read-only for the community. Instead you set the community to manager or to operator and specified the restricted or unrestricted option.

   The manager unrestricted community provides read-write access to all MIBs, manager restricted provides read-only access to all MIBs, operator unrestricted provides read-write access to all MIBs except the CONFIG MIB, and operator restricted provides read-only access to all MIBs except the CONFIG MIB.

   - Based on the table, which manager option is associated with read-only access and which with read-write access?

   - When might you want to use the operator-restricted options for the SNMP read-only community?

   The restricted keyword is associated with read-only access and the unrestricted keyword with read-write access. However, the exact read and write rights depend on the combination of user group (manager or operator) and restricted or unrestricted access. As you see in the table, the manager user group gives access to all or most rights.

You might want to use the operator-restricted options for the SNMP read-only community if you want to increase security by preventing the SNMP server from reading the SNMP community and SNMPv3 user names. The example configuration used this approach.

2. What steps did you have to complete on the Comware switches to set up login access for IMC?

   You enabled Telnet access. Because you were not using authentication in your environment, you also had to disable authentication on VTY interfaces and enable full privileges (level 3).

3. For the example environment, you allowed IMC Telnet access without authentication. How would you change the set up for the real world?

   You should configure passwords (or usernames and passwords) to secure Telnet access as you learned how to do in Module 4. You would then match those credentials when you set the IMC Telnet settings.

   You could use SSH for even more security, but the configuration for SSH is not covered in this study guide.

4. The basic auto discovery process has you enter a range of IP addresses to discover. When would this type of discovery be appropriate?

   This type of discovery works well when your switches all have IP addresses in a contiguous range.

5. Which form of discovery did you use? Why was this type of discovery appropriate for your network? How does it work?

   You used advanced auto discovery and selected ARP discovery. This type of discovery worked well for your environment because your switches have IP addresses on different subnets. IMC discovered the seed device and then the devices in its ARP table. In your network, the seed device has its OSPF neighbors in its ARP table. IMC then discovered those devices and the devices in their ARP tables, and so on, until all of your devices had been discovered.

   In the real world, all devices might not be in the ARP tables at all times. You might need to run discovery a few times or use other forms of discovery as well.

6. The switches in the example network have multiple IP addresses. Which does IMC use to manage the switches?

   When you have not taken steps to control the management address, IMC manages the switch at the IP address from which the switch replies to its discovery messages. This is the IP address that the switch uses to reach IMC, which depends on its routing table, and might be difficult for you to predict.

   You can set up a loopback interface on the switch. This is an interface that is always up and is not associated with a specific link. You can configure IMC to always use the loopback IP address to manage the switch. (Of course, you would need to set up routing to the loopback interfaces.)

## IMC users

You used the default admin user to log in to IMC. In the real world, of course, you would want to change the admin credentials. You might want to add more users with various privilege levels so that different managers can log in with the appropriate rights.

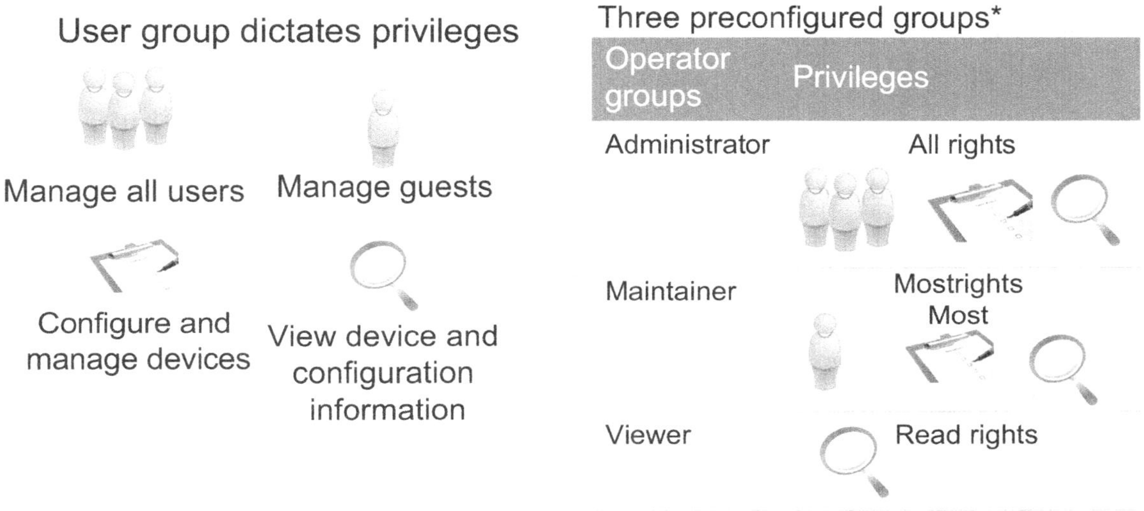


Figure 12-19: IMC users

IMC supports different types of operator groups (as shown in Figure 12-19), allowing you to grant each operator only the rights needed to complete certain tasks. IMC provides three preconfigured operator groups:

- Administrator group—Administrators have all rights to IMC. They can manage all users, including adding and deleting users. They can configure all network devices and view any information gathered and stored by IMC. Administrators have exclusive rights (rights that no other operator group has), including the following:
  - Operator management
  - Device group management
  - User group management
  - Login control template management
  - Password strategy management
  - System parameter settings

- Maintainer group—Maintainers have rights to configure network devices and can complete most configuration tasks that IMC supports. However, they cannot perform tasks that require the exclusive rights assigned to the administrator group, which are mostly related to user management. In addition, maintainers can view all information gathered and stored by IMC. This allows them to monitor key systems.
- Viewer group—Viewers have access to information but cannot perform management tasks. This allows them to monitor systems, although they must notify a maintainer or an administrator to remedy a problem.

You can add operators and then assign them to one of the existing operator groups, or you can create a new operator group and customize the rights operators in this group will have. IMC gives you granular control over the rights you assign each operator group and the devices they may access or manage.

Operators can log in with passwords that are stored on IMC. However, you can also authenticate users through a RADIUS or an LDAP server.

By default, IMC has one operator (or management user): **admin** is part of the administrator group. By default, this user logs in with a password, also **admin**.

## Example configuration: Use VLAN management in HP IMC

As you learned at the beginning of this chapter, IMC supports many monitoring, management, and configuration tools. In this section, you will learn about the VLAN Management service, just one of the many services that IMC provides. You will learn how to use this service to:

- View VLANs on your switches
- Add VLANs to switches
- Assign IP addresses to switches on the new VLAN
- Add switch ports to the VLAN

This section explains how to complete these steps in the example network shown in Figure 12-20. (This is the same configuration used in "Example configuration: Discover devices with HP IMC" on page 442.)

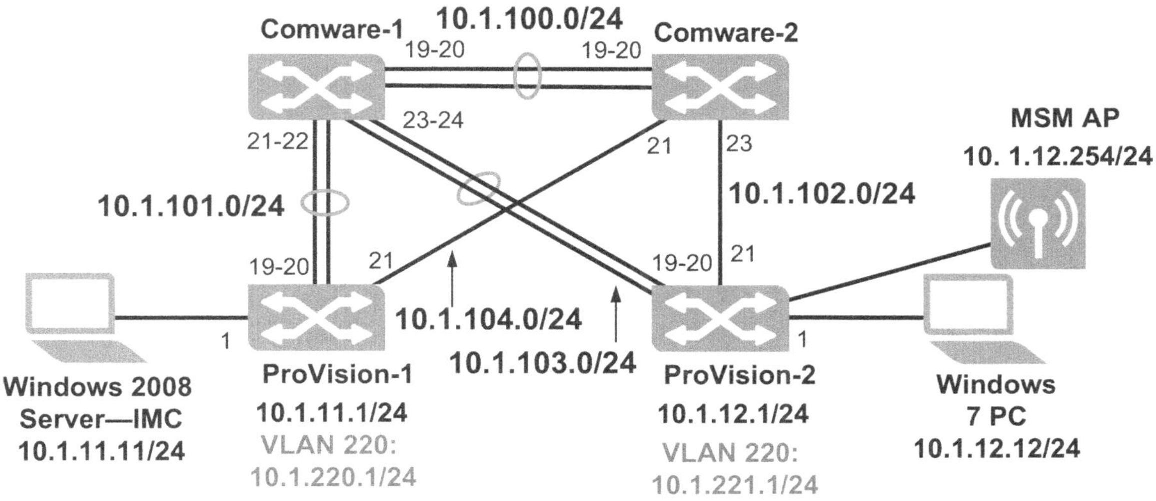


Figure 12-20: Example configuration: Use VLAN management in HP IMC

## Add a VLAN

You will use the VLAN Management service in HP IMC to add a new VLAN to ProVision-1 and ProVision-2. Actually, although the VLAN has the same ID on the two switches, it is a different VLAN on each switch. Think about why this is the case.

1. Access the Windows 2008 desktop.
2. Launch Internet Explorer or Mozilla and navigate to http://localhost:8080/imc.
3. Log in with username **admin** and password **admin**.
4. In the top navigation bar, navigate to **Service** > **VLAN Management** > **Global VLAN**, as shown in Figure 12-21.

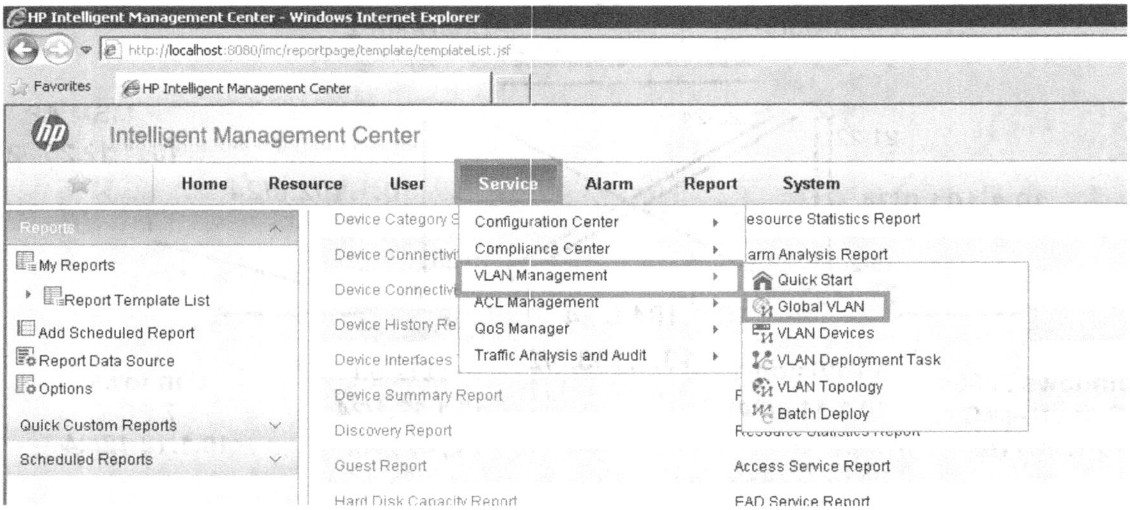


Figure 12-21: Access VLAN Management in IMC

5. You see a list of all VLANs on any switch that is managed by IMC.
6. Add a new VLAN. Click **Add**, as shown in Figure 12-22.

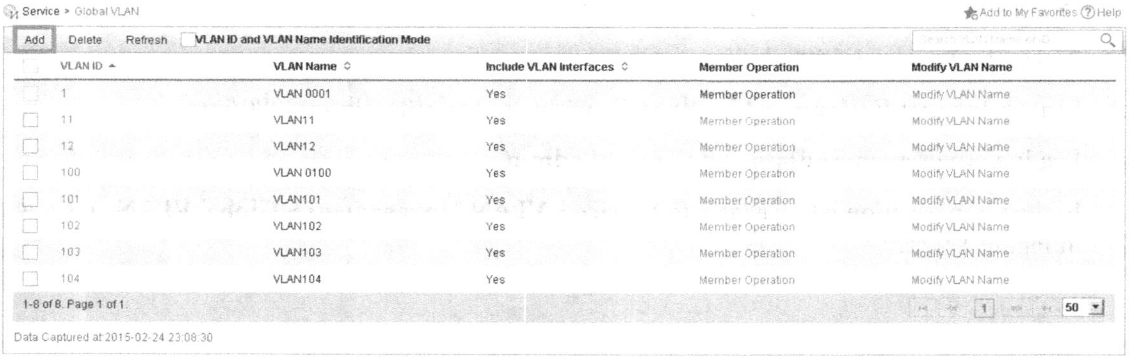


Figure 12-22: Add a new VLAN

7. Enter a VLAN ID of **220** and a VLAN Name of **Accounting**.
8. You will now choose the switches to which to deploy the VLAN. Under the VLAN Device List section, click **Add**, as shown in Figure 12-23.

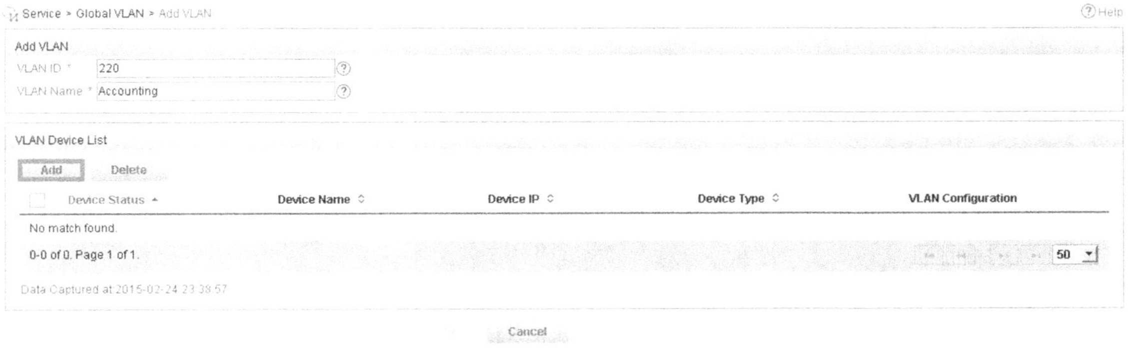


Figure 12-23: Add the new VLAN to switches

9. In the new window, select **ProVision-1** and **ProVision-2**.
   d. Click the **Device View** folder.
   e. Hold down **[Ctrl]** and click **ProVision-1** and **ProVision-2** in the **Devices Found** list, as shown in Figure 12-24.

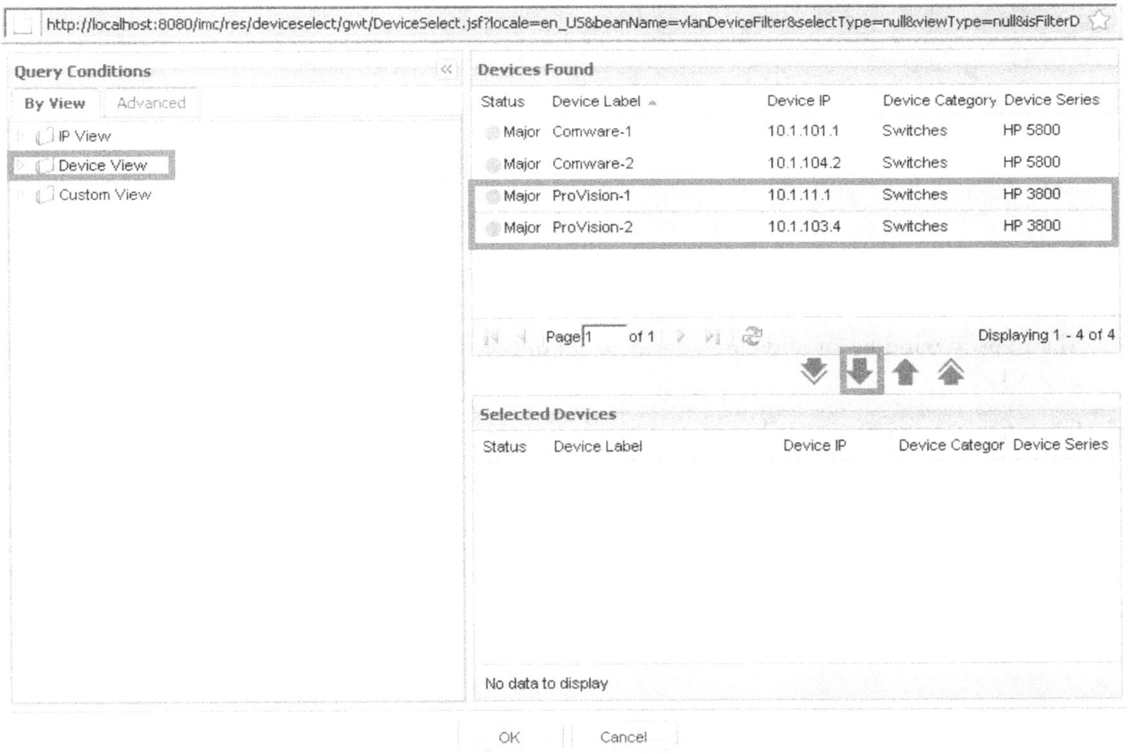


Figure 12-24: Add the new VLAN to switches

f. Click the down arrow.

g. ProVision-1 and ProVision-2 are now in the Selected Devices list, as shown in Figure 12-25. Click **OK**.

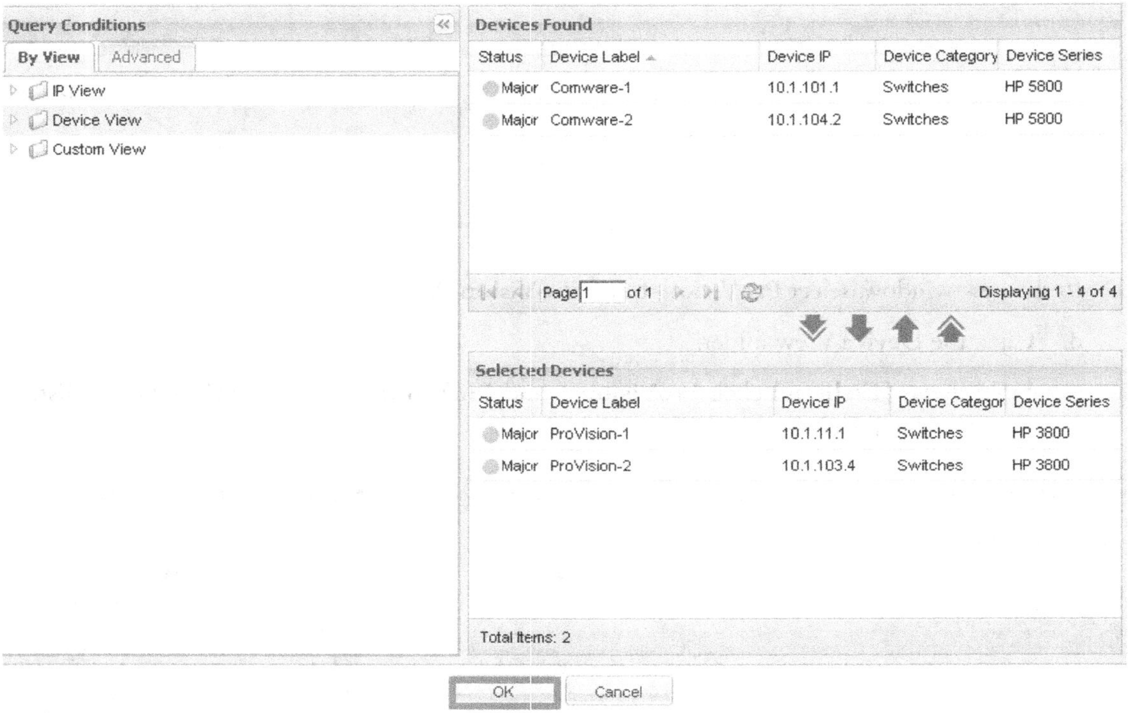


Figure 12-25: Add the new VLAN to switches

10. The pop-up window should close, and you should see that the switches are now selected for the new VLAN.

11. Click **OK** to activate your changes.

12. If you do not see VLAN 220 in the Global VLAN window, click **Refresh**, as shown in Figure 12-26.

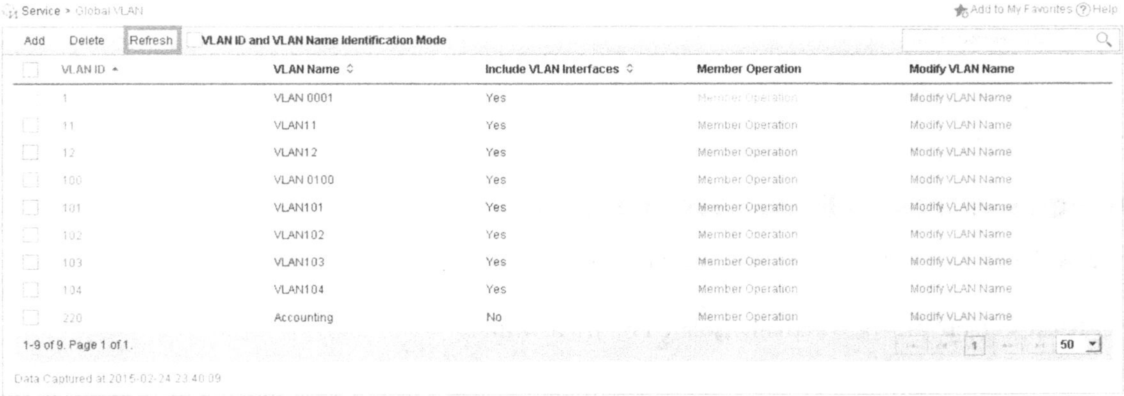


Figure 12-26: Click Refresh to see new configuration changes

The switches now support the VLAN. You will verify in the CLI.

13. Access a terminal session with ProVision-1 and view VLANs.

```
ProVision-1# show vlans
<-output omitted->
 VLAN ID Name                           | Status     Voice Jumbo
 ------- ------------------------------ + ---------- ----- -----
 1       DEFAULT_VLAN                   | Port-based No    No
 11      VLAN11                         | Port-based No    No
 101     VLAN101                        | Port-based No    No
 104     VLAN104                        | Port-based No    No
 220     Accounting                     | Port-based No    No
```

14. Access a terminal session with ProVision-2 and view VLANs.

```
ProVision-2# show vlans
<-output omitted->
 VLAN ID Name                           | Status     Voice Jumbo
 ------- ------------------------------ + ---------- ----- -----
 1       DEFAULT_VLAN                   | Port-based No    No
 12      VLAN12                         | Port-based No    No
```

```
102     VLAN102                          | Port-based No    No
103     VLAN103                          | Port-based No    No
220     Accounting                       | Port-based No    No
```

## Assign switches an IP address on the VLAN

You will now complete other configurations for the VLAN. First you will assign IP addresses to ProVision-1 and ProVision-2 on this VLAN. Each switch will have an IP address on a different subnet because each VLAN is actually isolated on each switch.

1. From the Global VLAN window, which is shown in Figure 12-27, click **220.**

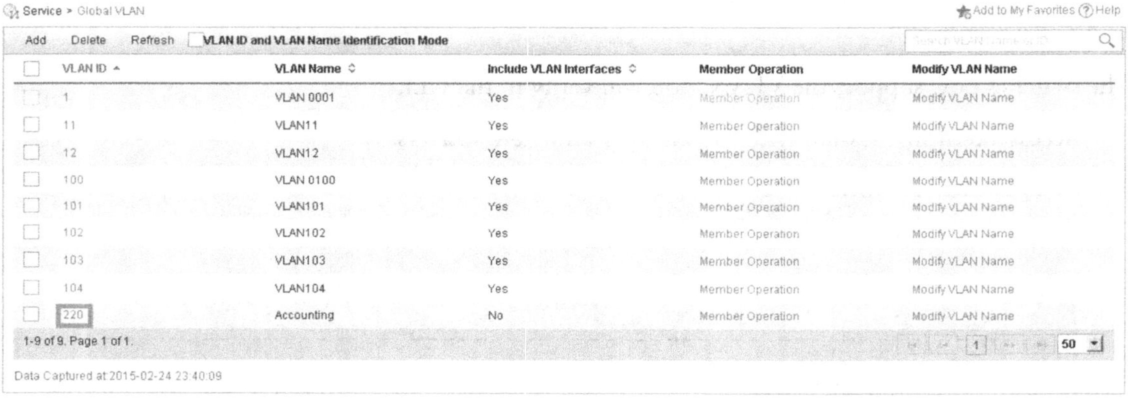


Figure 12-27: Click Refresh to see new configuration changes

2. The window lists the switches that support this VLAN. You should see ProVision-1 and ProVison-2.

3. Click the icon in ProVision-1's VLAN Configuration column, as shown in Figure 12-28.

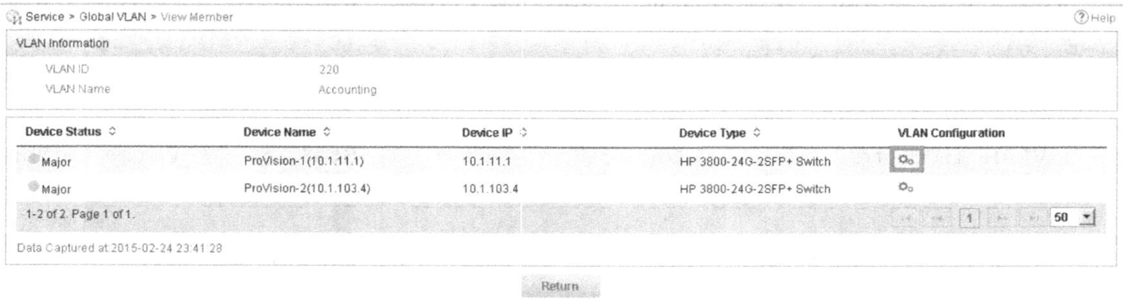


Figure 12-28: Click the icon in the VLAN Configuration column

4. From this window, you can configure VLAN settings on the switch. Click the **VLAN Interface** tab.

5. You see a list of all VLANs for which the switch has IP settings. You will now add IP settings for VLAN 220, as shown in Figure 12-29.

   a. Click **Add**.

6. For VLAN ID, select **220**.

7. For DHCP Enabled, select **No**.

8. Configure this IP address: 10.1.220.1 255.225.255.0.

9. Click **OK**.

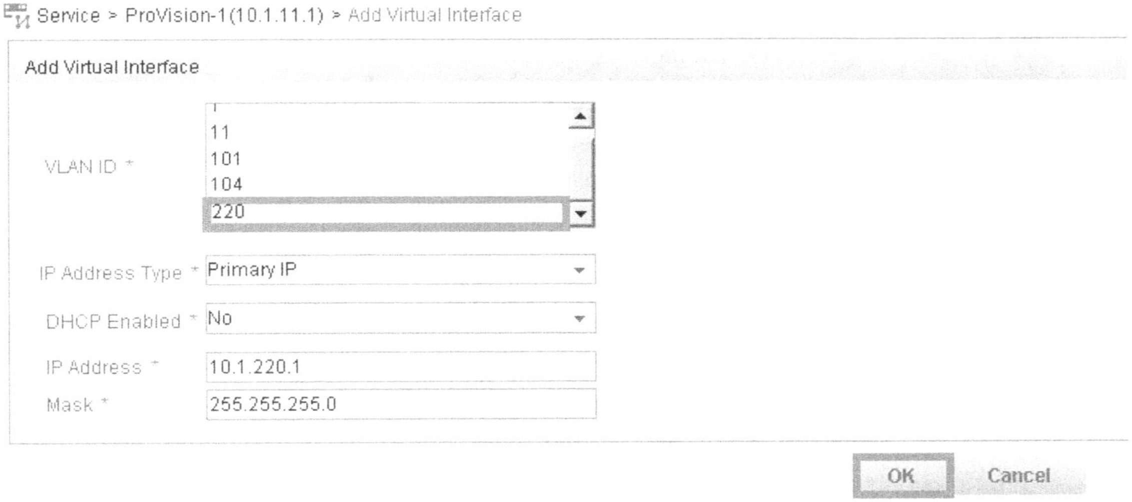


Figure 12-29: Configure an IP address for the VLAN

You will now configure an IP address on ProVision-2. First you must return to the list of devices.

10. In the left navigation bar, expand VLAN Management and click **VLAN Devices**, as shown in Figure 12-30.

Figure 12-30: Use the navigation tree to access VLAN Devices

11. Complete similar steps to assign this IP address to ProVision-2 on VLAN 220: 10.1.221.1 255.255.255.0.

The switches now have IP addresses on VLAN 220. Verify in the CLI.

12. View IP addresses on ProVision-1.

```
ProVision-1(config)# show ip

<-output omitted>

               |                                                Proxy ARP
  VLAN         | IP Config  IP Address    Subnet Mask     Std  Local
  ------------ + ---------- ------------ -------------- ----------
  DEFAULT_VLAN | Manual      10.1.1.3      255.255.255.0  No     No
  VLAN11       | Manual      10.1.11.1     255.255.255.0  No     No
  VLAN101      | Manual      10.1.101.3    255.255.255.0  No     No
  VLAN104      | Manual      10.1.104.3    255.255.255.0  No     No
  Accounting   | Manual      10.1.220.1    255.255.255.0  No     No
```

13. View IP addresses on ProVision-2.

```
ProVision-2(config)# show ip
<-output omitted>
                 |                                          Proxy ARP
  VLAN           | IP Config  IP Address  Subnet Mask    Std  Local
  ------------ + ---------- ----------- ------------- ----------
  DEFAULT_VLAN | Manual      10.1.1.4     255.255.255.0  No     No
  VLAN12         | Manual      10.1.12.1    255.255.255.0  No     No
  VLAN102        | Manual      10.1.102.4   255.255.255.0  No     No
  VLAN103        | Manual      10.1.103.4   255.255.255.0  No     No
  Accounting     | Manual      10.1.221.1   255.255.255.0  No     No
```

## Add ports to the VLAN

Finally, you will learn how to add ports to a VLAN using the VLAN Management tool.

1. Return to the Windows server desktop.
2. You should be in the ProVision-2 VLAN configuration window. Click the **Access Port** tab.
3. Click the check boxes for ports 3 and 4.
4. Click **Modify PVID**, as shown in Figure 12-31.

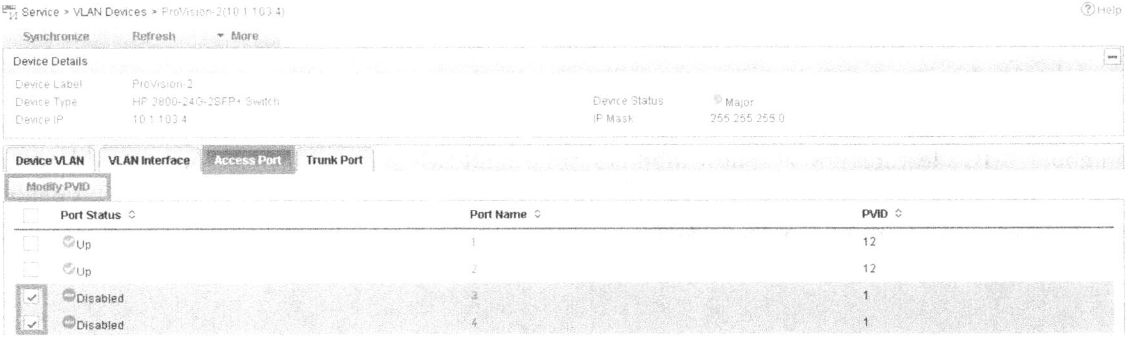


Figure 12-31: Add ports to the VLAN

**Note**

Access port and PVID are Comware concepts. When you set the PVID, you are setting the untagged VLAN for the ProVision port.

5. In the window that pops up, select **220**.
6. Click **OK**.
7. The pop-up window closes, and you should see that ports 3 and 4 are assigned to VLAN 220.
8. Verify the change in the CLI. View VLAN 220 on ProVision-2.

```
ProVision-2(config)# show vlan 220

 Status and Counters - VLAN Information - VLAN 220

  VLAN ID : 220

  Name : Accounting

  Status : Port-based

  Voice : No

  Jumbo : No

  Port Information Mode     Unknown VLAN Status

  ---------------- -------- ------------ ----------

  3                Untagged Learn        Down

  4                Untagged Learn        Down
```

## Summary

This chapter has introduced you to IMC, the HP networking management solution. You learned a bit about IMC's features and also how modules can extend IMC's functionality. You learned how to prepare your network infrastructure devices for discovery and how to discover them in IMC. Finally, you practiced using the VLAN Management service.

## Learning check 2

You will now answer some questions, based on what you have learned about configuring VLANs in IMC.

1. What are some advantages of using VLAN Management versus the CLI?

2. You configured a VLAN with ID 220 in ProVision-1 and ProVision-2. However, VLAN 220 on ProVision-1 and VLAN 220 on ProVision-2 are not actually the same VLAN in your network topology. Explain why this is the case.

3. Do you think that it is a good idea to use the same VLAN ID for a VLAN on ProVision-1 and on ProVision-2 in this network topology? What are the disadvantages? What are the advantages?

4. Describe the difference between a centralized and a distributed IMC deployment.

5. What are the default groups in IMC? (Select three.)
   a. Administrator
   b. Maintainer
   c. Manager
   d. Operator
   e. Viewer

6. You want to access the IMC interface. You open a web browser on the IMC server. To what URL should you navigate?

7. What are the default username and password on IMC?

## Learning check 2 answers

1. What are some advantages of using VLAN Management versus the CLI?

   You can configure a VLAN on multiple switches at once from a single location rather than having to log in to every switch separately. In the example configuration, you just configured the VLAN on two switches, but you can imagine how much quicker the configuration is if you have dozens of switches.

   You can configure the VLAN assignment on many ports at once.

   You can quickly see all of the switches that support a VLAN rather than having to log in to each switch and run `show` or `display` commands.

   You can use IMC to map the VLAN topology, a more comprehensive and intuitively understandable picture of where VLANs extend than `show` and `display` commands on individual switches.

2. You configured a VLAN with ID 220 in ProVision-1 and ProVision-2. However, VLAN 220 on ProVision-1 and VLAN 220 on ProVision-2 are not actually the same VLAN in your network topology. Explain why this is the case.

   A VLAN is a broadcast domain. In your network topology, broadcast domains are terminated on each switch.

   VLAN 220 exists on ProVision-1, and the broadcast domain extends over any ProVision-1 switch ports that are assigned to this VLAN. However, VLAN 220 is not carried on any switch-to-switch links. Instead, each switch-to-switch link is dedicated to a single VLAN that does not exist on any other link, terminating the VLAN 220 broadcast domain on ProVision-1.

   VLAN 220 also exists on ProVision-2, but this VLAN 220 is a different broadcast domain. No broadcasts from endpoints on ProVision-2 reach endpoints on ProVision-1. In effect, you have two VLANs that use the same ID.

   This is why you associated a different subnet with the VLAN on ProVision-1 and on ProVision-2.

   You know that many different private networks can use the same VLAN IDs, but endpoints in those different networks are not in the same VLANs as each other. Now you see that the same can hold true within a private network if you take steps to terminate the broadcast domains between routing switches.

3. Do you think that it is a good idea to use the same VLAN ID for a VLAN on ProVision-1 and on ProVision-2 in this network topology? What are the disadvantages? What are the advantages?

   The answer to this question depends on your perspective. There are both advantages and disadvantages to reusing VLAN IDs in different routed segments of a network.

A disadvantage could be accidentally extending the broadcast domain when you do not want to. For example, what happens if an admin mistakenly adds VLAN 220 to the links between switches? The VLANs would mix in an undesirable way. In addition, a broadcast storm could occur if the admin adds the VLAN to redundant switch-to-switch links.

You can make it less likely for this situation to occur. Rather than setting up Comware switch-to-switch links as access ports that are dedicated to a VLAN, you configure them as routing mode ports or routed link aggregation interfaces. Then you cannot add VLANs to the interface.

Reusing VLAN IDs could give you advantages in management. Perhaps you associate VLAN 220 with a particular type of user group. Then you always know that VLAN 220 is associated with that group no matter what the network segment. As you saw, it can be easy to add the VLANs to many switches at once even if you set a different subnet address in different areas. (Reusing the same VLAN ID can also make it easier to set up policies to assign users to dynamic VLANs, which were not covered in this study guide.)

4. Describe the difference between a centralized and a distributed IMC deployment.

   In a centralized deployment, the IMC platform and all components run on a single server. In a distributed deployment, you install IMC and its modules on a master server, but you deploy some components to different servers. A distributed deployment is typically recommended when you have added modules to IMC.

5. What are the default groups in IMC? (Select 3.)

   a. Administrator
   b. Maintainer
   c. Manager
   d. Operator
   e. Viewer

6. You want to access the IMC interface. You open a web browser on the IMC server. To what URL should you navigate?

   http://localhost:8080/imc

   Alternatively, you can navigate to https://localhost:8443/imc.

7. What are the default username and password on IMC?

   The username and password are both **admin**.

# 13 Practice Test

## INTRODUCTION

The HP Accredited Technical Professional (ATP)—FlexNetwork Solutions V3 validates your ability to configure and implement HP Networking products, focusing on the HP ProVision switches, HP Comware switches, and the wireless HP Multi-Service Mobility (MSM) access points. This certification covers implementations for small-to-medium sized businesses.

The intent of this study guide is to set expectations about the context of the exam and to help candidates prepare for it. Recommended training to prepare for this exam can be found at the HPE Certification and Learning website (http://certification-learning.hpe.com) and in books like this one. It is important to note that although training is recommended for exam preparation, successful completion of the training alone does not guarantee you will pass the exam. In addition to the training, exam items are based on knowledge gained from on-the-job experience and application and on other supplemental reference material that may be specified in this guide.

### Minimum qualifications

To achieve the HP ATP–FlexNetwork Solutions V3 certification, you must pass the HP0-Y52 or HP2-Z37 exam. Candidates should have a thorough understanding of HP switch and access point implementations in SMBs. To pass the exam, you should have at least six months experience deploying small to medium enterprise-level networks. You should also have an understanding of wired technologies used in edge and simple core environments and fundamental knowledge of wireless technologies.

Exams are based on an assumed level of industry-standard knowledge that may be gained from the training, hands-on experience, or other prerequisite events.

### HP0-Y52 exam details

The following are details about the HP0-Y52 exam:

- Number of items: 50
- Exam time: 1 hour 55 minutes
- Passing score: 67%
- Item types: Multiple choice (single-response), multiple choice (multiple response), and drag-and-drop
- Reference material: No online or hard copy reference material is allowed at the testing site.

## HP0-Y52 testing objectives

This exam validates that you can successfully perform the following objectives. Each main objective is given a weighting, indicating the emphasis this item has in the exam.

32% Identify, describe, and apply foundational networking architectures and technologies.

- Describe the basics of Layer 2 Ethernet, including broadcast domains and ARP messages, and describe basic wireless technologies.
- Interpret an IP routing table and explain default routes, static routing, and dynamic routing, including OSPF.
- Identify the role of TFTP, SFTP, FTP, Telnet, and SNMPv2 in managing HP network devices, and apply the appropriate security for these features.
- Describe Layer 2 redundancy technologies such as STP, RSTP, MSTP, and IRF, and recognize their benefits.
- Describe and apply link aggregation.
- Identify, describe, and explain VLANs.

13% Identify, describe, and differentiate the functions and features of HP Networking products and solutions.

- Identify and explain the components of a converged infrastructure using the HP FlexNetwork Architecture framework, which includes the following components: FlexFabric, FlexCampus, FlexBranch, FlexManagement, and Software-Defined Networks (SDN).
- Identify basic features and management options for HP products and solutions, including both wired and wireless.
- Compare and contrast HPE Networking solutions and features, and identify the appropriate products for an environment.

39% Install, configure, set up, and validate HP Networking solutions.

- Configure basic features on HP switches, including initial settings and management access.
- Configure Layer 2 technologies such as RSTP and MSTP, link aggregation, and VLANs on HP switches.
- Configure basic IP routing with static routes or OSPF on HP switches.
- Configure the management software and mange configuration files on HP switches.
- Validate the installed solution via debug technology, logging, and show and display commands.

6% Tune, optimize, and upgrade HP Networking solutions.

- Optimize Layer 2 and Layer 3 infrastructures via broadcast domain reduction, VLANs, and IRF.

10% Manage, monitor, administer, and operate HP Networking solutions.

- Perform network management according to best practices.
- Perform administrative tasks such as moves, adds, changes, deletions, and password resets for managed devices.

## Test preparation questions and answers

The following questions help you measure your understanding of the material presented in this book. Read all of the choices carefully, since there may be more than one correct answer. Choose all correct answers for each question.

### Questions

1. Which characterizes out-of-band management?
   a. Uses the same Ethernet link that the management station uses to transmit traffic onto the network
   b. Is the only secure method to manage switches
   c. Limits the network administrator's ability to configure switch features and functions
   d. Requires physical access to the switch

2. View the exhibit.

```
<Comware-1>display users
The user application information of the user interface(s):
  Idx UI      Delay    Type Userlevel
F 3   AUX 0   00:00:00          3
  29  VTY 0   00:00:00 TEL      2
```

   What rights would you have if you established an out-of-band connection to this switch?
   a. You could enter commands to diagnose network issues but could not enter commands to configure switch features.
   b. You could enter commands to maintain switch operations and to configure any switch feature.
   c. You could enter commands to maintain the system but could not enter commands to configure VLANs or interfaces.
   d. You could enter commands to configure VLANs and interfaces but could not enter commands to manage the file system or system access.

   For more information, see Chapter 3.

3. Which component of software-defined networking (SDN) implements and applies the actual policies for traffic flows?
   a. Application
   b. Control
   c. Infrastructure
   d. Management

4. Which interface identifier correctly specifies the second port on the second module of a ProVision 5400zl switch?
   a. 2/2
   b. A2/2
   c. B2
   d. 2/0/2

5. The following is a prompt from a Comware switch:

   ```
   <HP>
   ```

   Which Comware context is shown?
   a. System view
   b. Manager view
   c. User view
   d. Operator view

6. If a network manager enters the `include-credentials` command on a ProVision switch, what happens?
   a. Login credentials for switch access are stored in internal flash and are part of the configuration file.
   b. The switch sends login credentials to a network RADIUS server and grants access only to network administrators that the server approves.
   c. The local manager and operator usernames and passwords are stored in internal flash but are not part of the configuration file.
   d. The switch enforces 802.1X authentication for specified ports using the manager credentials configured on the switch.

7. What rights does operator-level access give network administrators on a ProVision switch?
   a. Viewing configuration information and configuring Layer 2 features
   b. Configuring and managing basic switch functionality
   c. Viewing switch configuration information (but not the configuration file itself)
   d. Debugging all switch features and managing users

8. Which Comware privilege level are you taken to when you log in to the console port of a Comware router or the auxiliary port of a Comware switch?
   a. Manager
   b. Monitor
   c. System
   d. Viewer

9. Which Comware privilege level would you assign a network administrator that should only monitor switch functions?
   a. Level 2
   b. Level 1
   c. Level 3
   d. Level 4

10. Which command correctly installs a new operating system from a USB drive on a ProVision switch?
    a. `upgrade usb flash KA.15.03.3004.swi primary`
    b. `copy usb flash KA.15.03.3004.swi secondary`
    c. `copy usb:/ KA.15.03.3004.swi flash:/KA.15.03.3004.swi secondary`
    d. `copy primary usb flash KA.15.03.3004.swi primary`

11. Which Comware command correctly identifies the configuration file that will be used upon the next reboot of the device?
    a. `startup saved-configuration`
    b. `startup-default`
    c. `boot system startup-config`
    d. `boot-file startup-config`

12. Which commands are used to remove a file and prevent a network administrator from recovering the file from a Comware flash file system? (Select two.)

    a. `erase <filename>`

    b. `remove <filename> permanent`

    c. `delete <filename>`

    d. `reset flash`

    e. `reset recycle-bin`

13. Which Comware configuration correctly configures a trunk allowing only VLANs 30 and 31, along with the untagged VLAN, based on the following output?

```
[Comware] display port trunk

Interface    PVID   VLAN passing
GE1/0/23     1      1, 100, 101
```

    a.
```
interface g1/0/23
   port trunk permit vlan 30 31
```

    b.
```
interface g1/0/23
   undo port trunk permit vlan 100 101
   port trunk permit vlan 30 31
```

    c.
```
vlan 30-31
   permit interface g1/0/23
```

    d.
```
vlan 100-101
   undo permit interface g1/0/23
   vlan 30-31
   permit interface g1/023
```

14. A PC is connected to the data port of a phone, and the phone is then connected to the port of a Comware switch. Neither the PC nor the phone support 802.1Q. Given this information, which port type should be configured on the Comware switch to place the two devices in different VLANs?

    a. Access

    b. Trunk

    c. Hybrid

d. Multihost

e. Multiple-mode

f. Untagged

15. Examine the network topology in Figure 13-1.

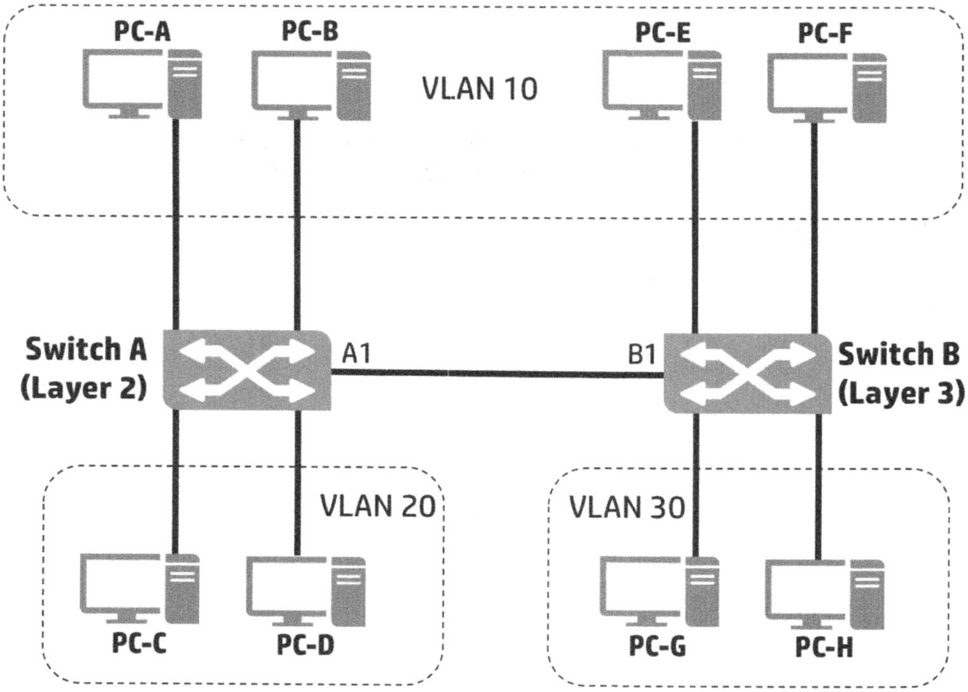


Figure 13-1: Question 15 network topology

SwitchA is a Layer 2 ProVision switch, and SwitchB is a Layer 3 ProVision switch. Based on this topology, what is the correct configuration for port B1 to ensure connectivity within VLAN 10 and inter-VLAN connectivity between all three VLANs?

a.
```
vlan 10 tagged B1
vlan 20 tagged B1
vlan 30 tagged B1
```

b.
```
vlan 10 tagged B1
vlan 20 tagged B1
```

c. `vlan 10 tagged B1`

   `vlan 30 tagged B1`

d. `interface B1`

   `port mode trunk 10 20 30`

16. Examine the network topology in Figure 13-2.

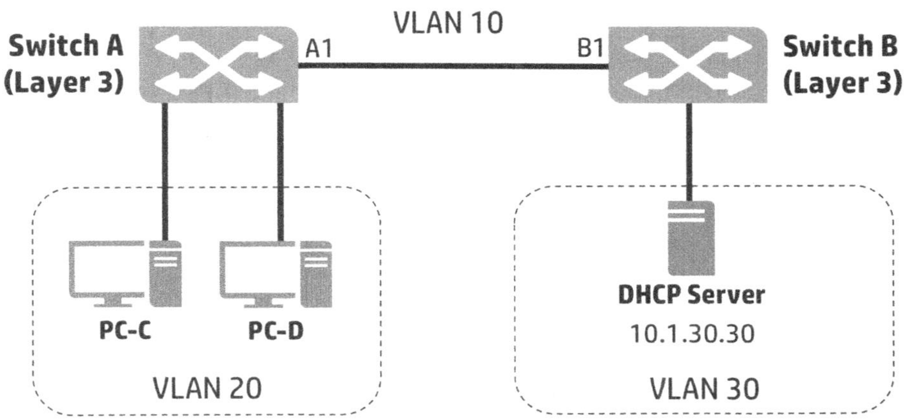


Figure 13-2: Question 16 network topology

Based on this topology, which configuration correctly sets up DHCP relay so that users in VLAN 20 can acquire addressing information from the DHCP server in VLAN 30?

a. `SwitchA(config)# vlan 20`

   `SwitchA(vlan-id 20)# ip helper-vlan 30`

b. `SwitchA(config)# vlan 20`

   `SwitchA(vlan-id 20)# ip helper-address 10.1.30.30`

c. `SwitchA(config)# vlan 10`

   `SwitchA(vlan-id 10)# ip helper-address 10.1.30.30`

d. `SwitchB(config)# vlan 10`

   `SwitchB(vlan-id 10)# ip helper-address 10.1.30.30`

17. View the network in Figure 13-3.

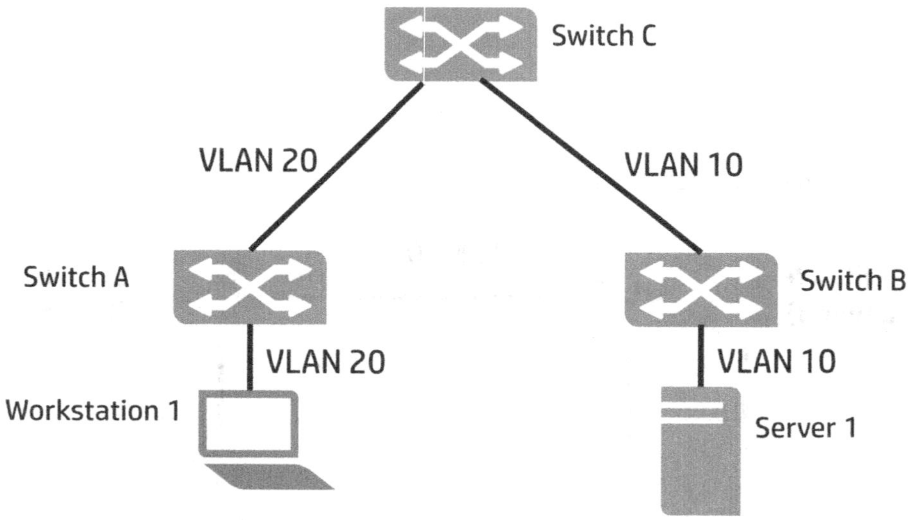


Figure 13-3: Question 17 network topology

Which devices are in the same broadcast domain?

a. Switch A, Switch B, and Switch C

b. Workstation 1, Switch A, Switch C, and Switch B

c. Workstation 1, Switch A, and Switch C

d. Server 1, Switch B, Switch C, and Switch A

18. View Figure 13-4.

The switches in this network are Comware switches. Switch C is the default router for all VLANs. The link between Switch A and Switch C is a trunk port that uses the default PVID and permits the VLANs shown in Figure 13-4.

Workstation 1 transmits a packet that is destined to Server 1. When Switch A sends the packet to Switch C, what VLAN assignment will the packet have?

a. Untagged in VLAN 1

b. Untagged in VLAN 20

c. Tagged in VLAN 20

d. Tagged in VLAN 10

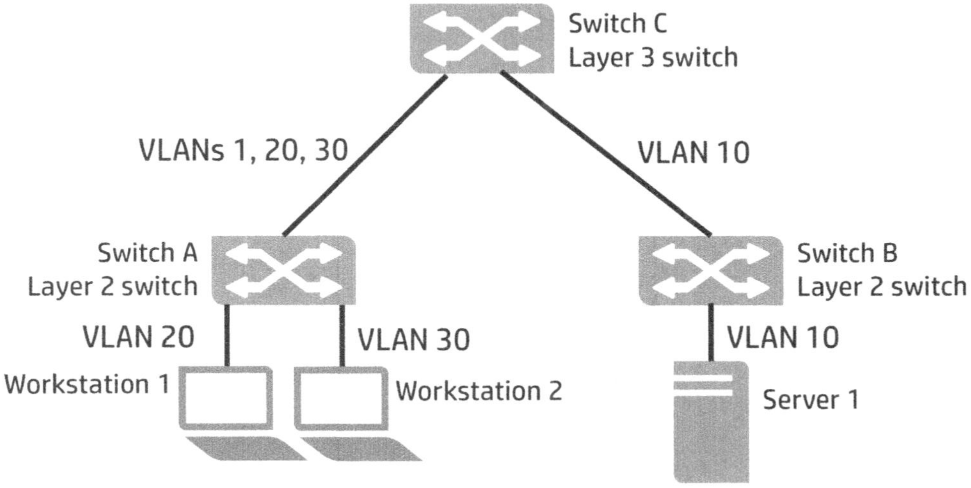


Figure 13-4: Question 18 network topology

19. View Figure 13-5.

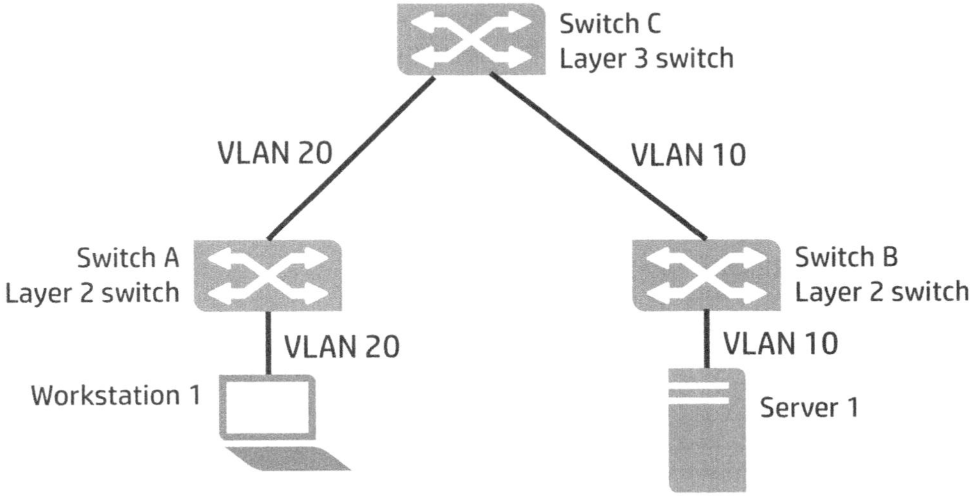


Figure 13-5: Question 19 network topology

In the sample network, Workstation 1 does not support 802.1Q. If Switch A is a Comware switch, how should the port that connects to Workstation 1 be configured?

a. Trunk port that has a PVID of 1 and permits VLAN 20

b. Untagged port in VLAN 20

c. Tagged port in VLAN 20

d. Access port that permits VLAN 20

20. Examine the network topology in Figure 13-6.

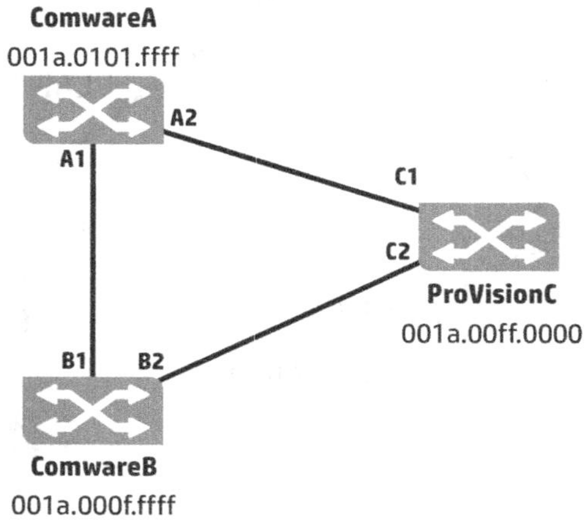


Figure 13-6: Question 20 network topology

ComwareA and ComwareB are Comware switches, and ProVisionC is a ProVision switch. All switches are at their factory default configurations with the exception of spanning tree being enabled on all three switches. Based on this information, which statements about this topology are correct? (Select two.)

a. ComwareA is the root switch.

b. ComwareB is the root switch.

c. ProVisionC is the root switch.

d. Port A1 is a designated port.

e. Port A2 is a designated port.

f. Port B2 is a root port.

21. A network administrator is configuring a DHCP server pool on a Comware switch. Which setting should the administrator generally include in the pool?

a. DNS server IP address

b. DHCP relay IP address

c. Excluded IP addresses

d. Backup DHCP server IP address

22. Examine the network topology in Figure 13-7.

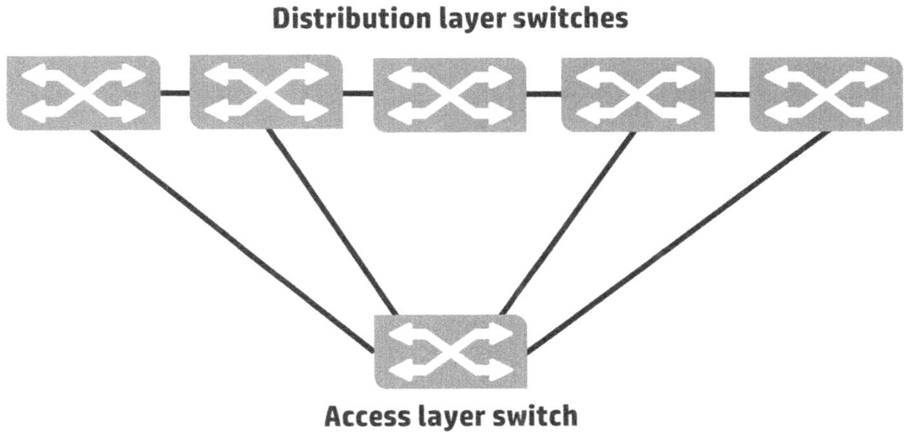


Figure 13-7: Question 22 network topology

This topology is using Multiple STP (MSTP) for spanning tree. Given this topology, how many instances would need to be created so that the access layer switch can implement load-sharing across all of its uplinks?

a. One

b. Two

c. Three

d. Four

e. Five

23. Put the steps in the DHCP process in the order in which they occur.

a. The server sends a DHCP ACK.

b. The client sends a DHCP Request.

c. The client sends a DHCP Discover message.

d. The server sends a DHCP Offer.

24. Examine the network topology in Figure 13-8.

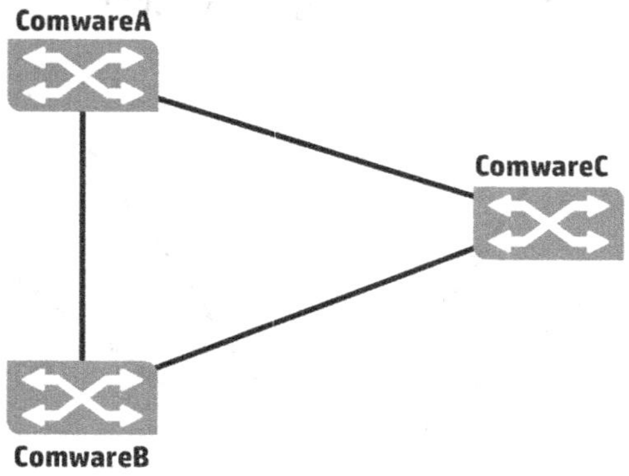


Figure 13-8: Question 24 network topology

Examine the commands executed on each switch from its factory default configuration:

```
[ComwareA] stp enable
[ComwareA] stp region-configuration
[ComwareA-mst-region] region-name HP
[ComwareA-mst-region] revision-level 1
[ComwareA-mst-region] instance 0 vlan 1,4-4094
[ComwareA-mst-region] instance 1 vlan 2
[ComwareA-mst-region] instance 2 vlan 3
[ComwareA-mst-region] active region-configuration
[ComwareA-mst-region] quit
[ComwareA] stp instance 0 root primary
[ComwareA] stp instance 1 root primary
[ComwareA] stp instance 2 root secondary

[ComwareB] stp enable
[ComwareB] stp region-configuration
[ComwareB-mst-region] region-name HP
[ComwareB-mst-region] revision-level 1
[ComwareB-mst-region] instance 0 vlan 1,4-4094
[ComwareB-mst-region] instance 1 vlan 2
[ComwareB-mst-region] instance 2 vlan 3
[ComwareB-mst-region] active region-configuration
[ComwareB-mst-region] quit
[ComwareB] stp instance 0 root secondary
[ComwareB] stp instance 1 root secondary
[ComwareB] stp instance 1 root primary
```

```
[ComwareC] stp enable
[ComwareC] stp region-configuration
[ComwareC-mst-region] region-name HP
[ComwareC-mst-region] revision-level 1
[ComwareC-mst-region] instance 0 vlan 1,4-4094
[ComwareC-mst-region] instance 1 vlan 2
[ComwareC-mst-region] instance 2 vlan 3
[ComwareC-mst-region] active region-configuration
[ComwareC-mst-region] instance 2 vlan 3-4
```

Based on the commands that were executed, which statement is true?

a. A Layer 2 loop exists.

b. ComwareC is the root switch for all three instances.

c. A spanning tree topology change will occur, with load-sharing no longer functioning on the uplinks from ComwareC.

d. No spanning tree topology change occurs, and load-sharing is functioning normally.

25. What is a valid configuration for logical Intelligent Resilient Framework (IRF) ports?

a.

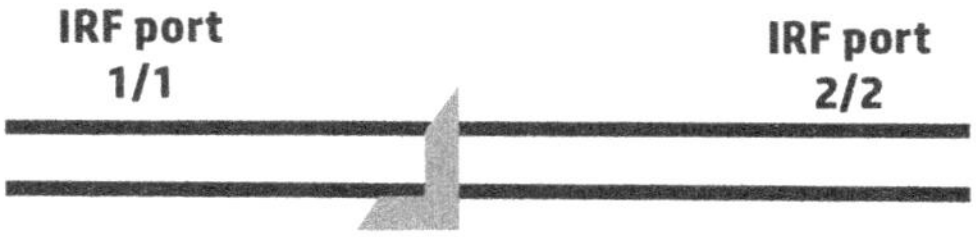


b.

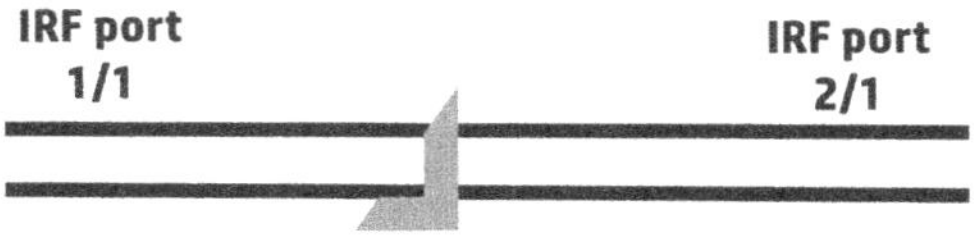


c.

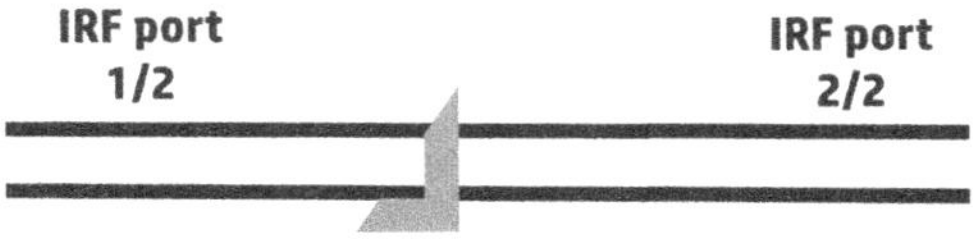


d.

26. Examine the network topology in Figure 13-9.

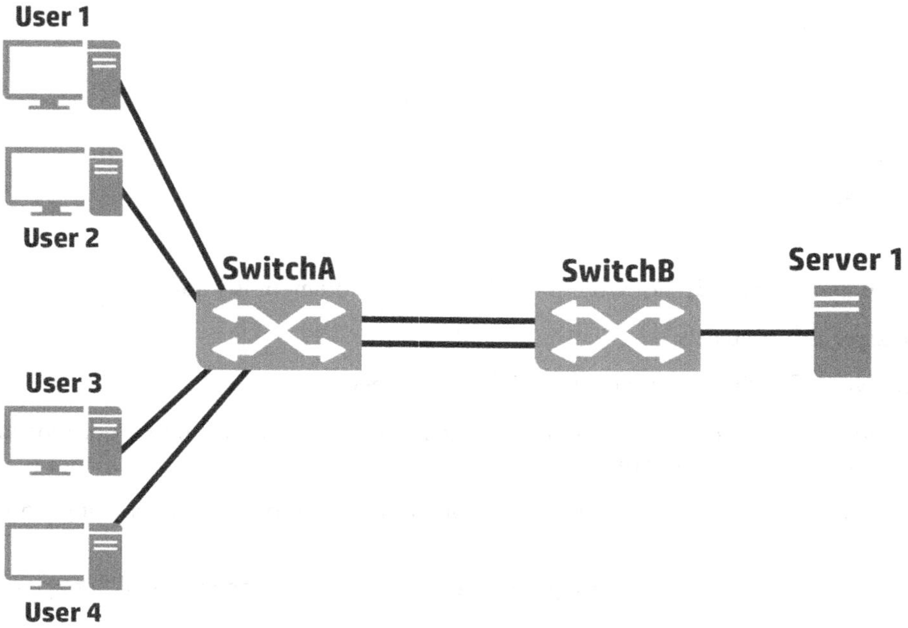


Figure 13-9: Question 26 network topology

The two ProVision switches are configured for link aggregation. Examine the output of SwitchA:

```
SwitchA# show trunk

Load  Balancing  Method:  L3-based  (Default),  L2-based  if  non-IP
traffic

Port   | Name     Type         | Group    Type

----- + -------- ---------- + -------- ----------

1      | Router   100/1000T  | Trk1     LACP

2      | Router   100/1000T  | Trk1     LACP

ProVision-2(config)# show lacp

                           LACP

     LACP    Trunk  Port             LACP     Admin  Oper

Port Enabled Group  Status  Partner  Status   Key    Key

---- ------- ------ ------- -------- -------  ------ ----

1    Active  Trk    Up      Yes      Success  0      562

2    Active  Trk    Up      Yes      Success  0      562
```

```
ProVision-2(config)# show interface display

                   Status and Counters - Port Counters

                                                            Flow  Bca*
Port   Total Bytes  Total Frames  Errors Rx  Drops Tx  Ctrl  Lim*
------ ----------- ------------- ---------- ---------  ----  ---*
1-Trk1     375,648        40,516          0         0   off   0
2-Trk2           0             0          0         0   off   0
<-output omitted->
```

Based on this configuration and output, which statement is true concerning the link aggregation configuration and load-sharing operation?

a. The two switches have failed to form an aggregated link.

b. Switch A is using destination MAC addresses for its load-sharing method, and Switch B is using source MAC addresses for its load-sharing.

c. Switch A is using source MAC addresses for its load-sharing method and Switch B is using destination MAC addresses for its load-sharing.

d. Switch A and Switch B are using source and destination MAC addresses for their load-sharing.

27. Examine the network topology in Figure 13-10.

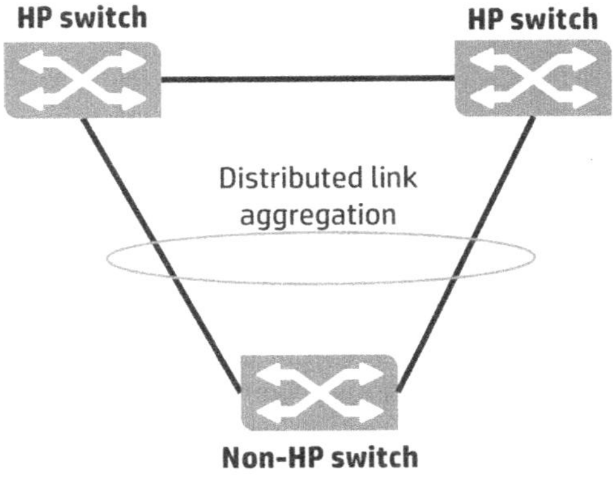


Figure 13-10: Question 27 network topology

Based on this topology, which technology should you deploy to successfully establish link aggregation from the non-HP switch to the two HP switches?

a. Comware using IRF

b. Comware using meshed stacking

c. Comware using dynamic LACP

d. Comware using a static aggregated link

28. A network administrator is setting up an LACP aggregated link between two ProVision switches. What must the administrator match so the switches set up the link successfully? (Select two.)

a. Media type

b. MSTP settings

c. VLAN settings

d. Duplex mode

e. QoS settings

29. Which ProVision switch command configures link aggregation to a connected device?

a. `link-aggregate`

b. `interface bridge-aggregation`

c. `interface trunk`

d. `trunk`

30. When configuring a static route on a Comware switch, which parameter changes the administrative distance of the route?

a. `preference`

b. `distance`

c. `administrative-distance`

d. `precedence`

31. A master is removed from an existing IRF virtual device. If the remaining members have the same IRF priority number, which will they use first to elect a new master?

a. Lowest IRF member ID

b. Longest system uptime

c. Highest bridge MAC address
d. Member designated as the stand-by

32. What advantages does IRF provide over multiple spanning tree protocol? (Select two.)
  a. Faster failover
  b. Link redundancy
  c. Device redundancy for Layer 3 functions
  d. Support for multiple VLANs
  e. Deployment at both access and core layers

33. Which are the responsibilities of only the master in an IRF virtual device? (Select two.)
  a. Run IP routing protocols
  b. Forward traffic
  c. Build MAC addressing table
  d. Learn MAC addresses on ports that connect to other devices
  e. Build new Layer 2 headers for forwarding routed traffic

34. View Figure 13-11.

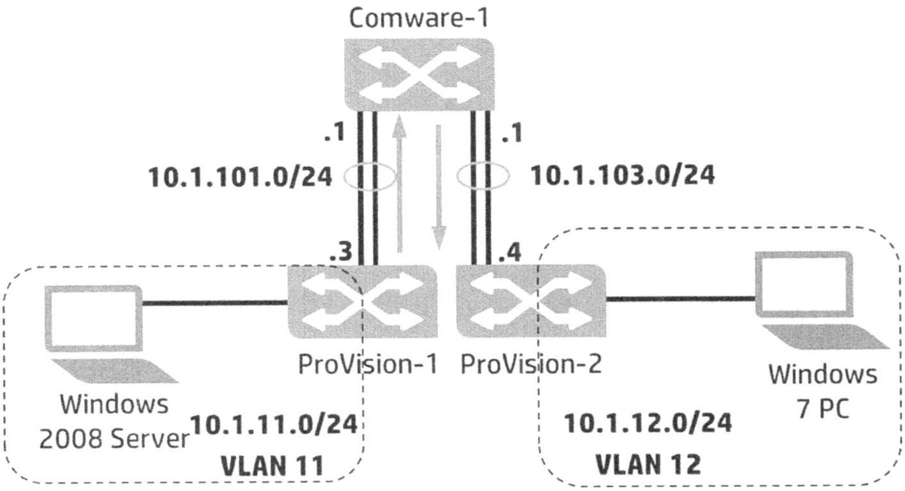


Figure 13-11: Question 34 network topology

In this network, Comware-1 is a Layer 3 switch, and ProVision-1 and ProVision-2 are Layer 2 switches. What type of route does Comware-1 typically need to route traffic to the 10.1.12.0/24 network?

a. Default route

b. Direct route

c. Indirect route

d. Connected route

35. Examine the following OSPF configuration on a Comware switch:

```
[Comware] interface vlan 10
[Comware-vlan-1] ip address 10.1.10.1 255.255.255.0
[Comware-vlan-1] quit
[Comware] interface vlan 11
[Comware-vlan-1] ip address 10.1.11.1 255.255.255.0
[Comware-vlan-1] quit
[Comware] interface vlan 12
[Comware-vlan-1] ip address 10.1.12.1 255.255.255.0
[Comware-vlan-1] quit
[Comware] interface vlan 13
[Comware-vlan-1] ip address 10.1.13.1 255.255.255.0
[Comware-vlan-1] quit
[Comware] interface vlan 14
[Comware-vlan-1] ip address 10.1.24.1 255.255.255.0
[Comware-vlan-1] quit
[Comware] ospf 1 router-id 1.1.1.1
[Comware-ospf-1] area 0
[Comware-ospf-1-area-0.0.0.0] network 10.1.8.0 0.0.3.255
```

Based on this information, which interfaces belong to Area 0?

a. VLAN 10 only

b. VLAN 10 and 11 only

c. VLAN 10, 11, and 12 only

d. VLAN 10, 11, 12, and 13 only

e. VLAN 10, 11, 12, and 13

36. Which ProVision command includes a network in a specific OSPF area?

a. `ip ospf area`

b. `ospf area`

c. `network`

d. `router ospf`

37. A ProVision switch receives two routes to the same network through OSPF. Which mechanism does the ProVision switch use to select the preferred route for its IP routing table?

   a. Administrative distance

   b. Metric

   c. Number of hops

   d. VLAN

38. View the table below, which lists the preference Comware gives to routes learned through different ways.

| Routing method | Comware |
|---|---|
| Static | 60 |
| OSPF internal | 10 |
| OSPF external | 150 |

   Which route will a Comware switch prefer?

   a. 192.168.115.0/24 10.1.1.2 cost 1 preference 10

   b. 192.168.115.0/24 10.1.2.2 cost 1 preference 60

   c. 192.168.115.0/24 10.1.3.2 cost 10 preference 10

   d. 192.168.115.0/24 10.1.3.3 cost 5 preference 60

39. Refer to Figure 13-12.

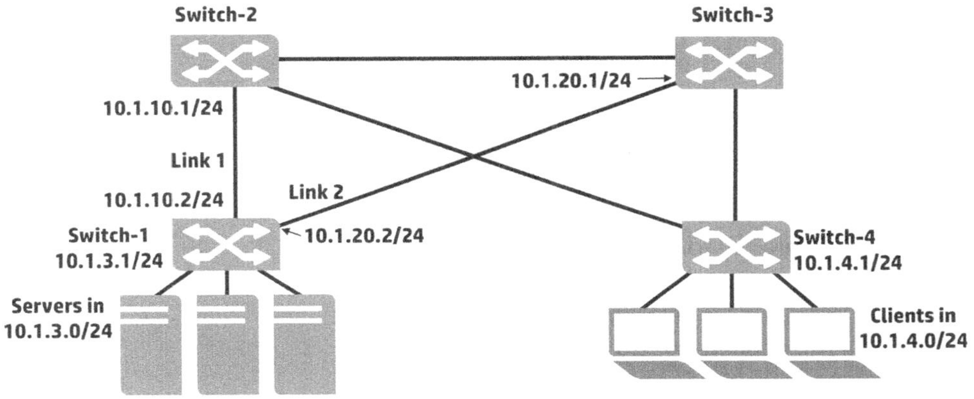


Figure 13-12: Question 39 network topology

```
<Switch-1> display ip routing-table
Routing Tables: Public
Destinations : 9 Routes : 9
Destination/Mask Proto Pre Cost NextHop     Interface
10.1.3.0/24      Direct 0   0   10.1.3.1    Vlan3
10.1.3.1/32      Direct 0   0   127.0.0.1   InLoop
10.1.4.0/24      Static 60  0   10.1.10.1   Vlan10
10.1.10.0/24     Direct 0   0   10.1.10.2   Vlan10
10.1.10.2/32     Direct 0   0   127.0.0.1   InLoop0
10.1.20.0/24     Direct 0   0   10.1.20.2   Vlan20
10.1.20.2/32     Direct 0   0   127.0.0.1   InLoop0
127.0.0.0/8      Direct 0   0   127.0.0.1   InLoop0
127.0.0.1/32     Direct 0   0   127.0.0.1   InLoop0
```

A network administrator wants to create a redundant IP route to 10.1.4.0/24 on Switch-1. The redundant route should use Link 2, and the switch should load balance traffic over both routes.

What is the correct setup?

a. An IP route through 10.1.4.1 with preference 1

b. An IP route through 10.1.4.1 with preference 60

c. An IP route through 10.1.20.1 with preference 1

d. An IP route through 10.1.20.1 with preference 60

40. Examine the following Comware configuration:

```
[Comware-1] display irf topology

                 Topology Info
-------------------------------------------------------

        IRF-Port1                IRF-Port2

Switch  Link   neighbor   Link  neighbor  Belong To

1       DIS    --         UP    2         0023-8929-54ad

2       UP     1          DIS   --        0023-8929-4f83
```

Based on this configuration, which statement is true?

a. This is a daisy-chain topology.

b. The IRF ports are misconfigured.

c. Switch 2 is the master.

d. Switch 1 has a higher priority in the IRF topology.

41. Which wireless mode is used to bridge traffic between two locations?

 a. Ad hoc mode
 b. In-cell relay mode
 c. Infrastructure mode
 d. Switch mode

42. For which standards does 802.11ac provide backward compatibility?

 a. 802.11b and g
 b. 802.11g and n
 c. 802.11a and n
 d. 802.11a and g

43. What is the default operational mode of an MSM access point?

 a. Autonomous
 b. Controlled
 c. Infrastructure
 d. Managed

44. Match the wireless security setting with the appropriate description.

 a. WEP with 802.1X
 b. WPA2-PSK
 c. WPA2 with 802.1X
 d. MAC-Auth to a RADIUS server

 _______i. Often used by small companies that do not have a backend authentication server

 _______ii. Not recommended for user authentication because it is time-consuming to configure and can be easily spoofed

 _______iii. Not supported with 802.11n or ac

 _______iv. Recommended for enterprises that want to secure employees' communications

45. Which group does the admin account belong to in IMC?

 a. Administrator
 b. Maintainer

c. Manager

d. Operator

46. Which method is not supported for auto-discovering devices in IMC?

a. ARP-based

b. IPsec VPN-based

c. LLDP-based

d. Point-to-Point protocol-based

e. Router-based

47. When importing devices into IMC, which item must you configure in IMC?

a. SNMP template

b. Telnet or SSH authentication credentials

c. Loopback interface

d. SNMP traps

48. A network administrator creates a port group on a Comware switch and shuts down interfaces in the port group using these commands:

```
[Switch] port-group manual myGroup
[Switch-port-group-manual-myGroup] group-member g1/0/1 to
g1/0/5
[Switch-port-group-manual-myGroup] shutdown
[Switch-port-group-manual-myGroup] quit
```

The administrator checks that the ports shut down correctly. Later the administrator enters these commands to enable the ports:

```
[Switch] port-group manual myGroup
[Switch-port-group-manual-myGroup] undo shutdown
```

However, when the administrator checks the status for the ports, the ports are still administratively shut down. What is the issue?

a. The port group is deactivated after five minutes, so the administrator must reactivate the port group before entering the `undo shutdown` command again.

b. The port group does not persist in the configuration, so it had no members when the administrator entered the `undo shutdown` command.

c. While the `shutdown` command is supported for use in port groups, the `undo shutdown` command is not.

d. The port group is using the default privilege level, which does not permit enabling interfaces.

49. A switch has ports that are assigned several different RSTP roles. Which type of port can immediately take over as a root port if the current root port fails?

a. Backup

b. Alternate

c. Designated

d. Disabled

50. A network administrator wants to configure a ProVision switch to use priority 16,384 for instance 0. What is the correct value for the `spanning-tree priority` command?

a. 4

b. 8

c. 2048

d 4096

## Answers

1. **D** is correct. For out-of-band management, you connect your management station directly to the switch's console port, using the serial cable that ships with the switch. You must, therefore, have physical access to the switch.

   **A** is incorrect because out-of-band management requires a direct connection, which is not used to transmit data onto the Ethernet network. **B** is incorrect because you can protect in-band management communications, using security mechanisms such as Secure Shell (SSH). **C** is incorrect because out-of-band management does not limit the network administrator's access to the switch features and functions.

   For more information, see Chapter 2.

2. **B** is correct. Out-of-band management is console access, which is the AUX interface on Comware switches. The exhibit shows that AUX has level 3 privileges. You could, therefore, enter commands to maintain switch operations and configure any switch feature.

   **A**, **C**, and **D** are incorrect because they limit privileges to configure the switch in some way.

   For more information, see Chapter 3.

3. **C** is correct. The infrastructure layer includes the underlying ports and forwarding hardware that move data across your network.

   **A** is incorrect because the application layer represents the actual applications in SDN. **B** is incorrect because the control layer presents an abstracted view of the infrastructure, allowing the network administrator to apply one or more policies across the network. The controller's job is to enforce these policies. A controller needs to communicate with the infrastructure but must also be able to communicate with applications. Answer **D** is incorrect because management is a nonexistent layer in SDN.

   For more information, see Chapter 3.

4. **C** is correct. The ProVision modular switches identify modules with letters and starting number ports at 1, so the correct answer is B2.

   **A** and **B** are incorrect because this is an invalid nomenclature for HP switches. **D** is incorrect because this nomenclature is used on Comware modular switches.

   For more information, see Chapter 2.

5. **C** is correct. On Comware switches, angle brackets (<>) indicate that you are at the user view.

   **A** is incorrect because system view is indicated by square brackets: [HP].

   **B** and **D** are incorrect because these are not valid views in the Comware operating system. (On ProVision switches, the enable context is sometimes referred to as the manager context; the basic context is sometimes referred to as operator context.)

   For more information, see Chapter 2.

6. **A** is correct. On ProVision switches, login credentials are not stored in flash by default. If you want to store the credentials in the internal flash and view them in the config, you should enter the `include-credentials` command.

   **B** and **D** are incorrect. The `include-credentials` command does not require the switch to send login credentials to an external RADIUS server or affect the 802.1X authentication process. **C** is incorrect because after the `include-credentials` command is entered, the login credentials for manager and operator are part of the configuration file as well as being stored in flash.

   For more information, see Chapter 3.

7. **C** is correct. With operator-level access on a ProVision switch, you can enter some `show` commands. You cannot, however, view the configuration file itself. In addition, you cannot configure switch features.

   **A**, **B**, and **D** are incorrect because operator-level access does not allow you to configure switch features or enter debug commands.

   For more information, see Chapter 2.

8. **A** is correct. When you log in to the console port of a Comware router or the auxiliary port of a Comware switch, you are immediately given Manager access, which is the highest privilege level of a Comware device.

   **B**, **C**, and **D** are incorrect. Even though they are valid privilege levels, they are not the default levels of access to the Comware router console port or the Comware switch auxiliary port.

   For more information, see Chapter 3.

9. **B** is correct. Level 1 is called the Monitor level. It gives you access to commands for system maintenance and service fault diagnosis.

   **A** is incorrect because Level 2 is the System level, which enables service configuration commands. **C** is incorrect because Level 3 is the Manager level, which enables basic operation of the system and configuration of the system support modules. **D** is incorrect because Comware does not support a Level 4 privilege level.

   For more information, see Chapter 3.

10. **B** is correct. The correct syntax to upgrade the ProVision operating system from a USB drive is: `copy usb flash <image-name> {primary | secondary}`.

    **A** is incorrect because the `upgrade` command is a nonexistent command. **C** and **D** are incorrect because the `copy` command is used for operating system upgrades, but these examples have incorrect syntax.

    For more information, see Chapter 4.

11. **A** is correct. Use the `startup saved-configuration` command to specify a startup configuration file to be used at the next startup of a Comware device.

    **B** is incorrect because this command is used to accomplish the task on a ProVision switch. **C** and **D** are incorrect because they are invalid commands.

    For more information, see Chapter 4.

12. **C** and **E** are correct. To permanently delete a file so that it cannot be recovered, use the `delete /unreserved` command. To empty the recycle bin so that the files cannot be recovered, use the `reset recycle-bin` command.

    **A** is incorrect because this is a nonexistent command. **B** and **D** are incorrect because these are invalid parameters.

    For more information, see Chapter 4.

13. **B** is correct. The `port trunk permit vlan` command adds VLANs to the existing trunk—you must manually remove the VLANs that you do not want from the trunk.

    **A** is incorrect because this configuration adds VLANs 30 to 31 to the trunk—it does not replace them. **C** and **D** are invalid configurations and, therefore, are incorrect.

    For more information, see Chapter 5.

14. **C** is correct. A hybrid port on a Comware device allows you to support multiple tagged or untagged VLANs on the same port.

    **A** is incorrect because an access port supports only a single untagged VLAN. **B** is incorrect because a trunk supports many VLANS, but only one can be untagged. **D** and **E** are incorrect because they are nonexistent port types. **F** is incorrect because this is a ProVision term.

    For more information, see Chapter 5.

15. **B** is correct. Because VLAN 10 extends between the two switches, it must be included on the trunk. And, to route traffic to VLAN 20, it also must be included on the VLAN trunk.

    **A** is incorrect because VLAN 30 is local and, therefore, does not have to be included on the VLAN trunk. Switch B will route traffic to/from it as necessary for the other VLANs. **C** is incorrect because VLAN 20—not 30—needs to be included on the trunk. **D** is incorrect because there is no such command as `port mode trunk` on the ProVision switches.

    For more information, see Chapter 5.

16. **B** is correct. The correct syntax requires that you specify the DHCP relay with the `ip helper-address` command for the VLAN where the users reside.

    **A** is incorrect because there is no such command as `ip helper-vlan`. **C** is incorrect because it does not specify the correct user VLAN. **D** is incorrect because it is configured on the wrong switch.

    For more information, see Chapter 6.

17. **C** is correct. Workstation 1, Switch A, and Switch C are members of VLAN 20. A VLAN defines a broadcast domain.

    **A**, **B**, and **C** are incorrect because devices in different VLANs are in separate broadcast domains.

    For more information, see Chapter 5.

18. **C** is correct. Because Switch A is a Layer 2 switch, it will forward the traffic in VLAN 20 to Switch C, which is responsible for routing the traffic. The packet will be tagged in VLAN 20 because Switch A's link to Switch C supports multiple VLANs, with VLAN 1 being the only VLAN that is untagged.

    **A** is incorrect because Switch A will not change the VLAN assignment to VLAN 1. **B** is incorrect because the link between the two switches supports multiple VLNAs, and the default VLAN, VLAN 1, is the only VLAN that is untagged. **D** is incorrect because Switch A is not routing traffic; it is operating at Layer 2.

    For more information, see Chapter 5.

19. **D** is correct. On Comware switches, ports that connect to endpoints should be configured as access ports that permit a particular VLAN—which is VLAN 20 in the example network shown.

    **A** is incorrect because the configuration would not allow the workstation to send untagged traffic in VLAN 20. There is no reason to configure the port as a trunk port, but if you did, the PVID would need to be 20. **B** and **C** are incorrect because Comware switches are not configured in this way; they are configured as a particular type of port (such as trunk, access, or hybrid). Access ports are then configured to permit a specific VLAN.

    For more information, see Chapter 5.

20. **B** and **E** are correct. ComwareB has the lowest switch ID (lowest MAC address) and, therefore, is the root switch. (The default priority on all HP switches is 32,768.) Comware switches use the older cost structure and, therefore, port A2 is the designated port since ComwareA has a lower cost path to the root than ProVisionC.

    **A** and **C** are incorrect because these switches have higher MAC addresses. **D** is incorrect because A1 is a root port. **F** is incorrect because all active ports on the root switch are designated ports.

    For more information, see Chapter 7.

21. **A** is correct. On Comware switches, you configure DNS server IP addresses as part of the DHCP server pool.

    **B** and **D** are incorrect because these settings are configured for DHCP relay; a DHCP server pool is configured when the switch functions as a DHCP server. **C** is incorrect because on Comware switches excluded IP addresses are configured globally, rather than as part of the DHCP server pool.

    For more information, see Chapter 6.

22. **D** is correct. Since there are four uplinks, you need to create four instances, separating out the VLANs appropriately, and configure a corresponding switch at the distribution layer as the root and the remaining distribution layer switches as secondary roots.

    **A**, **B**, and **C** are incorrect because this configuration would not utilize all four uplinks. **E** is incorrect because, even though there are five distribution switches, there are only four uplinks.

    For more information, see Chapter 7.

23. The correct answer is **c**, **d**, **b**, **a**. The DHCP process is outlined below.

    The client sends a DHCP Discover message.

    The server sends a DHCP Offer.

    The client sends a DHCP Request.

    The server sends a DHCP ACK.

    For more information, see Chapter 6.

24. **D** is correct. Because the `active region-configuration` command was not executed on ComwareC after changing instance 2's VLAN mappings, the change does not go into effect and, therefore, no STP recalculation occurs.

    **A** is incorrect because MSTP is configured and operating correctly. **B** is incorrect because ComwareA and ComwareB are the root switches based on their changed priorities. **C** is incorrect because the `active region-configuration` command was not executed after the `instance2 vlan 3-4` command on Switch C and, therefore, the change does not go into effect.

    For more information, see Chapter 7.

25. **A** is correct. When connecting IRF members, you must ensure that IRF port 1 on one member is connected to IRF port 2 on another member. For example, member 1 and member 2 can connect on ports 1/1 and 2/2 or on ports 1/2 and 2/1.

    **B** is incorrect because you cannot connect port 1 on member 1 (1/1) and port 1 on member 2 (2/1). **C** is incorrect because you cannot connect the two IRF ports on the same switch. **D** is incorrect because you cannot connect port 2 on member 1 (1/2) and port 2 on member 2 (2/2).

    For more information, see Chapter 10.

26. **B** is correct. Based on the `show lacp` command, both sides have successfully negotiated link aggregation. Since there are four users and one server, and only one link is being utilized, SwitchA is load-sharing based on destination MAC addresses and SwitchB is load-sharing based on source MAC addresses.

    **A** is incorrect because this contradicts the output of the `show lacp` command. **C** and **D** are incorrect because, if this was the case, traffic should be traversing both links in the aggregate connection.

    For more information, see Chapter 8.

27. **A** is correct. Comware using IRF allows you to set up distributed link aggregation from a neighboring device with a link to each individual HP switch.

    **B** is incorrect because meshed stacking is not supported on Comware switches. **C** and **D** are incorrect because these link aggregation technologies do not support an aggregated link with two switches on one side of the link.

    For more information, see Chapter 8.

28. **A** and **D** are correct. When using LACP to establish an aggregated link, ProVision switches check media type, media speed, and duplex mode. **B**, **C**, and **E** are incorrect. ProVision switches do not check these settings when establishing an LACP aggregated link.

    For more information, see Chapter 8.

29. **D** is correct. Use the `trunk` command to configure link aggregation on a ProVision switch.

    **B** is incorrect because this command is used to set up link aggregation on a Comware switch. **A** and **C** are incorrect because these are nonexistent commands on ProVision switches.

    For more information, see Chapter 8.

30. **A** is correct. To configure a static route on a Comware switch, use the following syntax:

```
[Comware] ip route-static <destination-network><mask>
<next-hop-ip-address> [preference <preference>]
[description <description>]
```

    The preference parameter defines the administrative distance, which defaults to 1 if omitted.

    **B**, **C**, and **D** are incorrect because these are nonexistent parameters.

    For more information, see Chapter 8.

31. **B** is correct. If all the members have the same IRF priority, they will first try to elect a new master by comparing system uptimes, and the member with the longest system uptime will win. If all the members have the same uptime (which is rounded to 10 minutes), the member with the lowest bridge MAC address wins.

    **A** is incorrect because IRF members never consider member ID in master elections. **C** is incorrect because members consider the bridge MAC address if all members have the same IRF priority and system uptime. In this case, however, the member with the lowest bridge MAC address wins. **D** is correct because IRF virtual devices do not designate a member as a "standby."

    For more information, see Chapter 10.

32. **A** and **C** are correct. IRF virtual devices have a much faster failover than MSTP, which was designed for legacy networks and can take as long as one second to reconverge. IRF virtual devices also provide device redundancy for Layer 3 functions, which MSTP does not offer.

    **B**, **D**, and **E** are incorrect because both IRF and MSTP provide link redundancy, support multiple VLANs, and can be deployed at the access and core layers.

    For more information, see Chapter 10.

33 **A** and **C** are correct. In an IRF virtual device, the master's responsibilities include running the IP routing protocols and building the MAC addressing tables. The master forwards the MAC addressing tables to the other members of the IRF virtual device.

    **B** and **E** are incorrect because all members in an IRF virtual device forward traffic and create new headers to forward routed traffic. **D** is incorrect because all members in the IRF virtual device learn MAC address on their ports that connect to other devices. The slaves forward this information to the master, which creates the MAC addressing tables.

For more information, see Chapter 10.

34. **C** is correct. In the network shown, the 10.1.12.0/24 network is not directly connected to Comware-1, so the switch requires an indirect route for that network.

    **A** is incorrect because Comware-1 would not have a default route to a switch that does not route. **B** and **D** are incorrect because the 10.1.12.0/24 network is not directly connected to Comware-1.

    For more information, see Chapter 9.

35. **B** is correct. A wildcard mask matches on a range of four addresses, starting on a power or a multiple of a power of 2, for which 8, in the third octet of the network number, qualifies. Therefore, addresses from 8 through 11 would match in the third octet, which includes 10.1.10.0/24 and 10.1.11.0/24.

    **A** is incorrect because it does not include 10.1.11.0/24. **C**, **D**, and **E** are incorrect because they include networks other than subnets 10 and 11.

    For more information, see Chapter 9.

36. **A** is correct. To associate networks to the areas, use the `ip ospf area` command under the VLAN context.

    **B** is incorrect because the command begins with ip. **C** is incorrect because the `network` command is used on Comware switches for this purpose. **D** is incorrect because the `router ospf` command creates the OSPF process only.

    For more information, see Chapter 9.

37. **B** is correct. A ProVision switch uses metrics to distinguish between routes that are learned in the same way, such as two routes learned through OSPF. Generally, these routes have the same administrative distance, so the ProVision switch selects the route that has the lower metric.

    **A** is incorrect because a ProVision switch uses administrative distance to select two routes that are learned through different ways, such as a static route and an OSPF route. **C** is incorrect because the switch uses metrics to differentiate between two OSPF routes, and OSPF metrics are not based on the number of hops to the destination network. **D** is incorrect because the VLAN is not a factor in selecting routes.

    For more information, see Chapter 9.

38. **A** is correct. A Comware switch uses preference (the term used for administrative distance on Comware switches) to select routes to the same network that are learned through different ways. Comware switches prefer internal OSPF routes over static routes. When choosing between two routes learned through the same way, Comware switches select the route with the lowest cost (the term used for metrics).

**B** and **D** are incorrect because the routes have a higher preference. **C** is incorrect because it has a higher cost than A.

For more information, see Chapter 9.

39. **D** is correct. To use Link 2, the route must go through 10.1.20.1, and to ensure that the switch will load share across the redundant routes, the second route must use the same preference as the first route: 60.

    **A** and **B** are incorrect because the new route cannot go through 10.1.4.1 on Link 2. **C** is incorrect because the new route must have the same preference as the existing route or the switch will not load share traffic across the two routes.

    For more information, see Chapter 9.

40. **A** is correct. Based on the CLI output "DIS," or disabled state, this is a daisy-chain topology.

    **B** is incorrect because the IRF ports are matched up 1 to 2. **C** and **D** are incorrect because there is not enough information to determine who the master is or who has the higher priority.

    For more information, see Chapter 10.

41. **B** is correct. In-cell relay mode is used to connect two or more network segments over a wireless connection.

    **A** is incorrect because only two devices communicate with each other directly. **C** is incorrect because an access point is used for devices to intercommunicate with each other. **D** is incorrect because this is an invalid mode.

    For more information, see Chapter 11.

42. **C** is correct. 802.11ac operates in the 5.0 GHz range. 802.11a operates in the 5.0 GHz range, and 802.11n can also operate in this range.

    **A** is incorrect because 802.11b and g operate in the 2.4 GHz range. **B** is incorrect because although 802.11n can operate in the 5.0 GHz range, 802.11g operates in the 2.4 GHz range. **D** is incorrect because although 802.11a operates in the 5.0 GHz range, 802.11g operates in the 2.4 GHz range.

    For more information, see Chapter 11.

43. **B** is correct. The default operational mode of an MSM AP is controlled.

    **A** is incorrect because you have to change the mode to autonomous and reboot for the change to take effect. **C** and **D** are incorrect because these are nonexistent modes.

    For more information, see Chapter 11.

44. Answers are listed below.

    b i. Often used by small companies that do not have a backend authentication server

    d ii. Not recommended for user authentication because it is time-consuming to configure and can be easily spoofed

    a iii. Not supported with 802.11n or ac

    c iv. Recommended for enterprises that want to secure employees' communications

    For more information, see Chapter 11.

45. **A** is correct. Operators in the Administrator group have all rights to IMC. They can manage all users, and they can add and delete users. They can configure all network devices and view any information gathered and stored by IMC. The default admin account is associated to this group, by default.

    **B** is incorrect because the Maintainer group has restricted access in IMC. Answers **C** and **D** are incorrect because these are ProVision privilege levels.

    For more information, see Chapter 12.

46. **C** is correct. LLDP is not a supported method for IMC auto-discovering devices.

    **A**, **B**, **D**, and **E** are valid methods of auto-discovering devices, are network-based, and, therefore, are incorrect answers.

    For more information, see Chapter 12.

47. **B** is correct. IMC needs CLI access to a device to perform management tasks that cannot be performed via SNMP. Either Telnet or SSH can be configured. You can predefine a template for this access or define the parameters during the auto-discovery or device import process.

    **A** is incorrect because you can manually define the SNMP parameters or use a template. **C** and **D** are incorrect because you can optionally specify these in IMC.

    For more information, see Chapter 12.

48. **B** is correct. When you configure a port group, it does not persist in the config. If you want to configure settings on a range of interfaces again, you must create the port group again.

    **A** is incorrect because port groups are not deactivated after 5 minutes. **C** is incorrect because you can enter the undo command in the port-group context. **D** is incorrect because the port-group does not use a privilege level. Users with the appropriate privilege level (system or manage) can configure interfaces.

    For more information, see Chapter 2.

49. **B** is correct. An alternate port offers the second (or the third, and so on) best route to the root device. If the root port fails, the alternate with the second best route can immediately take over.

**A** is incorrect because a backup port is connected to the same switch. The switch has a looped connection to itself or it has a loop through downstream devices that do not support spanning tree. **C** is incorrect because the designated port is the port closest to the root port. **D** is incorrect because a disabled port cannot participate in spanning tree.

For more information, see Chapter 7.

50. **A** is correct. You configure spanning tree priority in sets of 4096. If you enter 1, for example, the priority is 4096. If you want the switch to use a priority of 16,384, you enter 4 (4 × 4096 = 16,384).

**B** is incorrect because entering 8 would set the priority to 32,768 (8 × 4096). **C** and **D** are incorrect because you configure a priority in sets of 4096 (the number you enter × 4096 = the priority). You cannot enter a value of 2048 or 4096.

For more information, see Chapter 7.

# Index

## E

## P



## Q



## R

## S

## T

## X